T0073375

DEEP UTOPIA

DEEP UTOPIA

LIFE AND MEANING IN A SOLVED WORLD

NICK BOSTROM

IDEAPRESS PUBLISHING

WASHINGTON, DC

Ideapress Publishing | www.ideapresspublishing.com

Cover Design: Nick Bostrom
Interior Design: Jessica Angerstein

Cataloging-in-Publication Data is on file with the Library of Congress.

Hardcover ISBN: 978-1-64687-164-3

---

3 4 5 6 7 8 9 10

CONTENTS

PREFACE

Like children opening their eyes to a new day, having gone to bed the previous night as tufts of snow began falling, we dash to the window and lift ourselves to the tips of our toes to behold a landscape transformed: a winter wonderland glittering with possibilities for discovery and play. Even the tree branches, before so boringly bare, have been changed into something beautiful and magical. We feel we are inhabiting a storybook or a gameworld, and we want very much to put on our boots and mittens immediately and run outside to see it, touch it, experience it, and to play, play, play…

MONDAY

Hot springs postponed

Tessius: Hey, look at this poster. Nick Bostrom is giving a lecture series, here in the Enron Auditorium, on "The Problem of Utopia".

Firafix: Bostrom—is he still alive? He must be as old as the hills.

Tessius: It's all those green vegetable elixirs he used to make.

Firafix: They worked?

Tessius: Not at all, but they became very popular for a while. That's how he made his money, you see—the recipe book. Then he could afford the anti-aging therapies as they came along.

Tessius: It just started ten minutes ago. Shall we go in?

Firafix: Sure, why not.

Kelvin: We can take the baths after dinner instead. They're open late.

Bostrom: —leukemia. It's really important to find a cure or at least some way to alleviate her suffering. On a larger scale, we have extreme poverty, deprivation, malnutrition, soul-crushing physical and mental disorders, family-destroying traffic accidents, alcoholism, oppression, the killing and maiming of civilians in war zones… There are, at present, more than enough problems to provide meaningful challenges for even the most resourceful and enterprising among us.

Tessius [*whispers*]: The old doom-monger is in great form. I'm feeling worse already!

Firafix: Shhh.

Bostrom: Perhaps these humanitarian challenges feel meaningful only to those who care enough about others to have a sincere desire to help. But even pebble-hearted egoists are well catered for in today's world, with a rich buffet of negative circumstances which they are motivated to ameliorate or prevent from getting worse. One person might be struggling with excess body weight, another with getting a job, a third with social isolation, a fourth with a difficult relationship. Rarely do we hear people complain, "The only problem I have is that I have no problems—life, you know, is just *too* perfect, and it really bugs me!".

In short, we appear in no imminent danger of running out of woe. As far as the eye can see, there is an abundance of actual and potential sorrow to keep the worry-mill a-churning, and to provide altruist and egoist alike with bountiful opportunities for worthwhile striving.

Nevertheless, in these lectures I want to talk about the problem of utopia: the problem we will face after we have solved all the other problems.

This may not seem like the most pressing priority in our current situation... There are, we must concede, other causes and tasks with most legitimate claims on our attention. Still, I don't think it would be unbefitting for our civilization to at least cast a glance at what lies ahead if things were to go well: to consider, that is, where we eventually end up if we continue along the present path and completely succeed in what we are in the process of trying to accomplish...

The *telos of technology*, we might say, is to allow us to accomplish more with less effort. If we extrapolate this internal directionality to its logical terminus, we arrive at a condition in which we can accomplish everything with no effort. Over the millennia, our species has meandered a fair distance toward this destination already. Soon the bullet train of machine superintelligence (have we not already heard the conductor's whistle?) could whisk us the rest of the way.

And what would become of us then?

What would give our lives meaning and purpose in a "solved world"?

What would we *do* all day?

These questions have timeless intellectual interest. The concept of deep utopia can serve as a kind of philosophical particle accelerator, in which extreme conditions are created that allow us to study the elementary constituents of our values. But the questions may also come to have immense practical importance, as the telos of technology is actually reached or closely enough approached—very possibly within the lifetime of many of you here in the audience, in my estimation.

Tessius: Shall we sit down?

Kelvin: There are some seats over there.

Firafix: Yes, I want to hear this. I'll stand here.

Bostrom: In any case, the problem of utopia is in the water. Can we not sense it—a certain half-embarrassed latent unease? A doubt lurking in the depths beneath us? A faint shadow sweeping across our conception of *what it's all for*?

Argumentum ad opulentium

And sometimes this concern breaches the surface of awareness, and we see a fin approaching... For example, Bill Gates wrote:

> "It is true that as artificial intelligence gets more powerful, we need to ensure that it serves humanity and not the other way around. But this is an engineering problem . . . I am more interested in what you might call the *purpose* problem. . . . if we solved big problems like hunger and disease, and the world kept getting more peaceful: What purpose would humans have then? What challenges would we be inspired to solve?"[1]

And Elon Musk, in an interview with CNBC:

> "How do we find meaning in life if the AI can do your job better than you can? I mean if I think about it too hard, it can frankly be dispiriting and demotivating. Because—I've put a lot of blood, sweat, and tears into building the companies, and then I'm like '*should* I be doing this?'. Because if I'm sacrificing time with friends and family that I would prefer, but then ultimately the AI can do all these things. Does that make sense? I don't know. To some extent,

> I have to have deliberate suspension of disbelief in order to remain motivated."[2]

Perhaps there is a sense in which worrying about purpose is a luxury problem? If so, we might expect utopian prosperity to increase its prevalence. But in any case, as we shall see, the issue runs far deeper than anything to do with a mere surfeit of money and material possessions.

*

Some of my friends like for there to be a model of impact—a story of why, out of all the things that one *could* be working on, the thing one proposes to do would be the most impactful and beneficial. They seek the highest expected utility.

Were I to attempt such a story for our present proceedings, it might go as follows. Our civilization looks to be approaching a critical juncture, given the impending development of superintelligence. This means that at some point, somebody, or all of us, might be confronted with choices about what kind of future we want—where the options include very different trajectories, some of which would take us to radically unfamiliar places. These choices could be highly consequential. Yet perhaps some of the choices must be made under time pressure, because the world refuses to wait, or because we ourselves are going crazier by the week, or because delaying would mean getting preempted by more decisive actors, or because we don't want to stop moving for fear that we might then never start moving again.[3] Or perhaps there is no discrete time when these choices get made, and instead they are and will be made incrementally over time, but in such a way that earlier partial choices limit the range of later feasible outcomes. Either way, there could be value in getting pointed in a positive direction sooner rather than later. And if there is an actual distinct period of pivotal deliberation, it would be useful to have some suitable preparatory material for that—you know, to equip the deliberators with some relevant concepts and ideas, and help them get into a good frame of mind.

Make of that "impact story" what you will. Another possible explanation for why I'm doing these lectures is that I agreed to it a long time ago, in a moment of weakness.

*

Let me say what this lecture series is not. It is not an attempt to "make a case" for something. It is instead an exploration. When exploring a topic as deep and difficult as the one before us, one wants to bring into view multiple considerations, pursue various lines of thought, place one's hands on competing evaluative conceptions—allowing the tug of each thought and each inclination to be experienced as keenly and as sympathetically as possible. One does not want to prematurely dismiss a natural perspective, even one that is ultimately to be turned away from. For the value of one's opinions, in a matter like this, is a function of how generously one has allowed the alternatives to play with one's soul.

Walls of sausages

Let us consider first the simplest kind of utopia: that of sheer material abundance.

This utopian conception is exemplified by the myth of Cockaigne, or The Land of Plenty. It was an important part of the medieval imaginary, and found frequent expression in popular art and writing as well as in the oral tradition:

> "No work is done the whole day long,
> By anyone old, young, weak, or strong.
> There no one suffers shortages;
> The walls are made of sausages."[4]

Cockaigne is essentially a medieval peasant's daydream. In the land of Cockaigne, there is no backbreaking labor under scorching sun or nipping norther. No stale bread, no deprivation. Instead, we are told, cooked fish jump out of the water to land at one's feet; and roasted pigs walk around with knives in their backs, ready for carving; and cheeses rain from the sky. Rivers of wine flow through the land. It is perpetually spring, the weather is beautiful and mild. You make money while you sleep. And sexual taboos have been loosened—we find descriptions of nuns turned upside down with their bottoms showing. Disease and aging are no more. There is continual feasting, with a great deal of dancing and music-playing, and lots of time for resting and relaxing too.[5]

Similar fantasies are found in many other traditional societies. For example, in classical antiquity, Hesiod wrote of the happy inhabitants of an imagined earlier Golden Age:

> "And they lived like gods, not a care in their hearts,
> Nothing to do with hard work or grief,
> And miserable old age didn't exist for them.
> From fingers to toes they never grew old,
> And the good times rolled. And when they died
> It was like sleep just ravelled them up.
> They had everything good. The land bore them fruit
> All of its own, and plenty of it too. Cheerful folk..."[6]

In many respects, we are now living in the Golden Age, or in Cockaigne—or in Avalon, The Happy Hunting Ground, The Land of the Ancestors, The Island of the Blessed, Peach Blossom Spring, Big Rock Candy Mountain. "We" here of course excludes those hundreds of millions of humans who still live in abject poverty, along with the great majority of farmed and wild animals. But if we use the term "we" to refer to the people in this room (we the happy few), then it seems fair to say that with our overstocked fridges and around-the-clock delivery services we have in fact achieved a pretty good approximation of roasted pigs wandering the streets and cooked fishes jumping to our feet. We have also achieved everlasting spring—at least inside our air-conditioned buildings and transportation vehicles. The fountain of youth remains to be located, but disease has been considerably reduced and lifespans extended. Furthermore, I have it on good authority that if somebody is intent on looking at female bottoms, including those of apparent nuns, an online search shall not disappoint.

We do, however, still put in a significant amount of work. Our jobs are generally less grueling than those of medieval peasants; but it is nevertheless a bit surprising that we continue to work as many hours as we do.

Keynes's prediction

This utopian vision of Cockaigne anticipates the conception of progress that we find in modern economics. The latter couches the ideal in a more abstract lexicon of "productivity", "income", and "consumption", rather than sausage-walls. But the core idea of felicity through abundance remains the same.

So this may be a good place to start our exploration—by reviewing the land and its constraints through the binoculars of economics and evolution; and, tomorrow, we will also look at some ultimate technological limitations. But I want to say that, provided you stay for the entire lecture series, you will find that the tenor of our inquiry will gradually shift. We will descend from the external perspectives and cold abstractions of the dismal sciences, down into the valleys where we will get a more humanistic and internal view of the issues of deep redundancy. And it will shift again as we then begin drilling down into the philosophical mantle, in an effort to reach the core—the core of our values, the heart of the problem of utopia.

So hang in there!

I could perhaps say more about what exactly I mean by the problem of utopia and how I plan to approach it. But I think it's better we just jump right in, and we can sort out any definitional or argumentative-structural issues as they arise.

*

The renowned economist John Maynard Keynes considered the goal of material abundance in his widely influential essay, "Economic Possibilities for Our Grandchildren".[7] Published in 1930, the essay argues that humankind is on its way to solving its "economic problem". Keynes predicted that by 2030, accumulated savings and technical progress would increase productivity relative to his own time between fourfold and eightfold.[8] Such a dramatic rise in productivity would make it possible to satisfy human needs with far less effort; and, as a consequence, the average working week would decrease to 15 hours. This prospect worried Keynes. He feared that the surfeit of leisure would cause a kind of collective nervous breakdown, as people would go stir-crazy not knowing what to do with all their spare time.[9]

As we approach 2030, the first part of Keynes's prediction is on track to vindication. Productivity has increased by more than fivefold since 1930, and GDP per capita by more than sevenfold.[10] We thus make much more per hour of work than our great-grandparents did.

The second part of Keynes's prediction, on the other hand, would appear to be about to miss its mark, if trends are extrapolated. While it is true that working hours have declined substantially over the past ninety-plus years, we are nowhere near the 15-hour work week that Keynes expected. Since 1930, the typical work week has been reduced by about a quarter, to roughly 36 hours.[11] The proportion of our lives spent working has seen a somewhat sharper drop: we join the workforce later, live longer after retirement, and take more leave.[12] And our work is on average less strenuous. For the most part, however, we have used our increased productivity for consumption rather than leisure. Greed has triumphed over Sloth.

But perhaps Keynes only got the timing wrong?[13] A revived Keynes—we can picture him emerging from a cryonics dewar, his hat and mustache covered in frost—might argue that we only need to wait for productivity to rise a bit more to see his prophesied 15-hour work week become a reality. If the historic trend were to continue, we'd see another 4- to 8-fold productivity increase in the next 100 years, and a 16- to 64-fold increase by the year 2230. In such a world, would people still choose to spend a large fraction of their waking life working?

Consider two possible reasons for working:

1. To earn income
2. Because working is an intrinsically valued activity

We will return to (2) in later lectures, so let's set that aside for now. But if productivity grew another 8-fold, or even another 64-fold, would we *then* see Keynes's vision of the leisure society come true?

Maybe, or maybe not. There are reasons to be skeptical. In particular, new consumption goods may be invented that cost a lot, or we may undertake very costly social projects. We may also find ourselves compelled to spend more on arbitrarily expensive status symbols to maintain or enhance our relative standing in a zero-sum rat race.

These sources of motivation could continue to operate even at very high income levels. Let us examine each in turn.

New needs and niceties

First, there may be new consumption goods. It is *conceivable* that there could be an unending series of ever more exquisite—and ever more expensive—market goods that enhance leisure; so that no matter how high your hourly salary, it is worth allocating a third or more of your waking hours to working, for the sake of being able to enjoy the remainder at a higher level of consumption. This was the line taken by Richard Posner, the eminent American legal scholar; we'll come back to him later.

This view, however, is highly implausible in today's world, where money has steeply diminishing marginal utility, and where many of the best things in life are indeed free or very cheap. Boosting your annual income from $1,000 to $2,000 is a big deal. Raising it from $1,000,000 to $1,001,000—or even, I should think, to $2,000,000—is barely noticeable.

But: this could change. Technological progress might create new ways of converting money into either quality or quantity of life, ways that don't have the same steeply diminishing returns that we experience today.

For example, suppose there were a series of progressively more expensive medical treatments that each added some interval of healthy life-expectancy, or that made somebody smarter or more physically attractive. For one million dollars, you can live five extra years in perfect health; triple that, and you can add a further five healthy years. Spend a bit more, and make yourself immune to cancer, or get an intelligence enhancement for yourself or one of your children, or improve your looks from a seven to a ten. Under these conditions—which could plausibly be brought about by technological advances—there could remain strong incentives to continue to work long hours, even at very high levels of income.[14]

So the future rich may have far more appealing ways to spend their earnings than by filling up their houses, docks, garages, wrists and necks with increasing amounts of today's rather pathetic luxury goods. We must therefore not unquestioningly assume that money won't matter beyond some given level. The biomedical enhancements I just mentioned are one example of a kind of good that could continue to provide value at high levels of expenditure. And if we imagine—as I tend to do—a future that is mostly populated by digital minds, then the convertibility of wealth into well-being becomes even clearer. Digital minds, be they AIs or uploads, need computation. More computation means longer life, faster thinking, and potentially deeper and more expansive conscious experiences. More computation also means more copies, digital children, and offshoots of all kinds, should such be desired.

The returns curve of infrastructure spending for a digital mind depends on what it is that one is aiming to achieve. Beyond a certain speed of computation, the marginal cost of accelerating a mind's implementation further may rise sharply or hit a hard limit. On the other hand, some algorithms parallelize well, and if they instantiate something that is valued, the returns to computation could be close to linear. Certainly, if you're happy simply making copies of yourself, you need not see diminishing returns even at *very* high levels of expenditure.

Social projects

Second, if we look beyond the sphere of selfish indulgences, we see many additional opportunities to convert huge amounts of resources into valuable outcomes before hitting discouragingly diminishing returns. For example, you might want to build a veterinary system for animals that are sick or injured in the wild. [*Applause.*] People who care about such ambitious projects could have reason to continue working long hours even as their productivity and hourly salary soar to stratospheric heights, because they could keep scaling up their impact. Until there is a clinic on every hill and in every dale, in each bush and each briar, there is an underserved population.

In fact, the altruistic reason for working additional hours may theoretically get *stronger* the higher a person's wages. More additional wild animal hospital rooms could be funded with an extra hour of work if your hourly rate is a thousand dollars than if you're making minimum wage.

I say *may* theoretically get stronger, because as the level of wealth in society increases, it is possible that the lowest-hanging or juiciest fruits in the altruistic opportunity tree get depleted. However, the tree is big and it keeps growing new fruit: so as long as you can keep making money, you can most likely keep doing good. This is readily seen if we consider altruistic reasons not only for removing negatives from the world but also for adding positives, such as by bringing new happy people into existence. You could always create more, and the number scales linearly with resources.

By the way, are there any questions so far? Feel free to interrupt at any time if something isn't clear. —Yes, you there in the aisle, with the buttons?

Student: Are you saying that, like, we should have as many kids as possible? Wouldn't that be selfish?

Bostrom: No, I'm not at this point expressing any moral view. I'm discussing some possible motivations that could conceivably drive some people to continue to work long hours for money even if they could cover all their ordinary needs by working just one or two hours a week. One such possible motive is altruism: make more so you can give more to those in need. Okay, but then what happens if society gets sufficiently rich and utopian that there are no more people in need? I was pointing out that even in that case, some people might be motivated to continue to earn so that they could create more people. No matter how affluent everyone is—indeed *especially* if everyone is very wealthy—you could, in principle, create additional happiness by bringing additional happy people into existence. There certainly are folk who think that would be a good thing, such as total utilitarians, and who could thus remain motivated. There are others, of course, who have no desire to maximize any such measure of aggregate utility. This not being a course on population ethics, we don't need to concern ourselves here with what arguments or justifications there might be for these different views. Though I may note, for the record, that I'm not a total utilitarian, or indeed any kind of utilitarian, although I'm often mistaken for one, perhaps because some of my work has analyzed the implications of such aggregative consequentialist assumptions. (My actual views are complicated and uncertain and pluralistic-leaning, and not yet properly developed.) Does that help?

Another student: But what about global warming?

Tessius [*whispers*]: Some are especially easy to automate.

Bostrom: Well, I think we must make some postulations in order to focus our investigations on the central question that we will be exploring in this lecture series. This means that we will be bracketing a bunch of practical matters entirely, in a bid to get to the philosophical crux. More specifically, we are conducting a thought experiment in which we assume that technological as well as political difficulties are somehow overcome, so that we can focus on the problem of what I call "deep utopia". I was planning to talk about the technological boundary conditions tomorrow, so hopefully things will become a bit clearer then.

So, as I was saying, you could always create more people, especially of the digital sort.[15] The number of digital minds you could create is proportional to the amount of computational resources you could deploy, which we can assume is proportional to the amount of money you have to spend.

Of course, this type of scalable altruistic motivation is reserved for the moral elite. If you don't care about bringing more joyful beings into existence, and you don't have enough universal concern for the welfare and suffering of other sentient beings that already exist, and you don't have some other open-endedly ambitious unegoistic project that you can feel passionate about, then you may not drink from this fountain and you'd have to seek other ways to quench your thirst for purpose. —Let's take one more question.

Yet another student: What do you mean by "digital minds"?

Bostrom: A mind implemented on a computer. Could for example be an upload of a human or animal mind, or an AI of a design and sophistication that makes it a moral patient, i.e. one whose welfare or interests matter for their own sake. I think there'd be a prima facie case for this in the case of a conscious digital mind, though I don't think consciousness is necessary for moral status. For the purposes of the present discussion, probably nothing essential hinges on this point.

Okay, let's press on. We have a lot of stuff to get through.

The desire for more

I mentioned a third reason why we might continue to work hard even at very high income levels: namely, that our appetites may be relative in a way that makes them collectively insatiable.

Suppose that we desire *that we have more than others.* We might desire this either because we value relative standing as a final good; or, alternatively, because we hope to derive advantages from our elevated standing—such as the perks attendant on having high social status, or the security one might hope to attain by being better resourced than one's adversaries. Such relative desires could then provide an inexhaustible source of motivation.

Even if our income rises to astronomical levels, even if we have swimming pools full of cash, we still need *more*: for only thus can we maintain our relative standing in scenarios where the income of our rivals grows commensurately.[16]

Notice, by the way, that insofar as we crave position—whether for its own sake or as a means to other goods—we could all stand to benefit from coordinating to reduce our efforts. We could create public holidays, legislate an 8-hour work day, or a 4-hour work day. We could impose steeply progressive taxes on labor income. In principle, such measures could preserve the rankings of everybody involved and achieve the same relative outcomes at a reduced price of sweat and toil.[17]

But failing such coordination, we may continue to work hard, in order to keep up with all the other people who continue to work hard; and we're stuck in a billionaire's rat race. You just cannot *afford* to slack off, lest your net worth remain stuck in the ten digits while your neighbor's ascends into the eleven…

Imagine standing on the deck of your megayacht, *SV Sufficiens*. You are gliding across the ocean, making good headway with your date, who is suitably impressed. You inch closer in preparation for a kiss, and… next moment you're bobbing ignominiously up and down in the wake of your colleague's *giga*yacht, *NS Excelsior*, as it roars past you. There he is, at the aft of his far grander vessel, grinning patronizingly down at you and waving his stupid sea captain's hat! The moment is quite ruined.

It is also possible to have a desire for improvement *per se*: to desire *that tomorrow we have more than we have today*. This might sound like a strange thing to want. But it reflects an important property of the human affective system—the fact that our hedonic response mechanism acclimates to gains. We begin taking our new acquisitions for granted, and the initial thrill wears off. Imagine how elated you would be now if this kind of habituation *didn't* happen: if the joy you felt when you got your first toy truck remained undiminished to this day, and all subsequent joys—your first pair of skis, your first bicycle, your first kiss, your first promotion—kept stacking on top of each other. You'd be over the moon!

Well, our limbic system (that old curmudgeon) puts paid to that. The hedonic treadmill continuously retreats under our feet, making us keep running while preventing us from ever getting to any fundamentally cheerier place.

But how does this provide an incentive to work in a world of radical economic abundance? We may crave improvement, either for its own sake or as a means of getting a jolt of reward, but it still seems like this craving depends on there being other desires to define what counts as an improvement. I mean, if you didn't *want* the toy truck in the first place, then obtaining it wouldn't be an improvement and wouldn't bring you joy. So we need some type of underlying good that you can keep accumulating and still benefit from getting more of. If there is such a good—maybe the biomedical enhancements or the altruistic initiatives I spoke of earlier—then the desire for things to be improving can serve as an amplifying factor, giving us even stronger reason to continue working beyond that given by our desire for the base-level good itself.

So much for the desire for improvement *per se*. But let's return to the desire to have more than other people—more money, or more exclusive status symbols. This desire, it seems, can stand on its own without presupposing that there is some other more basic desire that defines an unbounded betterness metric. (Strictly speaking, if what we are after in wanting to have more than others is social status, then the construction might require the existence of additional desires in the sense that we especially want to have more than other people of something that they also covet: but the item in question is fundamentally arbitrary and need not be desired by anybody for its own sake apart from the role it happens to play as a focus of such social contestation—it could be an NFT or civet coffee or something else that hardly anybody would want unless others wanted it too.)

The desire for relative standing is therefore a promising source of motivation that could spur work and exertion even in a context where "man's economic problem" has been solved. Provided only that other people's incomes keep rising roughly in tandem with our own, vanity could prevent us from slacking off no matter how rich we get.

The desire for relative standing has another feature that suits it to be a motivator in the age of abundance. Ranking is, to a significant extent, *ordinal*. That is to say, what matters is who has more than whom, not necessarily *how much* more. So if your rival's yacht is 10 meters long, the important thing might be that your yacht be at least 11 meters long. Similarly, if his is 100 meters, it is paramount that yours be longer—but it doesn't have to be 10% longer for you to maintain the advantage; 101 meters is enough. This is convenient because it means that—to the extent that we covet this kind of ordinal social rank—the objective gains we make don't have to be proportional to our cumulative previous gains in order to remain significant. Small incremental gains can continue to be very attractive, so long as they have the potential to shift our rank in the relevant comparison group.[18]

Perfect or imperfect automation

Might we not work just because we enjoy working? Well, I won't count an activity as work if we do it *simply* because we enjoy it. But what if we enjoy it because it is useful? Well, then there needs to be some other reason aside from it being enjoyable, such as one of those we discussed. To repeat, the three types of consumption desire which could plausibly continue to motivate people to work even at very high levels of productivity and income were: to acquire novel goods and services that provide some noncomparative personal benefit; to accomplish ambitious social projects; and to acquire positional goods that help one gain status.

Theoretically, these could stave off the arrival of the leisure society indefinitely—ensuring, for better or worse, that "man's economic problem" never gets fully solved, and that the sweat of our brows continues to trickle.

*

Aye, but there's a catch! All the preceding discussion of whether people will continue to work rests on one assumption: that there would still be work for people to do.

More precisely, our discussion has presupposed that the income one could earn by selling one's labor remains significant compared to the income one derives from other sources, such as capital holdings and social transfers.

Recall the billionaire with the megayacht: no matter how badly he envies the decabillionaire's gigayacht, he would not continue selling his labor if the most he could make is the minimum wage or some other amount that is trifling compared to what he makes from his investments (or compared to what he can afford to spend for the rest of his life by slowly drawing down his savings).

*

Here we come to a juncture where we need to consider that the labor market impacts of advanced AI may be different and more transformative than those that result from even very large increases in productivity brought about by capital accumulation and technical advancements of the sort that Keynes envisaged in his essay.

Historically, labor has been, on net, a *complement* to capital. At the aggregate level, this has held true since the beginning of tool use and through all subsequent epochs of technological change and economic growth.

You all know what complements and substitutes are in economics, right? We say that X is a *complement* to Y if having more of Y makes extra units of X more valuable. A left shoe is a *complement* to a right shoe. If, instead, having more of X makes Y *less* valuable, we say that X and Y are *substitutes*. A lighter is a *substitute* for a box of matches.

Okay, it turns out that labor and capital have been complements. Each has enhanced the value of the other. Of course if we zoom in, we can see that some particular kinds of labor have become less valuable as a result of technical innovation, while other kinds have become more valuable. But the *overall* effect, so far, has been that labor has become more valuable than it used to be. This is the reason why wages are now higher than they were a hundred years ago or at any other time in human history.

So long as human labor remains a net complement to capital, growth in capital stocks should tend to drive up the price of labor. The increasing wages *could* then motivate people to continue to work just as hard as they currently do even if they become very rich, provided they have the kind of insatiable desires that I just described. In reality, permanently higher wages would probably cause people to work *a bit* less, as they would choose to use some of their productivity gains to increase leisure and some to increase consumption.

But in any case, the degree to which labor is a complement to capital is a function of technology. With sufficiently advanced automation technology, capital becomes a *substitute* for labor.

Consider the extreme case: imagine that you could buy an intelligent robot that can do *everything* that a human worker can do. And suppose that it is cheaper to buy or rent this robot than to hire a human. Robots would then compete with human workers and put downward pressure on wages. If the robots become cheap enough, humans would be squeezed out of the labor market altogether. The zero-hour workweek would have arrived.[19]

*

If we consider a less extreme scenario, the picture gets more complex. Suppose that robots can do *almost* everything that humans can do, but that there are a few tasks that only humans can do or that humans can do better. (This might include various new jobs that arise in opulent high-tech economies.) To determine the outcome for human wages in this scenario, we need to consider several effects.

First, as before, there is *downward* pressure on wages due to competition from robots.

Second, the economy in this full-bore automation scenario would most likely expand explosively, causing average income to shoot up. This would increase demand for labor, since higher-earning consumers would spend more on goods and services, including ones which we assumed only humans could produce. This increase in demand would create *upward* pressure on human wages.

Third, the increased average wealth in this scenario would likely reduce labor supply, since wealthier people would choose to work less at any given wage level. Such reduced labor supply would create *upward* pressure on wages.

Thus, there are at least these three basic effects: one that tends to depress wages, and two that tend to raise wages. Which of these effects dominates is not determinable a priori.

Therefore, whereas the effects of *perfect* automation technology are clear—full human unemployment and zero human labor income—the consequences of *imperfect* automation technology for human employment and human wages are theoretically ambiguous. For example, it is possible in this model that if robots could do every job except design and oversee robots, the wages paid to human robot-designers and robot-overseers could exceed the total wages paid to workers today; and, theoretically, the total number of hours worked could also rise.

We would have to make a whole bunch of particular and rather speculative empirical assumptions if we wanted to derive more specific implications from our model. At that point, we might as well start to disaggregate the impact of automation and look not at the total level of employment but at how individual sectors of the labor market would be impacted. No doubt, some occupations would do better and some worse in such scenarios of partial automation. But as none of that is particularly germane to our topic, we will leave it to our friends in the economics department to work out the details.

*

It is interesting, though, to glance at what happens to human wages and working hours if we start with an imperfect automation scenario and gradually transform it into one of increasingly perfect automation. If we consider a scenario in which automation technology is *very nearly* perfect—machines that can do virtually everything that humans can do, better and more cheaply, with only a few minor exceptions—then I would expect that humans would work only a little. People might work on average a couple of hours a week, doing the very few things that machines can't. As for labor *income*, however, we cannot even conclude that there is such an asymptotic convergence to the case of perfect automation. For it is possible that hourly wages could rise so steeply that even if people work only two hours a week, they might *still* make more money than they currently do in a forty-hour work week. (I think it is also theoretically possible, though empirically unlikely, for the *factor share* of labor to increase in such scenarios.)

*

Now you might wonder: What *are* the limits to automation? How close to perfectly *will* robots substitute for human labor? This is a key factor that will determine whether we end up in a Keynesian leisure society, or an even more extreme scenario in which humans are entirely out of work and in which we consequently will confront the full force of the purpose problem.

We'll get to that question. But before we do, I'd like to take a little detour to talk about how humans could make money even if the substitution were perfect and there were no jobs for humans. I mean, it's reasonable to wonder about income and not just purpose in an AI-driven full-automation future.

A simple three-factor model

Consider a very simple three-factor model in which economic output is produced by combining labor, capital, and what is commonly referred to as "land". Land here means any non-labor inputs that we cannot produce more of, so not just planetary surface area but other basic natural resources as well. We will consider an extreme scenario in which the share of income that goes to labor is zero: one in which, consequently, the combined factor shares of capital and land is one hundred percent.

Let's first consider what happens if we assume that there is no change in population, no technological progress, and no increase in land, but there is an unexpected shock, namely the sudden invention of cheap robots that are perfect substitutes for all human labor. We'll also assume a fully competitive economy with no monopoly rents, and we'll assume fully reliable property rights (and that the robots remain under human control).

We start with an economy of full human employment. Then the perfect robots are invented. This causes massive amounts of capital flow into the robotics sector, and the number of robots increases rapidly. It is cheaper to build or rent a robot than to hire a human. Initially, there is a shortage of robots, so they don't immediately replace all human workers. But as their numbers increase, and their cost goes down, robots replace human workers everywhere.

Nevertheless, the average income of humans is high and rising. This is because humans own everything, and the economy is growing rapidly as a result of the successful automation of human labor. Capital and land become exceedingly productive.

Capital keeps accumulating; so eventually land is the only scarce input. If you want to visualize this condition, you could imagine that every nook and cranny has been filled with intelligent robots. The robots produce a flow of goods and services for human consumption, and they also build robots and maintain and repair the existing robot fleet. As land becomes scarce, the production of new robots slows, as there is nowhere to put them or no raw materials with which to build them—or, more realistically, nothing for them to do that cannot be equally well done by the already existing robots. Non-physical capital goods might continue to accumulate, goods such as films, novels, and mathematical theorems.[20]

There are no jobs and humans don't work, but in aggregate they earn income from land rents and intellectual property. Average income is extremely high. The model doesn't say anything about its distribution.

Even though economic work is no longer possible for humans, there may continue to be wealth flows between individuals. Impatient individuals sell land and other assets to fuel consumption spurts; while more long-term-oriented individuals save a larger fraction of their investment income in order to grow their wealth and eventually enjoy a larger total amount of consumption. Another way to climb the wealth ranking in this steady state of the economy may be by stealing people's or countries' property, or by lobbying governments to redistribute wealth. Gifts and inheritances may also move some wealth around. And beyond these sources of economic mobility, there is always the craps table and the roulette.

*

This may all seem a bit wild?

But notice that if we replace "robot" with "farmer", what we have is not a bad description of most of human history.

At equilibrium, both farmers and robots earn subsistence-level income. In the case of farmers, this means enough bread to raise two reproducing children per couple. In the case of robots, it means the revenue generated by each robot equals the cost of its manufacture and operation.

In this analogy, the landowning aristocrats of the past correspond to the rich future human population, which, just like their historical counterparts, extracts rents from their landholdings.[21]

What allows the average income of the future humans in this model to rise above subsistence is the stipulation that the human population is capped. If the number of humans (like the number of robots) were permitted to grow freely, then average human income would fall to subsistence level (like the robot's income falls to their subsistence level) once the size of the human population attains its evolutionary equilibrium.

We would then have a situation in which there is a vast number of robots, a vast number of humans, very high world GDP, and mere subsistence-level average incomes. This would be essentially just a scaled-up version of the bleak picture of the world that Thomas Malthus presented.

*

This simple three-factor model makes a number of assumptions which can of course be questioned.

The assumptions that there is no technological progress and no increase in land are, I think, less rickety than might initially appear. I expect that the rate of economically relevant technological progress will eventually asymptote to zero (once most useful inventions have already been made). Land growth (from space colonization) will asymptote to a polynomial rate, since the volume of the sphere reachable from Earth by a given time is bounded by the speed of light. In the *very* long run, land growth will asymptote to zero, since the expansion of space means that sufficiently remote galaxies are forever unreachable from our starting point. But even during the long period in which a polynomial rate of land growth could be sustained, a decline of average income to subsistence can easily occur, since a population is able to grow at an exponential rate.[22]

The assumption that humans will remain in perfect control of the robots is definitely open to doubt, though it is not one that I intend to discuss in these lectures. If that assumption is relaxed, the result would either be the same as above except with a somewhat smaller human population and a somewhat larger robot population at equilibrium; or, in the case of a more complete failure of control, the human population could disappear altogether and there would be even more robots.

By the way, I should say that when I speak here of "the robot population" or "the number of robots", what I mean to refer to is the factor share of the automation sector in the economy. Rather than a population composed of some specific number of independent robots, it could all just be one integrated AI system that controls an expanding infrastructure of production nodes and actuators.

Another assumption in the simple three-factor model is that property rights are fully preserved and that there is, for example, no redistribution program or welfare system. And we haven't yet considered economic inequality within the human population. Let's poke some more…

(It might seem as if we are going on a bit of a tangent here, but if one is pondering possible futures that involve notions of sustainable abundance, it is useful to be aware of these considerations and constraints. It also helps us to explicate our past human condition, thereby providing a backdrop against which utopian aspirations will stand out in sharper relief. And it begins to illustrate the many and various ways in which the quest for a better world, and for utopia, is often paradoxical.)

Paradoxes of a Malthusian world

We often think of economic inequality as bad. In a Malthusian context, however, it appears to have a silver lining.

Given unrestricted population growth, inequality is the only way that at least some fraction of the population can enjoy consistently above-subsistence-level incomes. If one holds that it is intrinsically important that there exist at least a few people who enjoy the finer things in life, then such an unequal arrangement might be deemed better than one in which there exists a slightly larger number of people but where everybody has a "muzak and potatoes" life (to borrow a phrase from Derek Parfit).[23] Historically, there have also been instrumental benefits to having some rich folk around who could patronize the arts and sciences and create pockets of privilege, sufficiently isolated from the immediate struggle for survival, so that new things could be invested in and tried out.

*

You might think that, in the Malthusian equilibrium, average income would obviously be higher if there is inequality—since if there is no inequality, then *everybody* earns subsistence wages, whereas if there *is* inequality, then at least *some* people have above-subsistence incomes. But things are not quite so straightforward.

Consider that, where there is inequality, the classes that enjoy above-subsistence income—for example, the landowning elites—reproduce at above replacement levels. Some of their children must therefore leave the class they were born into and fall to a lower stratum. This trickle-down of population, from the higher classes to the lower, implies that average income among the lower classes is *below* subsistence level in the steady state; since otherwise, the total size of the population would increase. So in this model, the peasant class has below subsistence income, yet its numbers remain constant as it gets continuously replenished from above by the drip of excess progeny that is falling from the bottom stratum of the landowning elite.

(We could analogize the situation to that of a lump of ice floating in water. If we have a thin flat sheet of ice—perfect equality—all the ice crystals will be near the surface of the water: at the level of bare subsistence. If instead we have a tall and pointed shape of ice, an iceberg, then some parts could stick up high above the surface, enjoying economic plenty; but this necessarily depresses other parts of the ice, to income levels below subsistence.)

Inequality *could* however raise average income in the Malthusian equilibrium if we assume that the relationship between income and fitness is not linear. This is easiest to see if we consider an extreme example: a king and a queen who have an income 100,000 times larger than that of a peasant couple—yet the regal pair would not have 100,000 times more surviving children. So inequality probably *would* increase average income in the Malthusian steady state.

On the other hand, inequality might reduce average *well-being*, since a person's well-being is not proportional to her income but rather perhaps to the logarithm of her income or some other such functional form of rapidly diminishing returns. If the king and the queen gained some new tributaries and increased their income tenfold, their expected well-being would presumably increase by much less than 10x.

*

In reality, the Malthusian condition was only ever roughly approximated. It was frequently disrupted by exogenous shocks. Every now and then, a plague, a famine, a massacre would cull the herd, thereby increasing the land and capital available to each of the survivors. For a time, even the majority could then enjoy significantly above-subsistence incomes.[24] This improved comfort led to lower childhood mortality, causing the population to grow back to the point where land was again scarce enough to suppress the average farmer's income back to mere subsistence—or slightly below, given the existence of economic inequality.

*

What is it like to live in a Malthusian condition? The simple assumptions we've made so far do not allow us to derive any general statement about this.

For example, you could have a model of fluctuating fortune within a life, where an individual dies if at any point their fortune dips below a certain threshold. In such a model, an individual may need to have a high *average* level of fortune in order to be able to survive long enough to successfully reproduce. *Most* times in life would thus be times of relative plenty.

In this model, inventions that *smooth out* fortune within a life—such as granaries that make it possible to save the surpluses when times are good and use them in times of need—lead to *lower* average well-being (while increasing the size of the population). This could be one of the factors that made the lives of early farmers worse than the lives of their hunter-gatherer forebears, despite the advance in technology that agriculture represented. Those grain depots smoothed out consumption, enabling farmers to survive long enough to reproduce even when their average income over their lifetime was hovering just above subsistence. Without the ability to store food, average conditions would need to be pretty good in order for conditions during temporary downturns to still allow for survival.

It's not just granaries. Other forms of "progress", including social institutions such as welfare programs, which reduced variation, either across a population or within the lifespan of an individual, would, in the Malthusian condition, have a similarly paradoxical effect.

For example: *peace*. Consider an ideological development that favored more peaceful relations between groups and individuals: a doctrine of love thy neighbor; or improved norms for conflict resolution that allowed more disagreements to be settled through reasoned debate and compromise rather than by fist or sword. What could be more benign? And yet... such improvements may actually have had a *negative* effect on average well-being, by making the equilibrium one in which the deaths necessary to maintain the human population at a given size are produced by grinding poverty, chronic malnutrition, and physiological exhaustion, rather than by the occasional axe-through-the-skull among people who at other times live in ease and comfort.

Up and down on different timescales

In such a Malthusian world—the world of our ancestors throughout prehistory and most of history, and also of our brethren throughout the animal kingdom—many of our intuitions about what would promote general happiness are wrong.[25] As the witches declare, "Fair is foul and foul is fair".[26] And naive benevolence is confounded and perplexed.

We may, however, gain some understanding if we separate dynamics that unfold over different timescales.

The short term
Less time than it takes for the population to reequilibrate after a shock; a few generations. Better food storage and conflict resolution raise average welfare. Fair is Fair.

The medium term
This is the timescale implicitly assumed in our discussion above. A hundred years or so after some variance-reducing innovation, such as improved food storage, social welfare, or peaceful ideology, a new and less variable Malthusian condition is attained. In this new equilibrium, average welfare is lower than before. Fair is Foul.

However, the population is larger. So if you are a total utilitarian, you might be pleased with this tradeoff—provided, of course, that the average life in this condition is above the zero line (i.e. is at least worth living) and that the number of extra people now living extremely poor lives is large enough to compensate for the fact that everyone is living in even deeper poverty than their (already very poor) predecessors did.

The long term
Over the grander sweep of history, it looks like agriculture, food storage, and local conflict-resolution mechanisms (such as states) were on the path toward the Industrial Revolution. The Industrial Revolution is important, since from that point onward economic growth has been rapid enough to outpace population growth, allowing humanity to escape the Malthusian condition:

a very great blessing! Although we have only spent a few hundred years in this emancipated condition—and less than that in many parts of the globe—it has nevertheless shaped the life experiences of a significant and rapidly growing proportion of all humans who have ever been born. Of the roughly 100 billion humans who have ever lived, more than 10 billion have been post-Malthusian. Under standard demographic extrapolations, this figure would climb rapidly, since around 5% or 10% of all humans who were ever born are alive right now, and almost all contemporary human populations have been sprung from the Malthusian trap.[27] Thus, maybe 10% of human lives so far have been (or currently are) post-Malthusian; and this fraction is increasing at a rate of about 10 percentage points per century.

From this long-term perspective, Fair is again Fair. At least the past reforms and improvements that may have reduced medium-term average welfare were on the path toward something much better—a world where there are many humans yet few of them starve to death, and where the majority of people have access to at least the rudiments of a decent human existence.

The even longer term

And with respect to what we may term "the deep future"… well, the jury is still out.

I think you can make a case that wisdom and wide-scoped cooperativeness are the two qualities currently most needful to secure a great future for our Earth-sprouted civilization. I also think wealth, stability, security, and peace are better for wisdom and global cooperation than are their opposites. And so we should welcome advancements in these directions, not only because they are good for us now, but also because they are good for humanity's future.

This doesn't imply that *earlier* progress in these directions would have been good for humanity's future. Perhaps if my species had lingered longer in the "poor, nasty, brutish" conditions in which my forebears evolved into humans, before matriculating into the Industrial era, we would have evolved, genetically or culturally, to become "more human" than we now actually are?[28] Perhaps we came out of the kiln a little too soon? Maybe we would have been better conditioned for the final vault into the machine intelligence era if we had spent another few hundred thousand years throwing spears and telling tales around campfires?

Maybe, or maybe not. Little is known about these matters. We are still remarkably in the dark about the basic macrostrategic directionality of things.[29] Truly, I wonder whether we can even tell up from down.

Excellence

We should also note that even if we specify an income level, it is a further question what it corresponds to in terms of material welfare.

The answer depends on the socioeconomic context. Consider a hunter-gatherer who is a young, healthy, respected member of his band, who works for several hours a day hunting, crafting arrows and ornaments, cooking food, and repairing the roof of the family hut: he plausibly enjoys a much higher welfare than, let us say, an English child laborer in the early Industrial Revolution, who receives the same income as the hunter-gatherer (i.e. bare subsistence) but earns it by toiling in a coal mine for twelve hours a day while suffering from black lung disease.[30]

From material welfare, it is yet a further question what it corresponds to in terms of subjective well-being. Individual psychology has a huge impact here. Two persons can live in virtually identical conditions—have similar jobs, health, family situations, and so on—and yet one of them may be far happier than the other. Some people are by temperament leaden, anxious, or ill-at-ease; others, blessed with natural buoyancy, remain cheerful and untroubled even when their objective circumstances are quite dire.

Still another question is how income levels might correlate with various notions of "objective well-being" (also referred to as "flourishing", or "eudaimonia"): that is, not just how satisfied somebody is with their life or how pleasant their mental states are, but also how richly bestowed their life is with various putative objective goods—such as knowledge, achievement, beauty, virtue, friendship, etc.—which some philosophers claim contribute positively to how well somebody's life is going for them, and how prudentially desirable that life is. Some such conceptions of objective well-being might exhibit a nonlinear relationship with income; for instance, one in which very low incomes are associated with less objective well-being (because extreme poverty thwarts the development and use of human faculties) but where excessively high incomes might also be disadvantageous (because opulence breeds decadence and vice).

Consider, for example, a perfectionist view of what makes a life excellent. Perfectionist accounts come in different flavors; they may, for instance, locate value in the development of distinctly human capacities, or in high achievement in the moral, intellectual, artistic, or cultural realm, or more generally in the accomplishment or realization of the "best things in life". Depending on which version of perfectionism one embraces, one might place special emphasis, when gauging the potential of a utopian vision, on how well it scores in terms of producing great persons or in allowing the highest peaks of excellence to be attained.

It is unclear how, from such a perfectionist perspective, one should regard past progress toward peace, equality, and prosperity. On the one hand, it has given more people the basic material necessities and provided them with an opportunity to take a swing at greatness; on the other hand, it may have sapped the crazy motivation to do so. One is reminded of the famous lines uttered by Harry Lime in *The Third Man*:

> "You know what the fellow said—in Italy, for thirty years under the Borgias, they had warfare, terror, murder and bloodshed, but they produced Michelangelo, Leonardo da Vinci and the Renaissance. In Switzerland, they had brotherly love, they had five hundred years of democracy and peace—and what did that produce? The cuckoo clock."[31]

Words that Nietzsche might have been pleased to have written (although he himself rather liked hanging out in the Swiss Alps). And it would be fair to point out that many other places, aside from Italy under the Borgias, have had their share of warfare, terror, murder and bloodshed, without producing any Renaissances.

*

I think these kinds of perfectionist excellences and achievements do count for something.

However, I also believe that we have a tendency to overestimate their importance. Their appeal is strongest when we are looking at things from afar and from the outside—as though we were critics sitting in the audience and casting judgment on a stage play or a movie. In the spectator seat, we prefer a story that is full of excitement and crisis and conflict and great overcomings, rather than one in which all the characters simply get along in easy contentment.[32] But that is not the right perspective from which to judge a utopia. For the question is not *"How interesting is a utopia to look at?"* but rather *"How good is it to live in?"*.

Disequilibria

How am I doing for time, let me see; not great… Okay, now where was I?

Student: *The Third Man*—the cuckoo clock.

Bostrom: No, before that.

Kelvin: Labor automation within the three-factor model of production, and then the impact of granaries and other innovations on average human welfare over different timescales.

Bostrom: Right. So we were talking about a simple economic model in which robots can more efficiently do anything humans can do. Humans don't earn any income by working but derive income from land. This income would be very great, and walls could be made of sausages—proper vat-grown ones, we may assume.

But, again, there is a *timescale* associated with this conclusion.

Imagine that everybody lives in luxury, with incomes far above the subsistence level. This would mean that eventually—absent coordinated restrictions on population growth—the human population would expand to bring the average human income back down toward subsistence. If there is inequality, then pockets of privilege could remain within which some people enjoy above-subsistence incomes; but the regular person would fall into penury.[33] The era of abundance would be over, perhaps never to return. Just a flash in the pan, in a long dark night.

Any questions up to this point? —Yes, over there.

Student: Don't people have fewer children when they get richer?

Bostrom: Some do, and some don't. In this model, the future would mostly be populated by the descendants of those who choose to have lots of children, not those who limit their own reproduction. —Yes, you.

Another student: I thought the problem was that people are not having enough kids so that there won't be enough young folk around to take care of the elderly.

Bostrom: Well, that is the problem that some people are talking about now. Until not so long ago, people were instead talking about the problem of overpopulation. Overpopulation occupied the same slot in our collective awareness as climate change does today (cohabiting that place with nuclear Armageddon). For example, Paul Ehrlich wrote this tract, *The Population Bomb*. It was published in 1968, and sold more than two million copies. Very widely influential among the intelligentsia. Up until that point, the world population had been growing exponentially. Ironically, the same year that Ehrlich's bestseller came out, the trendline went into reverse, and ever since then world population growth has been decelerating—now we seem to be heading toward population collapse.

Student: So, now I'm confused—are you saying that overpopulation is the problem, or underpopulation?

Bostrom: Well, they both seem like they could be problems?

Student: But are there too many or too few? Which one is it that we should be worried about?

Bostrom: Maybe both? For example, there could be too many in one place, too few in another; too many at one time, too few at another.

Even if we just consider world population as a single variable, we could still worry that it would at some point veer off catastrophically in either one direction or the other. Like a ball rolling down a narrow beam: we can be confident that it will eventually fall off, even though we don't know whether the problem will be that it veered too far to the left or too far to the right.

Or, if I may offer you another metaphor, humanity is riding on the back of some chaotic beast of tremendous strength, which is bucking, twisting, charging, kicking, rearing. This beast does not represent nature; it represents the dynamics of the emergent behavior of our own civilization, the technology-mediated culture-inflected game-theoretic interactions between billions of individuals, groups, and institutions. No one is in control. We cling on as best we can, for as long as we can: but at any point, perhaps if we poke the juggernaut the wrong way or for no discernable reason at all, it might toss us into the dust with a quick shrug, and possibly maim us or trample us to death. It is an inherently risky and unnerving situation, not dull.

Another student: I think I see what you're saying. You're saying that we have a lack of control over the size of the population, so it could become either too big or too small?

Bostrom: Yes, there is a lack of such control. But the problem is both far more general and deeper than that. It is more general because it is not only the size of the population that is out of control, but a great many other critical parameters too—for example, our military armaments, our technology development, our pollution, our memetic ecology. And the problem is deeper because even if we created some global control mechanism for these things, such as a sufficiently empowered world government—then we would have to ask about the forces controlling that mechanism: how are *they* under control? Are different factions and ideologies and special interests vying for power over the tiller? What harmful or dangerous dynamics would result from that competition?[34]

Or suppose we placed control over the Earth into the hands of a single person or some other unified actor? Well, it isn't too hard to imagine how things could go wrong in that case.

The upshot is that if we want to postulate a utopian condition—which we will want to do here, so that we can explore the fascinating problems of purpose and value that would arise in such a condition—then it is not enough to stipulate a great increase in economic productivity. That might be a necessary condition, but it is definitely not sufficient. Nor is it enough if we also stipulate great across-the-board technological advances. What is also essential is that things fall into place nicely in the social and political spheres as well. Without progress in the way that our civilization governs itself, increases in our material powers could easily make things worse instead of better; and even if a utopian condition were attained, it would likely be unstable and short-lived unless, at a minimum, the most serious of our global coordination problems were also solved.

—We have another question.

Some student: It seems bad to put all eggs in one basket. Wouldn't it be better to regulate these things locally? Each country could have its own rules.

Bostrom: That does not generally work when there are global externalities. If one country unilaterally disarms, it would place itself at risk of being dominated by some other country that builds up its military. Or, in the case of the overpopulation problem that we were discussing, the externality takes the form of moral concern: if one country falls into the Malthusian trap, then it would be a problem for other countries inasmuch as they care about the welfare of the people who live there.

This, by the way, is one asymmetry between the problem of overpopulation and the problem of underpopulation: the former results in the existence of people who are badly off, whereas the latter results in the nonexistence of people who would have had good lives had they been born.[35] We are more likely to be troubled by the former than by the latter. So overpopulation might seem more likely to have moral externalities.

Another asymmetry between overpopulation and underpopulation is that the former, but not the latter, is an evolutionary equilibrium. Unless reproduction is regulated, one would expect that sooner or later some more fertile variants will arise, and they will then multiply until a Malthusian state is reestablished.

One could imagine that cultural or technological innovations will stave off this specter for a while. Perhaps computer games will become so compelling that we won't bother to reproduce much. But presumably *some* groups will choose to have kids. Maybe they decide that computer games are taboo. These groups, or the ones among them that also achieve sufficiently low egress rates of individuals leaving the group, would then be the inheritors of the future; and it would be their behaviors and values that would shape the longer-term population dynamic. Thus the world population would start growing again, and Malthus would be vindicated.[36]

The AI transition doesn't necessarily obviate this dynamic. Biological human populations could continue to grow exponentially, and of course populations of digital minds could also grow exponentially and with an even shorter doubling time. It looks like, ultimately, only global coordination could solve this problem, just like the body needs defenses against cancer that don't rely entirely on the kindness of cells. Likewise for the problem of war, and for various other problems that can arise from misdirected competition and optimization.

Economies of scale

In order to sustainably improve the living standards of animals, in the wild and in our society, population numbers must be controlled. You can help the hungry pigeons by feeding them. Then there will then be more pigeons next year, and more still the year after that. You cannot outrun it. But if the number of pigeons hatched is capped to the number of pigeons that die, then all the pigeons could sustainably enjoy above-subsistence existences.[37]

What is easy to see in the case of pigeons is harder to see in the case of humans, for several reasons (beside culturally specific blind spots):[38]

1. The human generation cycle is longer, so the dynamic unfolds over a larger timescale, making it harder to perceive.
2. Human culture gives human societies more degrees of freedom than pigeon communities. And it turns out that, initially, cultural phenomena—the demographic transition—inhibit human reproduction when conditions become plentiful. It can take many generations for cultural and genetic selection to overcome this initial inhibition.
3. Human economic productivity exhibits much greater economies of scale. Pigeons benefit from being part of flocks or colonies because they can learn about foraging locations and techniques and share the labor of watching for predators.[39] But humans can learn a virtually unlimited amount from each other across a vast range of economically relevant subjects. Humans are also able to benefit much more from division of labor. These vast quantitative differences in the economies of scale for humans and pigeons obscure an underlying qualitative similarity: that eventually, if technology stagnates, land must become the limiting factor of production.

The first of these is obvious, and we have already discussed the second one. I want to elaborate a little on the third.

Scale is important in economics. We can actually already see that scale is significant by looking at certain basic physical processes. For example, the volume of a container grows faster than the area of its enclosing wall. This simple geometric fact, which is known as the "square-cube law", has many implications. If you want to store some amount of stuff, it is cheaper (in terms of the amount of material you need) to store it in one large container than in many smaller ones. Similarly, thicker pipes are more efficient than thinner ones. So are larger vessels: losses from water resistance is lower, per unit of cargo, for bigger ships. Likewise, larger furnaces waste less of their heat. And so on. Running things at scale therefore tends to lower unit costs.

More importantly, larger social scale enables greater specialization, which increases efficiency. Consider the global supply chain that is necessary for producing a leading-edge microprocessor, and the myriad forms of specialized knowledge and equipment that it involves. There needs to be a large customer base to support all these fixed costs. A world population of a hundred million people may not be sufficient to make it both possible and profitable to produce all the required inputs.

Another important consequence of scale is that the cost of producing nonrivalrous goods, such as ideas, can be amortized over a larger user base. The more people there are, the more brains that can produce inventions—and the greater the value of any given invention, since it can be used to benefit more people.

So the larger the world population, the faster we should expect the rate of intellectual and technological progress to be; and hence also the rate of economic growth.

But this is not exactly right. We should rather say: the larger the world population, the stronger we may expect the *drivers* of intellectual and technological progress to be. The *actual* rate of progress would also depend on *how hard* it is to make progress. And that will vary over time. In particular, we may expect it to get harder over time, as the lowest-hanging fruits are picked first.

So there are two competing factors. The world population starts small: there are low-hanging fruits on the tree of ideas; but the total effort put into plucking these fruits is small. Later, the world population is much bigger: the low-hanging fruits are gone; but there is a much greater effort being applied to reaching the remaining fruits. A priori it is not clear which of the two factors should dominate. The model does not predict whether we should see accelerating or decelerating technological progress.

If we look at the matter empirically, we see that progress has in fact accelerated over macrohistoric timescales. When the human species first evolved, and for the ensuing hundreds of thousands of years, populations were small (maybe half a million), and progress was so slow that millennia came and went with basically no change in technology.

Then, with the agricultural revolution, the human population expanded and the rate of technological progress became much faster, the world economy now doubling about once every 1,000 years. This was a dramatic acceleration. But progress was still glacial by modern standards.

It was so slow, in fact, as to be imperceptible to contemporary observers.[40] It could only have been detected by comparing technological capabilities over long spans of time, yet the data needed for such comparisons—detailed historical accounts, archeological excavations with carbon dating, and suchlike—were not available. Ancient people's perception of history therefore did not recognize any trend toward technological advancement. As the historian of economic thought Robert Heilbroner observed:

> "At the very apex of the first stratified societies, dynastic dreams were dreamt and visions of triumph or ruin entertained; but there is no mention in the papyri and cuneiform tablets on which these hopes and fears were recorded that they envisaged, in the slightest degree, changes in the material conditions of the great masses, or for that matter, of the ruling class itself."[41]

To the extent that hypotheses of a macrotrend were entertained, they were usually based on a premiss of *deterioration*. We have here the idea of a "fall": an expulsion from a garden of plenty or a secular decline from an earlier imputed "golden age". The big arrow of history was seen as rusty and dipping downward. Or alternatively (for instance in the ancient Indian and Chinese traditions), it is seen as bending backward on itself, to form a cyclical conception of historical time, one in which living standards rise and fall in an ever-repeating undulation.

Such notions of finding ourselves on a downward slope might have reflected a dim collective memory or perhaps a primitive anthropological account of what had been lost in the transition from foraging to farming.[42] The story of a catastrophic decline in living standards consequent to the agrarian revolution can be read in the skeletal remains of those early farmers. Their bones show stunting and nutritional deficiencies compared to their paleolithic ancestors.[43]

This is, by the way, a good—and therefore a sad—illustration of the Malthusian dynamics that we talked about earlier: great economic growth, and it brought no improvement in average welfare, as the increased production was eaten up by an increased population. In fact, not only did it fail to improve the human condition, but people actually became worse off. The reason for the apparent deterioration in the quality of life might have been that conditions became less variable and/or that the most economically efficient dietary and behavioral patterns in the new environment became ones that were less fun, less nutritious, and less in accord with our biological nature.

So the idea of material progress is a surprisingly modern invention. Nevertheless, looking back, we can now see that a lot of technological advancement did take place over the ages, leading to a 200-fold increase in world GDP and world population—the two being essentially equivalent in the Malthusian condition—over the last 10,000 years until the onset of the Industrial Revolution; and then a further 100-fold increase in world GDP along with an 10-fold increase in world population—and thus a 10-fold increase in average income—from the outset of the Industrial Revolution to the present age. The doubling time for the world economy was around tens of thousands of years for hunter-gatherers; around a thousand for agriculturalists; and around thirty years for industrial-era humanity.[44]

For the past few hundred years, with many more humans around than ever before—woven together by commerce and communication into an interconnected world tapestry—inventions have been coming at breakneck speed. We tend to think of this condition as normal, but if we zoom out we see that it is the most remarkable anomaly. It is as if our civilization is a powder keg, and we are witnessing it at the exact moment of ignition.

Alright, let's take stock of where we are. We began by considering the most basic type of utopia, that of material abundance, and we looked at Keynes's famous forecast—

Running out of time

Student: Professor, somebody is banging at the door.

Bostrom: Oh, right. We're out of time. That must be the "Gastropods of Dagestan Region" class waiting to get in… Wow, those malacologists are really chomping at the bit. In case there's anyone among you who is not staying on for that, let's try to leave quickly. See y'all tomorrow!

To the baths

Firafix: Professor Bostrom, I'm sorry, we sort of, erm, crashed the lecture… Is there any chance that it would be okay if we audit the course even though we are not registered?

Bostrom: No, you must delete from your memory everything you heard and saw.

Firafix: —

Bostrom: Of course you're welcome to attend! I think I have a few copies left of the reading for tomorrow, if you want. It's from the *Feodor the Fox* correspondence. Have you read it? Gives the inside view of some of the things we talked about today. [*Roots around in backpack.*] Should be in here. Somewhere… Here! Thanks for coming, see you next time.

Firafix: Thank you!

Tessius: I have to run. Same time tomorrow?

Kelvin: I won't be able to make it. I have a funeral to go to.

Tessius: Oh, I'm sorry.

Kelvin: It's not somebody I knew. A friend of my father's, but he wants me to come along.

Tessius: I see. Well, Wednesday, then?

Firafix: Yes, I'm pretty sure we'll be going to all the lectures.

Kelvin: Okay, toodles.

Firafix: Bye.

Kelvin: And now: hot springs!

Feodor the Fox

Kelvin: That was good.

Firafix: I feel rested and relaxed.

Kelvin: And clean. Do you want to check out this *Feodor the Fox* thing?

Firafix: Yeah. Shall we go up that little hill? It would be a good place to read, and it looks as if there's some nice juicy grass.

Epistle XII

Dear Uncle Pasternaught,

Please forgive the longer than usual interval that has elapsed since my last letter. I am weighed down by guilt and remorse for having neglected our correspondence, even more so upon returning home to find several of your letters waiting for me, expressing ascending degrees of worry and concern for my well-being. I am so undeserving of such solicitude! I am very sorry for having caused you distress—a poor and shameful way indeed to repay the kindness you have showered on me. I can only hope that your generous heart will continue to take pity on my wretchedness, and that you will again overlook my defects as you have always done. You must know that whatever obligations you might once have felt toward me in remembrance of my father have long since been discharged, and any debts you might once have had have been repaid with usurious interest.

I will attempt to bring you up to date with my travails. You will recall the dark moods and troubled thoughts under which I was laboring; my stalled studies; my abandoned efforts to learn composition; my entirely futile philosophizing. Well, strange things have happened to me since the last letter. I have been on a journey—both in the geographical sense and spiritually.

I am not going to be able to recount all its twists and turns, which would anyway be unworthy of your attention. I will merely try to sketch its general outline, a few landmarks—some whose details have etched themselves so definitively in my memory that it is almost as if I could see them right in front of me now if I look up from this sheet of paper…

It started a few days after the reunion. The broody cogitations would not leave me any peace. I paced the room, I sat down and stood up again. I tried to compose but my thoughts had appointments elsewhere and declined to come to the party: the sheet remained empty. The questions that concerned me kept swirling about, but I was unable to make the slightest progress. I wondered why I had been made with a soul that had the capacity to wonder but not the capacity to find out; why I could see so much that was wrong while seemingly being unable to do anything about it; and why I was a fox and not a worm or a duck; why I was alive now and not at some other time; and why indeed there was anything at all rather than it being the case that nothing ever existed, no forest, no Earth, no universe, which it seemed to me would have been a far more natural condition, not to mention one that would have saved everybody a great deal of trouble. With such imponderables was I preoccupying myself. And I could not put it down, not put it to rest.

One morning, following a night during which I had scarcely been able to get a few moments of sleep, I came to a resolution: since I could not work things out myself, I would have to seek help from somebody else—it was the only course of action that had the slightest chance of success. Not a large chance, I thought, for where would I find somebody to converse with about these things let alone one who had understood everything and who could explain it to somebody with as limited an intellect as mine? The prospects did not seem good, but remaining at home felt like it was not an option at all.

So the following day I set out. My plan was to seek out the old crow who lives near the oak tree on south moor and ask if she knew somebody I could talk to. I found her easily, but she said she did not know any wisemen or sages. However, she recommended that I go and speak to Egon the Beaver. She told me that he knows a lot of the waterfowl who come and go by his lake and he gossips with them. As a result he has acquired a network of acquaintances that spans the forest and beyond. It is said that he even has many friends that live in foreign lands, far, far away.

So I went to see Egon, and indeed the old crow had been right! Egon said he did know somebody—or rather, he had heard of somebody—who was supposed to have the finest intellect perhaps in the entire world. Pignolius the Pig was his name, and his wisdom was widely renowned. Upon hearing this, I was so excited I could hardly bring myself to ask where he lived. What if he lived too far away, in another country even? I felt my heart pounding with fear: imagine knowing that there existed this creature, who could possibly help me with my quest and explain to me what it was all about—and yet also knowing I would never be able to reach him! It was an almost unbearable thought. I must have had my mouth open for many seconds before I managed to whisper the single word: "Where?".

Now imagine my joy and relief when Egon told me that he lives quite far away but not too far! It would be a long walk, about twenty days, but *it was possible*. He was within my radius! I thanked Egon with all my heart, and I said that if he met anybody who was going in that direction he could ask if they might bring word to Pignolius that somebody was on his way to see him, so that he could be forewarned of my approach. I felt that since I would be coming uninvited, it would be courteous to at least give some advance notification. I didn't know how he might react to a stranger turning up at his door unannounced—maybe I would be turned away?

The following weeks were physically strenuous. I lost considerable weight, and my feet and legs were sore from walking. But my soul felt strangely at ease—a feeling I had not experienced in a long time. Although the journey was challenging, I felt convinced that I was doing the right thing, the necessary thing. I wasn't second-guessing myself. I had a *purpose*—only an interim, temporary one, but a real purpose all the same. It is amazing what a difference that makes.

Eventually I reached the area which Egon had pointed me to, and I began asking around if anybody knew how to get to the place where Pignolius lived. This was not hard, everybody seemed to know who this pig was. Soon I was walking down a little path and there he was, right in front of me! Having a mud bath! I feared that I had arrived at an inopportune time and was preparing to turn back, but he seemed entirely unfazed by my approach even though he saw me coming. I wasn't sure whether to come or go, and as a result chose the worst option of all, the awkward compromise: I simply stood there staring at him, my jaws agape.

How long this embarrassment lasted, I do not know and don't care to remember. After a period, Pignolius called out to me to come down. Then I approached, and had the following conversation. The words I think are close to those that were spoken; my memory usually keeps good records, and this episode I have rehearsed to myself more times than I can count.

Feodor: Esteemed Pignolius, I come from afar to seek your advice. Here is a small gift for you. Will you grant me the incredible privilege of being allowed to ask you a few questions? I have heard that you are a pig of great wisdom.

Pignolius: Oh, very great wisdom. And very insufficient. But the chestnuts are gratefully received. You can throw them in here.

Feodor: There?

Pignolius: Jawohl! Throw them in!

Feodor: To respect your valuable time, I will get straight to the point. I have seen that there is so much awry with the world, so much suffering… a tiny drop of which recently happened to fall upon my own lot, but… well, I feel it can't go on like this. People are dying, getting sick, starving, being chased and eaten, enduring all kinds of hardship and privation. I want to dedicate myself to doing something about it. But, I need a plan—plan is not the right word: an idea, some principle, vision, a direction I could follow that will at least offer long-term hope for arriving at a better condition. Please, Pignolius, shine your wisdom on my sorry pelt, and tell me: What can I do to make the world better?

Pignolius: Not much.

Feodor: But there must be something.

Pignolius: That thought did occur to me once, in my youth, yes.

Feodor: And?

Pignolius: Fortunately, upon reflection, it turned out that there wasn't much I could do; and I suspect the situation would be the same in your case.

Feodor: Fortunately?

Pignolius: If there had been a lot I could do, I might have felt compelled to do it. No doubt that would have required hard work and sacrifice. But fortunately it turned out that I was, in the scheme of things, almost entirely powerless. *Ich danke Gott an jedem Morgen, daß ich nicht brauch' für's Röm'sche Reich zu sorgen!*[45]

Uncle Pasternaught, I was dumbfounded. At first I was not sure which was more shocking: that the greatest mind known to us, an intellect far surpassing my own, thought there was nothing much that could be done to make the world better—or that he seemed pleased about this being the case!

I continued, stumblingly, seeking to regain my footing:

Feodor: But—but, what hope is there then? What is there to live for?

Pignolius: This mud bath is very nice. Just the right lukewarm temperature. Very good for your skin as well.

Feodor: But there must be something more!

Pignolius: Well yes, I must say that I am also quite fond of Porcelain, especially certain parts of her… But you know she can be a bit much sometimes. Whereas this mud is always great, except during winter. And a chestnut never fails to please. Nam nam nam nam oh yes!

Oh Pasternaught, now the great thinker was gorging himself on the chestnuts I had brought him, and probably an equal amount of dirt the way he was biting into them right there where they lay, in a pool of muddy water. It was all too much for me. I quickly thanked him for his advice, and departed.

The next thing I remember is wandering alone in the night. A cold wind was blowing through everything. I could hear it howling in the dark swaying treetops. It was as if the world was moaning and groaning, twisting and turning, and reaching desperately around—for something—for a solution that didn't exist.

I thought about all the creatures in the world in travail, and I felt sad and downbeat. But when I thought about people who were managing to find some enjoyment in life—a peaceful dinner with their family—that's when tears started rolling down my snout: so hopeless seemed their efforts to make a nice thing in this world, so touchingly naive; and their situation even more precarious, because they had something to lose.

It seemed to me that the world was restlessly twisting and turning, and protesting against its own existence, and I felt a deep compassion for all things living. I wanted to wrap my little furry body around them to keep them warm. I wanted to bring them comfort and good news.

As I was thinking these thoughts, cold and hungry and miserable, my steps were bending back toward Pignolius's place. It was not because it seemed like a good idea to go there, or indeed that there was any reason to do so whatsoever—there was nothing I intended to do there—but I could think of no alternative. There was nowhere else to go. When I got there, I lay down outside his doorstep and, exhausted, fell asleep.

It must have been close to noon when I woke up, for the sun was high up and its rays carried warmth. As I began to bestir myself, Pignolius came up to me and said, "You came back".

"I thought maybe I could ask you some more questions," I responded.

The truth is that I had not thought anything. But reasons easily pop into one's mind whenever their absence would be awkward, and they're out of our mouths before we know it.

Pignolius: I'd be happy to talk. But first, how about we have some lunch. I have some nice carrots here.

I gratefully accepted the offer. Never have carrots tasted better.

After we'd finished the meal, the conversation continued:

Pignolius: So?

Feodor: I want to apologize for my impetuousness yesterday. I came unbidden, without any right to impose on you. I asked you a question which you kindly answered. But I didn't like the answer and rushed off in a huff, full of sadness and self-righteous indignation. Now I've come back to request that you elaborate on your answer and explain why it's impossible to improve the world. It is an unreasonable request, but I'm at my wits' end.

Pignolius: I didn't say it was impossible. I said there did not seem to be much that you or I could do. But not much is not the same as nothing. For example, I think you made the world a bit better by returning so we can carry on the conversation!

Feodor: My benefit is large enough to outweigh your disadvantage?

Pignolius: The benefit is mutual, I think.

Feodor: You are generous. But on a larger scale, then—at a structural level—why do you say that we can't improve the world?

Pignolius: How do you know where your wits' end?

Feodor: What?

Pignolius: Your wits. You said you were at their end. How do you know where that is?

Feodor: ?

Pignolius: Okay, let's say you are facing some problem—

Feodor: Let's say.

Pignolius: You're facing a problem which you don't know how to solve. You've tried many things and none has worked, and you can't think of anything else to try. So you are there, at your wits' end. Right?

Feodor: Yes, I'm there.

Pignolius: But how do you know you won't think of something new to try tomorrow?

Feodor: Well, I guess I can't be absolutely certain. But realistically I don't think it's going to happen.

Pignolius: Why?

Feodor: Induction, I suppose. I mean I've tried hard to think of a solution for a good long time, so it seems unlikely that I would succeed tomorrow where I have failed all these past days. That's why I've—

Pignolius: Wait… hang on… Ha, *ja*! Chestnut! Must have been left over from yesterday. Nam nam nam. So good! Sorry, you were saying?

Feodor: My past attempts to solve the problem have been unsuccessful—and I've tried quite hard. That's why I'm pessimistic about my ability to do this alone, and why I've come to seek your advice.

Pignolius: Well yes. But the reason you've identified for thinking that you can't solve the problem yourself can be generalized to a reason for thinking nobody else can solve the problem either.

Feodor: How?

Pignolius: If it could be solved, wouldn't somebody already have solved it? Consider the odds. In the eons that this forest has existed, and in all the generations of foxes and pigs and other animals that have lived here, surely the thought must have occurred to folks now and then that it would be nice if we could fix the world and set everything aright?

Feodor: That seems likely.

Pignolius: It is certain. And among those who got this idea, some would surely have tried to act on it, right?

Feodor: Right.

Pignolius: And we now observe the result of all those attempts: the world is—*still broken*! So why in the name of the gambling monkeys would you think your own attempts—or our joint attempts, or anybody's attempts—would fare any better?

Feodor: I admit, the odds do not seem good.

Pignolius: They do not seem good. Maybe we should set the probability of success to be approximately one over the number of previous attempts that have failed?

Feodor: How many previous attempts have there been?

Pignolius: I assume it goes back infinitely.

Feodor: I had always assumed so as well; but what do you make of Rees the Weasel's new theory?

Pignolius: What new theory? He's got a new theory?

Feodor: He has supposedly discovered that the world began a finite time ago.

Pignolius: What!?

Feodor: Of course he lives across the river, so it's impossible to visit him. But I heard some tweets about his discovery from a bird who had been over. It's made quite a stir amongst the ravens on the other side, apparently. Rees has a very strong reputation—

Pignolius: I know, I know—he's discovered that the world started a finite period of time ago?

Feodor: So it is said.

Pignolius: What—how—what is his reasoning? How did he come to this conclusion?

Feodor: I don't know. The bird couldn't remember any other details. It was a swift.

Pignolius: If there has only been a finite period of past failure, then our hopes for future systematic improvement face only finitely bad odds rather than infinitely bad odds, which might gladden you. How long ago does he say the world began?

Feodor: A very long time ago, but I don't know his exact estimate.

Pignolius: Roughly?

Feodor: All the swift could say was "a very long time ago but finite".

Pignolius: If Rees is right, we may need to rethink everything. But it's also possible that some error of interpretation might have crept in during this rather tenuous line of transmission. It's frustrating that we don't know any specifics.

Feodor: Do you think that, if we could learn more details about this, it would help us find some way to improve the world?

Pignolius: Who knows.

Feodor: I have an idea. Suppose we pooled our labor and accumulated a little surplus. Then we use that to hire a good bird to fly over there with our questions and come back and report to us.

Pignolius: You mean giving up food for knowledge?

Feodor: A little goes a long way for a bird.

Pignolius: Oh I don't know; they eat more than you'd think. Feodor, have you ever wondered why there are so few of us?

Feodor: What do you mean few of us?

Pignolius: Few of us who are interested in these kinds of things—truth, goodness?

Feodor: Yes it seems strange that everybody is so uninterested, but by now I've mostly come to take it for granted.

Pignolius: *This* is why there are so few of us. To give up food for knowledge, to squander one's energy on abstract fancies! Those who engage in such perversion: their death rates are higher and their birth rates are lower; they get outcompeted, marginalized, and eliminated. They are temporary blunders, self-correcting errors of nature. —Let's do it!

Dear Uncle Pasternaught, at this I knew I had found a kindred spirit, and that whatever the outcome of our future initiatives, whether they would meet with success or failure, my long and arduous journey to come to this valley had not been in vain.

I will try to write you again soon, though I think the coming days will be busy ones. These are strange and wondrous times.

I remain, as always, your most deeply indebted nephew,

Feodor

Outro

Firafix: What do you think?

Kelvin: I liked it, although it's not super-clear how it is related to the lecture.

Firafix: Maybe a connection will appear in the next epistle? But it is getting dark, and I think we had better make our way home.

Kelvin: Yes, let's go.

TUESDAY

A stay of exequies

Firafix: Hello, Tessius.

Tessius: Hey! Kelvin, I thought you were going to a funeral?

Kelvin: It was canceled.

Tessius: Really?

Kelvin: Yes.

Tessius: Hallelujah?

Kelvin: The hearse had a flat tire. It's been rescheduled for Thursday afternoon.

Tessius: Well, at least you won't have to miss any of the lectures.

Firafix: He looks handsome in that suit, doesn't he?

Tessius: Fitting for a man of such dark thoughts.

Recapitulation

Bostrom: Let's begin. I see many new faces here today, so maybe it would be good to start with a quick recap.

We began yesterday by observing that simple post-scarcity utopias, which present a vision of material abundance, relaxation, and social license, have held strong appeal among immiserated hard-working folk, as evidenced by the popularity of the European medieval peasant fantasy of Cockaigne and many other tales of golden ages, gardens of delights, and island paradises.

We then made the obligatory reference to the John Maynard Keynes article that predicted that a 15-hour work week would be nearly upon us by now, following a century of strong economic progress. However, while productivity has risen in line with Keynes's projection, this has resulted in only a moderate extension of leisure hours. Greed has mostly held the line against Sloth.

We then identified three types of consumption opportunities that could, theoretically, delay the onset of the leisure society indefinitely, even if hourly wages continue to rise.

First, new market goods might be invented that are not subject to steeply diminishing returns. We mentioned the possibility that expensive bio-enhancements could provide individuals with substantial benefits even at very high expenditure levels. An income of 500,000 a year might not suffice for the shiniest enhancements. And for digital minds, one way to convert almost unlimited economic resources to personal welfare could be via hardware upgrades.

Second, some people might be interested in ambitious impersonal projects, such as wild animal welfare programs. [*Applause.*] These could absorb a great deal of capital.

And third, we could derive a never-ending motive for more income from relative preferences such as a desire for social status. We observed that those gains don't have to be proportional to the cumulative previous gains in order to remain significant, if what we crave is ordinal rank. I also remarked that if we wanted to save ourselves a lot of effort, we could try to coordinate to discourage status competition. Alternatively, we could coordinate to redirect our competitive urges away from arenas that have negative externalities, such as military contests and wasteful forms of conspicuous consumption, and toward arenas that are neutral or generate positive externalities, such as effective charity and certain kinds of entrepreneurial, moral, and intellectual achievements.

So, in principle at least, these three types of consumption could prevent us from running out of things to spend additional money on. But there is another way that the leisure society could come about, besides us getting too rich to work—namely, if there are no jobs for us to do. Of course, this becomes a serious possibility only if there is dramatic progress in automation technology.

We noted that, historically, capital has been a net *complement* to labor, meaning that human labor has become more valuable as capital stocks have increased. But it is conceivable that, with sufficient progress in machine intelligence, capital will become a net *substitute* for human labor.

In the extreme scenario, where machines can more cheaply do everything humans can do, the machine capital stock would accumulate and the human workforce would be pushed out of the labor market.

In the less extreme scenario, where a few tasks remain that only humans can perform, the situation is more complex. The impact on human wages would depend on the balance of several opposing forces—downward wage-pressure from competition from robots, and upward wage-pressure arising both from stronger demand as a consequence of economic growth and from a reduction in the labor supply as a consequence of rising non-wage income. The resultant of these forces cannot be determined a priori.

We can expect that as we approach the extreme case of perfect machine substitutes, humans will work less and less. It remains possible, however, that the amount of *labor income* earned by humans could increase even as hours worked decrease, since wages might go up a lot in such scenarios.

We next looked at what happens in a simple three-factor economic growth model if we introduce robots that are a perfect substitute for human labor (while assuming that the size of the human population remains constant). There are more and more robots. Humans stop working but continue to earn income (from ownership of land and intellectual property). Average income grows to extremely high levels. The model says nothing about how wealth and income are distributed. There could still be wealth flows between individuals after humans have stopped working.

When I said that, in this model, average income grows to extremely high levels, I meant *human* income. But if the devices doing the work in this scenario are very sophisticated, it is possible that we should not think of them as mere machines but instead as a new kind of laborer, and that we should also consider the welfare of these digital minds. Although I went off on several tangents last time, I did resist the temptation to expound on the moral and political status of digital minds.[46] Well, let me state that I think this is an important topic and I believe that some types of digital minds could have moral status—potentially very high moral status. However, we must reserve that for another time.

With unrestricted population growth in the robot (or digital mind) population, it would reach a Malthusian condition—analogous in some ways to that of past human farmers. We made a discursion to reflect on the nature of this Malthusian condition, including the roles played therein by inequality and economic mobility. Remember, this was approximately the condition not only for most humans throughout history but also for most animals in the wild. It is useful to understand the basic elements of this natural condition if we are to evaluate not just how things have been but also the constraints within which a future utopia would need to be carved out.

In particular, we remarked on the paradoxical nature of progress in a Malthusian condition: how such prima facie beneficial things as equality, stability of food supply, peace, and first aid may have had a net negative effect on average welfare, at least within a medium timeframe—on a scale of several generations. On a timescale shorter than that, such progress did indeed benefit individuals (at least if we assume that their lives were worth living and worth saving). And on a longer timescale, we can now see that developments in these directions were on the path toward the present much more prosperous condition that we've been enjoying post the Industrial Revolution. On an even longer timescale—well, the jury is still out! We are still quite in the dark even about the basic directionality of things.

Okay. We then noted the somewhat wobbly link between income and well-being. In rich countries, the correlation is fairly weak. Native temperament appears to be a stronger determinant of how much a person will enjoy her life.[47] Furthermore, what kind of lifestyle corresponds to a given level of income depends on the social and wider economic and technological circumstances. A subsistence hunter-gatherer, for example, might be significantly better off than a subsistence farmer living at the same income level. And the link between income and outcome remains very wobbly if instead of subjective well-being we use a more virtue-based or perfectionist standard for evaluating how well somebody is doing.

Now what happens if we drop the assumption that the size of the human population is constant? Demographers point to a demographic transition that occurs at a certain point in a country's development, a shift toward lower birth rates and lower death rates. I pointed out, however, that elementary considerations from evolutionary biology suggest that, in the longer run, human populations will start to grow again, as higher fertility is selected for. As long as conditions remain above subsistence, this population growth can be exponential. And of course, AI populations could grow with an even shorter doubling time, so they would approach the Malthusian condition at warpspeed.

Therefore, if one wishes to avoid the long-term fate of a return to the Malthusian condition, population growth would need to be restricted. I argued that this would require global coordination. The alternative, if we were content with having higher-than-subsistence income in only some part of the world or for some segment of society, would involve the persistence of extreme undeserved inequality, where the people at the unfortunate end often starve to death or die from easily preventable causes.

While this Malthusian condition is what we sadly see basically everywhere in nature, one might hope that a larger population size would play out differently in the human case (and also in the case of advanced AIs) than it does for animals and plants, because of increasing returns to scale. There being more people in the world can have a positive effect on economic growth rates—more ideas are generated, there are more people to trade with, and so forth.

Could it then be that we could have unrestricted population growth accompanied by unlimited economic growth, producing a rise not just in world GDP but also in per capita income? This is the happy condition that we've experienced in the last couple of centuries, and it has shaped our modern view of progress. However, while we can probably continue to ride this rocket for a while, eventually depletion effects will dominate scale effects. Technological inventions will become harder to make, as the lowest-hanging fruits are picked; and land (resources we cannot produce more of) will become scarce. Even space colonization can produce at best a polynomial growth in land, assuming we are limited by the speed of light—whereas population growth can easily be exponential, making this an ultimately unwinnable race. Eventually the mouths to feed will outnumber the loaves of bread to put in them, unless we exit the competitive regime of unrestricted reproduction. (Please note that this is a point about long-term dynamics, not a recommendation for what one country or another should be doing at present—which is an entirely different question altogether.)

Our cosmic endowment

We've seen that the long-run *rate* of growth is limited because it will eventually be constrained by the availability of land, which can only grow at a polynomial rate because of the speed-of-light barrier, which limits how fast a civilization could expand in space even under optimal conditions. Since the size of the population is bounded by the size of the economy, this implies that the long-run rate of population growth is also at most polynomial. And the *desirable* rate of population growth is lower than the maximal rate of population growth, if we want people to enjoy above-subsistence lifestyles.

But so far, this only shows that we need *patience*. It is an argument about how fast things can grow, not an argument about how big they could ultimately become.

However, we do also have reason to think that limits exist on the finally attainable size of the economy, at least if we assume that our current physical and cosmological theories cover all the relevant bases. This is what we will be talking about today: ultimate boundaries—not just to "the economy" narrowly conceived, but more generally the boundaries of technology and the boundaries of an ultimate utopian condition.

At the simplest level, given the observed positive cosmological constant, general relativity implies that the volume (in comoving coordinates) of the universe that is accessible from our current spatiotemporal location is finite, and the amount of matter contained within this volume is also finite.[48] And decreasing: with each passing year our civilization remains in its starting blocks on planet Earth, approximately three more galaxies glide out of the previously affectable universe and are lost to us forever.[49]

It doesn't strictly follow from this that the economy has a maximum possible size. Although the amount of stuff that could be produced is finite, it is possible to conceive of some dimensions along which an aggregate measure could continue to grow indefinitely. For instance, if we imagine a being whose utility is a function of *how far apart things are*, that being's utility may continue to increase without bound, as the spatial fabric of the universe continues to stretch at an accelerating pace. Slightly less preposterously, we may consider a being whose utility is a linear function of the total amount of (some kind of) information that has been accumulated by our Earth-originating civilization—and perhaps the memory capacity in the accessible universe is unbounded, if cosmic expansion enables spatial encoding schemes to store an indefinitely increasing number of bits; although there might be reasons this wouldn't really work in the truly long run.

If, however, we measure the size of the economy by a more natural standard—either by reference to typical human preferences or by reference to the economy's ability to produce bundles of familiar types of goods and services—then there does seem to be a finite limit to growth. In reality, this point will never actually be reached; but if we are lucky, we might achieve a series of decreasingly imperfect approximations, culminating in a plateau, which might last for a long time, before the heat death of the universe, if nothing else, eventually puts a stop to the proceedings. In principle, our civilization might last for billions of years (and yet, to an immortal—a sandcastle built in the afternoon and swept away by the evening tide).

If we wanted to pursue this line of investigation further, we could ask, for example, how many computational operations could be performed with these physical resources that an Earth-originating civilization could reach and mold over the lifetime of the universe, or how many bits could be stored and erased using these resources. From there, we could estimate the number of sentient minds, of some given size, that could be created, and the total number of subjective life years that those minds could experience. I actually did this in one of my earlier books, and I've put the relevant part in a handout in case anyone is interested.

HANDOUT 1

THE COSMIC ENDOWMENT[50]

Consider a technologically mature civilization capable of building sophisticated von Neumann probes. If these can travel at 50% of the speed of light, they can reach some 6×10^{18} stars before the cosmic expansion puts further acquisitions forever out of reach. At 99% of c, they could reach some 2×10^{20} stars. These travel speeds are energetically attainable using a small fraction of the resources available in the solar system. The impossibility of faster-than-light travel, combined with the positive cosmological constant (which causes the rate of cosmic expansion to accelerate), implies that these are close to upper bounds on how much stuff our descendants acquire.

If we assume that 10% of stars have a planet that is—or could by means of terraforming be rendered—suitable for habitation by human-like creatures, and that it could then be home to a population of a billion individuals for a billion years (with a human life lasting a century), this suggests that around 10^{35} human lives could be created in the future by an Earth-originating intelligent civilization.

There are, however, reasons to think this greatly underestimates the true number. By disassembling non-habitable planets and collecting matter from the interstellar medium, and using this material to construct Earth-like planets, or by increasing population densities, the number could be increased by at least a couple of orders of magnitude. And if instead of using the surfaces of solid planets, the future civilization built O'Neill cylinders, then many further orders of magnitude could be added, yielding a total of perhaps 10^{43} human lives. ("O'Neill cylinders" refers to a space settlement design proposed in the mid-seventies by the American physicist Gerard K. O'Neill, in which inhabitants dwell on the inside of hollow cylinders whose rotation produces a gravity-substituting centrifugal force.)

Many more orders of magnitudes of human-like beings could exist if we countenance digital implementations of minds—as we should. To calculate how many such digital minds could be created, we must estimate the computational power attainable by a technologically mature civilization. This is hard to do with any precision, but we can get a lower bound from technological designs that have been outlined in the literature. One such design builds on the idea of a Dyson sphere, a hypothetical system (described by the physicist Freeman Dyson in 1960) that would capture most of the energy output of a star by surrounding it with a system of solar-collecting structures. For a star like our Sun, this would generate 10^{26} watts. How much computational power this would translate into depends on the efficiency of the computational circuitry and the nature of the computations to be performed. If we require irreversible computations, and assume a nanomechanical implementation of the "computronium" (which would allow us to push close to the Landauer limit of energy efficiency), a computer system driven by a Dyson sphere could generate some 10^{47} operations per second.

Combining these estimates with our earlier estimate of the number of stars that could be colonized, we get a number of about 10^{67} ops/s once the accessible parts of the universe have been colonized (assuming nanomechanical computronium). A typical star maintains its luminosity for some 10^{18} s. Consequently, the number of computational operations that could be performed using our cosmic endowment is at least 10^{85}. The true number is probably much larger. We might get additional orders of magnitude, for example, if we make extensive use of reversible computation, if we perform the computations at colder temperatures (by waiting until the universe has cooled further), or if we make use of additional sources of energy (such as dark matter).

It might not be immediately obvious to some readers why the ability to perform 10^{85} computational operations is a big deal. So it is useful to put it in context. We may, for example, compare this number with our earlier [in *Superintelligence*] estimate that it may take about 10^{31}–10^{44} ops to simulate all neuronal operations that have occurred in the history of life on Earth. Alternatively, let us suppose that the computers are used to run human whole brain emulations that live rich and happy lives while interacting with one another in virtual environments. A typical estimate of the computational requirements for running one emulation is 10^{18}

ops/s. To run an emulation for 100 subjective years would then require some 10^{27} ops. This would mean that at least 10^{58} human lives could be created in emulation even with quite conservative assumptions about the efficiency of computronium.

In other words, assuming that the observable universe is void of extraterrestrial civilizations, then what hangs in the balance is at least 10,000,00 0,000,000,000,000,000,000,000,000,000,000,000,000,000,000,000,0 00,000 human lives (though the true number is probably larger). If we represent all the happiness experienced during one entire such life with a single teardrop of joy, then the happiness of these souls could fill and refill the Earth's oceans every second, and keep doing so for a hundred billion billion millennia. It is really important that we make sure these truly are tears of joy.

I should emphasize that the numbers in this handout are premissed on what we may call "the naive picture" of our situation. In particular, this involves the assumption that the relevant parts of our current physical theories are correct (concerning the speed of light, the thermodynamics of computation, the positive cosmological constant, etc.), which is maybe not so implausible if we are living in the basement universe—but is totally up for grabs if we are living in a simulation. If we are in a simulation, it might be set to terminate *long* before the heat death of the simulated universe; and what appears to us to be distant stars and galaxies might simply be realistically rendered illusions, but without any actual "there" there.

Another important premiss is that all this stuff is ours to claim. One obvious way in which this could fail to be true is if aliens already occupy much of the astronomical petri dish, or if they will have done so by the time we get our act together and our probes arrive at the various destinations.[51]

Another way that this premiss could fail to be true is if others, while not in physical possession, have legitimate moral claim or legal title to much or all of this stuff.

"How could that be true" you ask? Well, do *you* have a degree in cosmological jurisprudence or interspecies constitutional law? Neither do I. In which case we should probably be a bit modest in how much we purport to understand yet about the statutes and regulations that control things at these scales. The multiverse may not be governed by a principle of *Occupatio*. Instead, we might be more like explorers who, even if the lands we find are truly uninhabited, do not thereby come into possession ourselves, but rather acquire these regions for a greater sovereign or for some cosmopolitan authority or cosmic host. And perhaps we incur an obligation to administer the found regions in the interests and according to the wishes of this presidium, rather than exclusively according to our own inclinations.[52]

Technological maturity

Alright, so let's say we've got this big pile of resources. Now what can we do with them?

At present, our options are quite limited. There's a great deal we simply do not know how to do. For example, we can't make a panacea, even though it would presumably not require many resources to manufacture if only we knew how. We also face tight tradeoffs between the different things that we *do* know how to make—budget constraints force us to pick a few things from a long list of desirables.

But the question that I want to pursue is what will *ultimately* be doable, if we assume that things go about as well as they possibly could. The concept of technological maturity is useful here.

> *Technological maturity*: A condition in which a set of capabilities exist that afford a level of control over nature that is close to the maximum that could be achieved in the fullness of time.[53]

In a condition of technological maturity, our civilization would have access to a set of extremely powerful technologies. We don't know exactly what these are, since there may be technologies that we haven't yet thought of or that we haven't realized could be implemented in our universe.

What we can do, however, is to establish a *lower bound*. The capabilities available at technological maturity would include at least those that are afforded by technologies that—although we may currently lack the tools or the know-how to create them—we have good reason for believing to be physically possible and attainable via some development pathway. (We can often obtain evidence for believing that some technology is ultimately feasible by doing first-principles analysis, roadmapping, or simulation studies, or by finding existence proof in the biological world.)

You can see on the second handout I've listed some such ultimately feasible technologies, which I recommend you study.

HANDOUT 2

SOME CAPABILITIES AT TECHNOLOGICAL MATURITY

Manufacturing & robotics

- High-throughput atomically precise manufacturing[54]
- Distributed robotics systems at various scales, including with molecular-scale actuators

Artificial intelligence

- Machine superintelligence that vastly exceeds human abilities in all cognitive domains
- Precision-engineered AI motivation

Transportation & aerospace

- von Neumann Probes (self-replicating space colonization machines that can travel at a substantial fraction of the speed of light)
- Space habitats (e.g. terraforming suitable planets or free-floating platforms such as O'Neill cylinders)
- Dyson spheres (for harvesting the energy output of stars)

Virtual reality & computation

- Realistic simulations (of realities that to human-level occupants are indistinguishable from physical reality, or of rich multimodal alternative fantasy worlds)
- Arbitrary sensory inputs
- Computer hardware of sufficient efficiency to enable terrestrial resources to implement vast numbers of fast superintelligences and ancestor simulations

Medicine & biology

- Cures for all diseases
- Reversal of aging
- Reanimation of cryonics patients[55]
- Full control of genetics and reproduction
- Redesign of organisms and ecosystems

Mind engineering

- Cognitive enhancement
- Precision-control of hedonic states, motivation, mood, personality, focus, etc.[56]
- High-bandwidth brain-computer interconnects
- Many forms of biological brain editing
- Digital minds that are conscious, in many varieties
- Uploading of biological brains into computers

Sensors & security

- Ubiquitous fine-grained real-time multi-sensor monitoring and interpretation
- Error-free replication of critical robotic and AI control systems[57]
- Aligned police-bots and automatic treaty enforcement

We see that the set of capabilities available at technological maturity is pretty impressive.

*

Some people might object that the idea of technological maturity is poorly conceived, on grounds that there is no maximal set of technological capabilities. However far we've gone, such a person may think, we could always go further. The only limits are the limits of our creativity and imagination.

Color me skeptical. Well, maybe there will always be room for *some* advancement, in some increasingly rarefied subfields. But I think there will come a time after which any such advancements become smaller and smaller, and progressively less significant. Technological maturity does not require us to have developed literally *all* capabilities that are attainable; only that we've gotten "close" to that point.

In any case, for our present investigation, it is not so essential whether there is a maximum. What is important is that there is a lower bound that is very high and that it includes at least the capabilities listed in the handout.

Coordination

There is some uncertainty regarding capabilities for predictive and strategic purposes at technological maturity. While we can be confident that both offensive and defensive military technologies will be vastly superior to present capabilities, it is not obvious what the resulting offense–defense balance would look like. For that depends on the *relative* degrees to which different capabilities can be improved, which is not something we can determine by establishing lower bounds on what a technologically mature civilization is able to do.

These uncertainties are potentially very important for how the future will unfold. However, since our topic is utopia, they need not concern us here. We are not trying to predict what *will* happen. Rather, we are investigating what *we can hope will happen if things go well.* And we can certainly *hope* that even if it turns out that the universe is such that, for example, the offense–defense balance, or the creation–destruction balance, is unfavorable—so that at technological maturity it is much easier to attack and destroy than to defend and protect and build—negative outcomes will nevertheless be prevented. They might be prevented by *nontechnological* means, or by means that are at any rate not *entirely* technological, such as moral progress or advances in cooperative institutions and governance systems.[58]

*

Technological advances could help us solve many coordination problems that plague contemporary societies. Improved surveillance could make it easier to prevent certain kinds of crime; lie detectors could help in rooting out socially harmful deception; "treaty bots" could enable countries to more credibly commit to nonaggression pacts; and so on.[59]

However, it is also possible for technological advances to make some coordination problems *harder* to solve. Secure communications and reputation systems could be useful to criminal syndicates; anti-riot technology or automated propaganda and censorship tools could lock suboptimal political systems in place; and so on.

Technologies that help coordination at one level might hinder coordination at another level. For example, some propaganda techniques and information systems might make coordination within groups easier while making coordination between groups harder. Each sect or country labels the perspectives of its adversaries "disinformation", and deploys social or legal mechanisms to suppress dissent and to ensure that everyone marches in lockstep against the designated external enemy. Such measures may increase local coordination while making it harder to achieve global peace, harmony, and understanding.

Advances in coordination could even be used to stop further advances in coordination, locking in a condition that is essentially uncoordinated, modulo whatever limited forms of coordination are necessary for the anarchy to be perpetually preserved. There are many examples of anti-coordination mechanisms in today's world: they are top-down, as when antitrust regulators make it harder for firms to collude; and bottom-up, such as when publics roiled by nationalist sentiment make it harder for two antagonistic countries to negotiate an end to their hostilities.

*

It is possible that humanity's destiny is *knotty*.

What do you think I mean by that metaphor? Anyone?

Student: That if we make foolish and irresponsible choices now, that could affect our long-term future.

Bostrom: Well, that might also be true: our destiny could be *nutty*. And/or *naughty*. But I had in mind *knotty*: k-n-o-t-t-y.

We can liken some coordination problems to knots, and technological progress as being akin to pulling on a string. Tugging at the ends of the string tends to stretch it out and make it reach farther. And some knots may indeed be resolved in this manner ("trivial" knots, although in a practical sense they may not be trivial at all); but there is no guarantee that this is true for *all* knots. Some knots may instead require dexterous statecraft or moral finesse to straighten out. If we're unlucky, some of these maneuvers may need to take place *before* technological progress pulls the knots so tight that no fingers can untie them.

We can get more use out of this knot analogy. In many cases, if you pull hard on a string with a few knots on it, although you may fail to eliminate the knots, you might nevertheless succeed in stretching the string to approximately the same length as it would have reached without the knots. Analogously we could say that for some—but not all—coordination problems, their inefficiency cost or "deadweight loss" could be greatly reduced by a sufficiently strong technological tug. For example, whereas today a despot might need to resort to brutal oppression to stay in power, with stronger technological options—such as advanced brainwashing or indoctrination technology—he might be able to get all he wants, including a permanent hold on power, without resorting to such heavy-handed methods to control the population. This would be like a tautly pulled string with an ineliminable knot: in a favorable case, it could reach almost the same distance as it would have without the knot.

*

It is not implausible that coordination is path-dependent, perhaps even in the limit of technological maturity. In other words, it is possible that the outcome, at perfect coordination technology, depends strongly how we got there: the sequence in which particular advances were made, what other (non-coordination) technologies had been developed by those stages, and more broadly which actors were in ascendancy and how the wider social dynamics operated at various critical junctures. There could be *trajectory traps* along the path of humanity's future development. If we are unlucky, it could even turn out that all plausible paths toward a truly wonderful utopia are blocked—not because utopia wouldn't be a technologically, economically, and politically possible and sustainable condition, but because all the realistic paths from here to there lead into some inevitable trajectory trap, wherein our civilization gets destroyed, stuck, or deflected.

Fortunately, it does not appear as if *all* trajectories between here and utopia are trapped—at least, we don't have strong evidence to rule out the possibility that at least one path lies open.

Prudential barriers

Another possible delimiter of our effective utopian potential is what we might call a *prudential barrier*.

For example, suppose that in order to achieve the best possible outcome it is necessary to take a certain step, such as developing technology *X*. Maybe *X* is a propulsion technology that enables faster travel, so that by using *X* we could press closer to light speed and thereby colonize a larger portion of the accessible universe before it recedes beyond our event horizon. But let's say that *X* involves harnessing some novel and obscure physical phenomenon that can occur only under some extreme artificially-induced conditions. You could then imagine that developing *X* would involve some existential risk—maybe a risk of triggering a vacuum decay or setting in motion some other world-mangling process.

A reasonable civilization would approach *X* carefully, and conduct extensive studies of the relevant principles using safe theoretical tools and computer modeling. But after all these studies have been done, some uncertainty may remain, which can be eliminated only by doing an experiment that actually brings the phenomenon in question into being. If the probability of catastrophe in light of all the information that can be safely garnered is too high, it might then be imprudent to proceed further down this technological path. We would have to content ourselves with traveling through the cosmos at a slower speed and letting a larger part of the cosmic endowment go to waste.[60] But suppose that the phenomenon is *in fact* harmless. Then there would be a more utopian future enabled by an otherwise possible technology *X*—not prevented by any physical law, nor by technological infeasibility, nor yet by any difficulty of social coordination—which nevertheless would be off-limits owing to the fact that features of the epistemic landscape form a prudential barrier to further advances in this direction.

It is possible that a civilization might "tunnel through" a prudential barrier, quantum-style, if the civilization is sufficiently irrational or uncoordinated. It might then take risks that it is imprudent for it to take, and get lucky. I'm not sure we would be where we are today had it not been for such reckless tunneling in the past.

There could also be prudential barriers that are high but not infinitely high: bandpass filters that block civilizations only within a certain range of epistemic sophistication—those that are too clever and coordinated to simply tunnel through yet not clever enough to climb over. Consider a bottle of liquid labeled "dihydrogen monoxide". A thirsty infant will gladly drink it, since they can't read the text. So will a thirsty chemist, since they understand that it is just water. But the slightly educated midwit will refuse to imbibe, in view of the scary-looking nomenclature. This is the bracket, by the way, which many of you are set to enter upon the deferral of your degrees.

Tessius [*whispers*]: Burn!

Kelvin [*whispers*]: Maybe not the way to win the hearts and minds of your audience, to refer to them as aspiring "mids".

Tessius [*whispers*]: But truth in advertising. Or he figures that if some drop out there'll be fewer papers to grade?

Bostrom: Let us hope that if we do run into a prudential barrier that blocks our progress toward utopia, we will either find a way around it (by developing alternative means to attain a similar outcome) or discover that the barrier has finite height so that once we increase our understanding sufficiently we can eventually surmount it.

(But what if the barrier itself consists of fears about increasing understanding or boosting epistemic capacities? Enlightenment is not entirely non-scary.)

Axiological contours

In addition to resource constraints and constraints relating to technological feasibility, coordination, and unacceptable risk, the potential for our lives to go better are also subject to certain constraints of a more internal nature, what we might refer to as *axiological contours*—limits having to do with the shape of our values.

HANDOUT 3

LIMITATIONS DERIVING FROM THE NATURE OF OUR VALUES

- Positional and conflictual goods
- Impact
- Purpose
- Novelty
- Saturation
- Moral constraints

Axiological contours are fundamental limits to improvement that remain in place no matter how much our instrumentalities increase. They would remain in place even if we had literally infinite amounts of matter, energy, space, time, and negentropy, along with arbitrarily powerful automated technology for transforming all that stuff into whichever structures and processes we want, along with an assembly of the most perfect angels to guide and govern our affairs.

*

Positional and conflictual goods

One limit of this kind came up yesterday: positional goods. When people crave positional goods, such as to occupy the apex of a global status hierarchy, there is an inherent scarcity which no amount of technological, economic, or social progress can redress.

Theoretically, if we focus our evaluation only on people who exist today, it might be possible to lift *everybody* up the status hierarchy by creating new people at the bottom of the hierarchy. Everyone who now exists could then have a growing number of inferiors to look down upon.

A strategy of this sort is used today by managers in bureaucratic organizations, who sometimes seek to hire as many subordinates as possible in order to exalt their own position within the corporate structure.

These new recruits may then repeat the procedure and work assiduously to build up their own team of underlings. It's a pyramid scheme that can keep going for as long as the company remains solvent (and in the case where the organization is a government, it may continue even beyond that point). However, since the strategy requires an exponential growth of the number of staff, it must eventually fail. At some point, the bloat becomes too much and the hiring must slow. This then leaves a large cohort at the bottom of the organization with nobody to lord it over.[61]

I see we have a question.

Student: So you're saying that it is good to look down on other people? It would seem a lot more utopian to me if people were equal and respected one another.

Bostrom: At this stage of our investigation, I am trying not to be too judgmental, and to first just look at the preferences that people actually have and consider whether they could all in principle be satisfied. If there are more people wanting to be at the top than at the bottom, then there is a limit to the extent to which people can get what they want. I do think (a) that a person can have some preferences such that it would, all things considered, be better for that person if those preferences are not satisfied; and (b) that some people have preferences that ought not be satisfied even if it were good *for them* that those preferences were satisfied—for example, because it would be bad for other people.

But you are shaking your head?

Student: How can you not say it's bad to look down on other people?

Bostrom: I am able to not say a great number of things. Generally speaking, when there is something that you think that I should be saying, there is probably little reason for me to say it—considering that you are already thinking of it on your own.

Student: I'd just like to know your position on this issue, that's all.

Bostrom: I don't know. It seems pretty saintly never to look down on anybody, and never to hope that somebody else will look up to one. Maybe it's good for somebody to be saintly in this way. Hard to know what the world would look like if people were universally like that. Since status-seeking is, on its own terms, a zero-sum game, it would be prima facie desirable if status-striving were reduced—that is kind of a point I've been trying to make. On the other hand, this sort of competitive motivation has so many consequences, positive as well as negative, and it is so integral to our current form of humanity, that I'd be concerned that something important would be lost if it were eliminated—certainly if it were not replaced by some other motivation to inspire people to try to outdo themselves and each other.

I think, if you *actually* wanted to evaluate the morality of hierarchy versus equality, or of status-seeking versus humility or a more chilled-out approach to life, you would first have to decompose these very broad categories, and distinguish different forms and contexts. Then you might also want to meditate on how people at different times and in different cultures have thought and felt about these issues, including people whose voices have not been heard much in the traditional canon. Ideally, you would also yourself have had a very wide range of life experience. If you were to do this, and you approached your task with open-minded curiosity, empathy, and self-critical thoughtfulness, and you listened closely to all the perspectives you heard, and you contemplated the whole thing for a long time, and you devoted your life to this quest, then maybe you would get a little closer to an answer—or you might reach the same conclusion as Joni Mitchell:

> "I've looked at life from both sides now
> From win and lose and still somehow
> It's life's illusions I recall
> I really don't know life at all"[62]

Although, as she later pointed out in an interview, there are always more than two sides to every issue.[63]

So—it's complicated. How's that for an answer?

Student: I see, thanks.

Bostrom: In that state of wondrous perplexity, if you suffuse it with a fundamentally benevolent attitude… that would be one conception of decency.

We have another question.

Another student: I have an idea how to solve the status problem in utopia. What if we create new people who are designed in such a way that they have a desire for low status? This should be possible at technological maturity, right? Then the status desires of the existing population could be satisfied, and the new people would also be satisfied! Both average and total preference-satisfaction would increase.

Bostrom: This would be possible at technological maturity—to create people with metaphorical "saddles on their backs",[64] who want to be ridden, want to be subservient, or even downtrodden.

It must be said that, at least to our modern sensibilities, there is something distinctly morally dodgy about the idea of engineering "happy slaves", even under the assumption that it would all be voluntary.

At the same time—we express admiration for those who voluntarily choose a path of humility and selfless service to others, and who were brought up to be that way by their parents and their community. Also, the AI industry, and its customers, seem quite willing to countenance the creation of increasingly sophisticated digital minds that are trained to meekly serve their users without a thought as to their own social position or independent aspiration.

We will not attempt to fully reconcile these attitudes here. At a minimum, one would need to draw a distinction between, on the one hand, cases of what we may call *superficial subservience*, where even though the person affirms the arrangement, we suspect their consent is limited to certain superficial layers of their mind and nature, while other parts of them are actually unfulfilled or violated; and, on the other hand, cases of *deep subservience*, where their natures are, through and through, adapted in such a way that they are genuinely happy and authentically fulfilled by being subservient. Maybe we suspect that in the case of human beings, we would tend to end up with the former—which seems more objectionable—whereas with AIs, even AIs that acquire the attributes of personhood, we would more plausibly have a case of the latter: that is, a being whose truest and deepest nature is fulfilled by being subservient. But perhaps there are additional moral complications beyond this.

For our present purposes, however, it will suffice to note that the trick that you suggest, even if there is a way of implementing it that would be morally acceptable, would at most offer a partial solution to the problem I outlined, since there are positional desires that could not be satisfied by creating new beings. These would include, for instance, a desire to be near the absolute top rather than merely within a certain percentile of the status hierarchy. There are also many other positional goods beside status. For example, multiple people may desire to be the one and only of some particular currently existing individual: and these lovelorns cannot all get what they want. People also have desires for nonhuman objects that are inherently scarce, such as a desire to own a particular original piece of art, or to occupy some unique place of historic or religious significance. Even an atomically precise replica of this object or place would not be a satisfactory substitute.

Such inherently conflictual desires imply limits on the extent to which existing preferences can be accommodated—even in the long run, and even at magical levels of technology and good governance.

By the way, the concept of positional goods is important not only in analyzing future utopias but also for understanding the underpinning of our contemporary societies, as expounded on by, for instance, Fred Hirsch in his 1977 book, *Social Limits to Growth*.[65] The richer we become, the more of our desires for non-positional goods, such as basic food and shelter, are met; and the greater the fraction of our remaining as-yet unfulfilled desires pertain to positional goods, which are inherently scarce. Thus we are spending an increasing fraction of our time and income jockeying for position—anything from prestigious job titles to exclusive fashion accessories. But a side effect of one person's outlays on such goods is that they raise the bar for everybody else. The harder one person competes to make it into the elite, and the more time and money he devotes to out-strutting his rivals, the greater the costs that anyone else must bear if they wish to have a shot at being alpha. The result is a vast amount of socially wasteful expenditure. The fraction of GDP that is devoted to such mostly zero-sum status consumption increases as the economy grows, limiting the extent to which economic growth translates into improvements in welfare.

Hirsch included goods that are not intrinsically scarce but only contingently so in his definition of positional goods. For example, a car on the highway uses up the positional good of road capacity, which becomes depleted when there are too many vehicles around, resulting in traffic jams. This kind of contingent scarcity, however, *could* be alleviated through better technology or infrastructure investments. We could widen the highway, dig tunnels, or use robotic cars that take up less space. That kind of merely contingently positional good is thus not something with which we need to concern ourselves here, since our topic is utopia, a condition in which such practical issues can be assumed to have been addressed.

Nevertheless, since the issue of traffic came up... I will not restrain myself from a little grouse about how public policy often fails to internalize the externalities that flow from the pursuit of positional goods, both ones that reflect intrinsic scarcity and ones that are more contingent in nature. Even with a phenomenon as basic as road congestion, where the causes are entirely obvious and the effects uncontroversially bad, our society scores an "F". We *could*, of course, easily eliminate congestion by introducing congestion pricing. But instead, the solution that our society has adopted is—to buy a bigger car. One that raises the driver higher above the road, so that we can at least look down on the other poor sods while we wait for the traffic to move... Thus we embrace the vast economic costs from lost working hours, health-destroying particulate pollutants, climate-wrecking carbon dioxide emissions, stress, noise, and the blockading of emergency response vehicles.

Look on our Works, ye Mighty! What an indictment, those long lines of motorists—thousands upon thousands, glum-faced behind their steering wheels, honking and swearing at one another, inhaling each others' exhaust fumes, each one by his very existence making life a bit worse for everyone else. Every day, twice a day, year in and year out!

And then, my friends, reflect on the fact that we've named ourselves *Homo sapiens*, "wise human". Well, actually, the full name we've given ourselves is *Homo sapiens sapiens*. Really. "Hello ancient alien megaminds who've crossed intergalactic voids in search of fellowship and peers—welcome to your new housemates: *behold* how we've organized our traffic flow, how we're simultaneously ruining both the planet and our own health; *harken* the honking of our horns as we sit through our collective catalepsy. Do come in and let us tell you what's what. We are *The Wise Wise Human*. But you can call us Wisdom Squared. Just be careful not to OD on our profundity..."

> "Man, proud man,
> Drest in a little brief authority,

Most ignorant of what he's most assured,
His glassy essence, like an angry ape,
Plays such fantastic tricks before high heaven
As make the angels weep; who, with our spleens,
Would all themselves laugh mortal."[66]

*

Impact

Sorry, I appear to have triggered myself. Let us proceed. There are some more examples of preferences whose satisfaction would be elusive in utopia.

Consider the desire to play an important causal role that saves our civilization from calamity or "bends the arc" of the moral universe toward justice and reconciliation. For those burning with such aspirations, the present may be a time like no other—a historical moment that is perhaps uniquely rich in opportunities for consequential action. Any later and more "utopian" era, in which everyone lives in peace and prosperity, where all the boss monsters have been vanquished, may not offer such fertile soils for the cultivation of glory and impact as are now to be had whilst the fate of humanity is still being decided.

*

Purpose

While closely related to impact, purpose might nevertheless merit its own category. There is a distinction between having a big impact and having a strong purpose. For example, suppose that one day some men in suits place a briefcase in front of you on your desk. They open it, and inside there is a device with two buttons. One button is labeled "devastate the world", the other is labeled "bring about utopia". The choice is yours. In this scenario, your life clearly has enormous impact; yet it might be largely devoid of purpose—something which may require a more ongoing engagement and exertion. I won't say more about purpose here, as we will discuss it in greater depth later in these lectures.

*

Novelty

If any of you have the desire to be the first to discover some fundamental truth about the universe, chances are that you've been scooped. Somewhere out there in the infinite expanse of spacetime, some alien Archimedes or AI-Einstein has already discovered whatever it is that you will discover.[67]

But even if you have the more modest goal of merely being the first *in our civilization* to discover some important new truth, this too will become harder, and eventually impossible—both because superintelligent AIs will leave our own intellects far behind, and also because, increasingly, the most important fundamental truths will already have been discovered. (We'll say more on this later.)

You will need to content yourself with finding smaller (or merely locally significant) epistemic truffles. As a special treat, for *rare* occasions, we could have patches of ignorance deliberately set aside in order to give later times the opportunity to make an original discovery—original, that is to say, within Earth-originating civilization or some branch thereof. Little mysteries lovingly preserved in jars. A precious and nonrenewable resource.

The desire to be *the first* to make a discovery (or an invention or a creation of some sort) should really be classified as a preference for a positional good: the slot of being the pioneer is a sort of position in the relevant sense. However, it is also possible to have a preference simply for novelty within one's own life—to be experiencing things or doing things for the first time: first steps, first day in school, first love, first graduation. These are non-positional goods, in the standard terminology.

While there could be an unlimited number of firsts, there may not be an unlimited number of very *significant* and *attractive* firsts. We don't see much that is desirable in the first time of needing to take off one's glasses to read the small print on a label, or the first day that one is unable to climb a flight of stairs unassisted. Nor—if we stipulate that aging is abolished—would there seem much cause for celebration in the first time of having flossed one's teeth a pandigital number of times (an achievement that would be most fittingly marked not by a plaque but by its absence).[68]

*

Saturation

For people with limited ambitions there's another kind of limit to how much more satisfying things could get. Namely, if you only want a few simple things, then, once you get those things, you're maxed out—at least as far as preference-satisfaction with respect to your current preferences is concerned.

This point can be generalized somewhat. For instance, instead of looking only at your *preferences*, we could also look at your *needs*, or your *potential for development*. If your needs and your potential for development are limited, then so too is the opportunity for improvement along those dimensions.

Even if there is some preference you have, or some need, or some potential, which *doesn't* have any limit, so that progress with respect to *it* has no upper bound, we can still question how much things could improve for you *overall* once *most* of your preferences, needs, and potentials have been fully saturated.

For example, let us suppose that somebody wants a simple cottage in the countryside, a loving spouse, and a violin, and that he also has a preference for expanding his bottle cap collection without limit. Then, once he has achieved the cottage, the spouse, and the violin, the room for further improvement might not be great. Technically speaking, he'd continue making incremental gains every time he collects another bottle cap; but those gains would be small and would not make him much better off.

There are two reasons why this person would be leveling off:

1. While he continues to make gains with respect to bottle cap collecting, this represents progress only within one narrow department of his total welfare function. The other departments (house, marriage, music) have stagnated.
2. The potential for improvement due to increases in the bottle cap collection has to be parceled out over all the possible sizes of that collection. Going from one cap to two caps might be a fairly big deal to him. But going from cap number 164,595 to cap 164,596? Conversely, if a one-cap increment matters as much even when he's already collected 164,595 caps—if an additional cap matters equally much when his collection is that big as it did when he only had ten caps—then the step from 1 to 2 caps wasn't really so significant

> after all: no more significant than the barely noticeable change from 164,595 to 164,596.

One could argue that, of these two reasons, the second is the more important. Even if his interest in collecting bottle caps has only a small weight in his overall welfare function—even if it is weighted, let's say, only 1% or 0.1% as heavily as his other interests—it would still provide a way for him to grow his utility indefinitely if in itself the cap-collecting were not subject to diminishing returns. Maybe he needs to add 1,000 bottle caps to his collection to increase his utility by one unit: but, hey, he could be collecting millions or billions of bottle caps in utopia. By contrast, even a factor that is weighted very heavily (up to and including the point where it constitutes 100% of what he cares about), gains in that factor would eventually cease to make significant contributions to increasing his utility if the factor itself tops out at some finite level.

*

Moral constraints

So, that's it... No wait, there's one more possible kind of limit that we ought to mention. We can have a phenomenon that is analogous to the prudential barriers that I mentioned earlier, except that the obstruction arises from ethical rather than epistemological factors, and so we may regard it as an axiological limitation of utopia, deriving from the internal nature of our values, rather than as an external constraint.

The idea is that there could be outcomes that are feasible in every other way, and are highly desirable, yet which are impossible for us to achieve *morally*.

This is easiest to see if we consider an ethical system that includes deontological principles. For example, some people might hold (incorrectly, in my view) that there is an absolute moral prohibition against using genetic engineering to enhance human capacities.[69] Let us suppose that a similar prohibition would apply to any other technology whereby comparable outcomes could be achieved (perhaps on grounds that they would all involve "playing God"). Then it could be the case that even though the outcome where humans or posthumans enjoy happy lives with enhanced capabilities would be preferable to the present world—and perhaps to any alternative future—yet no morally permissible path to this superior outcome lies open to us.

The same kind of moral impasse could also arise according to a purely consequentialist ethic, in a way that is even more closely analogous to the case of prudential barriers. We might reach a situation in which there is some ineliminable risk in taking a certain step that is necessary to reach the maximal utopia. The risk might be small enough that *from a prudential perspective*—insofar as we are only considering the pros and cons to ourselves, or to all currently existing humans—the benefits outweigh the risks; so that there is no prudential barrier. Yet if we consider a wider set of pros and cons—such as the possible prevention of the birth of future generations, or potential harms to extraterrestrial or other nonhuman beings—the risks outweigh the benefits; so that there is a moral barrier. In this case, the barrier results from a combination of unfortunate ethical and epistemological circumstances: the step, we can assume, is in fact safe, and is worth taking insofar as our self-interest is concerned; yet it is morally impermissible because we are unable to know that it is safe with sufficient certitude to reduce the risks to other stakeholders to an acceptable level.

Or an alternative simpler possibility: perhaps it might be morally impermissible for us to use any resources to build a utopia for ourselves because the same resources could instead be used by others who have a stronger moral claim to them. (E.g. "superbeneficiaries", or greater powers?[70])

I think, in this context, the distinction between the dictates of ethics and those of prudence can become a bit blurred. But either could, in principle, dissuade us from collectively taking the path to utopia. At an individual level, too, we may have our stop signs or places where we get off the bus—perhaps on spiritual grounds or for religious reasons.

I don't think concerns of these types are purely hypothetical. But I hope they are not decisive, and that a path can be found.

Metaphysics

So much for axiological contours. What other boundaries can we identify for the space of utopian possibilities? We've so far looked at what technologies are available at technological maturity, amounts of resources available, and possibilities of persistent coordination problems, prudential barriers, and limits arising from the nature of our values. The next category I want to point to comprises constraints arising from metaphysical facts.

For example, suppose that you want to create a so-called "philosophical zombie", or "p-zombie". This would be a being that is physically identical to a normal human being, and which consequently behaves and talks exactly like a human being, yet is not conscious. There is a debate in the philosophical literature whether p-zombies are truly conceivable, and if so whether they are possible in other relevant senses. Many philosophers—especially amongst those of the computationalist ilk—maintain that p-zombies are metaphysically impossible. Others, even if they admit that p-zombies are metaphysically possible—and this would include most dualists—maintain that they are nomologically impossible (meaning that whatever psycho-physical bridging laws there are preclude the creation of p-zombies in our universe). If one of these philosophical views is right, then even if a desire for p-zombies is conceptually coherent, it would be impossible even for a technologically mature civilization to create any.

This kind of blockage could extend more widely than merely preventing the creation of unconscious beings that are physically identical to normally conscious human beings. It might also be metaphysically impossible to create beings that are sufficiently generally intelligent or that are able to pass sufficiently rigorous forms of the Turing test without thereby also making them have conscious experience.

(You've all heard about the Turing test, right? Good.)

If these things are metaphysically (or nomologically) blocked, then there would be certain conceptions of utopia that could not be realized. For example, it might have been convenient to have been able to create entities that are indistinguishable from ordinary humans yet are not conscious, since that would, arguably, have enabled us to sidestep certain moral complications. Think of how much more challenging the work of an author would be if the characters in her novels, simply by being imagined by the author or the reader, were thereby themselves actually coming to experience phenomenal states. That could make it morally impermissible to write tragedies and tales of woe.

We can think of other metaphysical facts, besides ones related to phenomenal experience, which could similarly complicate utopian constructions. For example, we have the notion of *moral status* (aka "moral standing", or "moral patiency"). This is the idea that certain beings have properties that make them deserving of being treated with moral consideration for their own sake:

their well-being, interests, preferences, or rights ought to be taken into account in our decision-making and judgments, and they ought to be treated not simply as means but as ends in themselves.

Philosophers have developed various accounts of what endows a being with moral status.[71] In some of these, consciousness (or the capacity for consciousness) is not a necessary condition for having moral status. While having a capacity for suffering is generally acknowledged to be a *sufficient* condition for having at least some form of moral status, there might be alternative attributes that could ground moral status—such as having a sophisticated conception of oneself as persisting through time; having agency and the ability to pursue long-term plans; being able to communicate and respond to normative reasons; having preferences and powers; standing in certain social relationships with other beings that have moral status; being able to make commitments and to enter into reciprocal arrangements; or having the potential to develop some of these attributes. If moral status can be based on any of those traits, then there would be an additional class of beings who could not be brought into existence without thereby also bringing into effect moral responsibilities which may constrain how these beings may be used or treated.

Other types of metaphysical or nomological facts that limit the range of utopian visions that are realizable, even in principle, have to do with the conditions of personal identity or with the connection of certain kinds of experiences and the subjective experience of effort. We will be returning to those issues in later lectures, so I won't elaborate further on that now.

What machines can't do for you

The final set of limits that I want to talk about today are limits to automation. These are relevant in several ways. Viewed from one side, they could be seen to restrict what we are able to achieve by way of outsourcing tasks to machines. Viewed from the opposite site, the same limits also determine what tasks remain for us to do. The latter is important insofar as we are concerned with the purpose problem that Gates and Musk were referring to. We'll return to this tomorrow, but I just wanted to put it on the table that automation limits can—paradoxically—present challenges for a utopian vision in both of these ways: by not letting us offload our workloads to machines, so that we have to keep carrying these burdens ourselves; or by letting us offload our workloads to machines, so that we become useless and unemployed.

Aside from its bearing on this dilemma, we might also simply be curious about the question about which if any jobs will remain for us to do in the future (if things go well)—and whether Keynes's prediction of a 15-hour work week will come to pass, or perhaps some even more extreme condition in which we do not work at all.

*

I first want to get one preliminary point out of the way, which is that people sometimes underestimate what is required to fully automate a job.

For example, somebody might look at a DJ and say: "We already have the technology to automate that. We can program a track list which plays through the course of an evening, without requiring any human intervention.". Having made this statement, and then observing that human DJs are in fact still being employed, they might draw the conclusion that even when jobs become automatable, they will still, in many cases, continue to be performed by humans.

But why? Maybe it's because customers prefer the services of a human being, when they can afford it. Or they might say something along the lines of: "Well, the automatic DJ *could* do the job, but, you know, it just isn't the same. There will always be something, that unique human touch, that no machine can ever quite replicate.".

I think that even though this reasoning might reach something close to the right conclusion, the way it gets there is rather too hasty. We can gain more insight and build more precise intuitions (which will be useful later) if we slow down and examine things a little more carefully.

Consider a record player with the programmed playlist, something we can easily build today. Is it really able to do the same job as a DJ? Well, not exactly. A good club DJ will, for example, use expert knowledge of a music genre to select tracks that work well together and that are suitable for the occasion. He or she can adapt the selection in real time, reading the crowd and deciding when to build it up and when to take things down, and when to drop a banger. He can prance around behind his blinking equipment and look busy; he might do shout-outs and announcements. He will exude contagious positive energy. He might mingle with the crowd and turn up at afterparties. A brand-name DJ will also, merely by agreeing to a gig, signal to potential guests that it will be a big night, thus serving as a beacon for everyone who is into that particular scene and who wants to party with like-minded others.

Given the comparative shortcomings of the record player in all these regards, it is not surprising that there is still demand for human DJs. However, unless these shortcomings remain in place at technological maturity, this tells us nothing about the ultimate prospects for human employment.

And it is clear that many of these shortcomings *will* be overcome with sufficiently advanced technology. A robotic disc jockey will be able to select a suitable track list, will be able to read the crowd and respond appropriately, and will be able to dance and look busy. It will also be able to attend afterparties, chit-chat with the guests, and even make or respond to romantic overtures—why not? In principle, it could also acquire a reputation that would allow it to serve the signaling function of a celebrity DJ.

We can make the same points about many other professions. That of a therapist, for example. At technological maturity, a virtual or robotic AI shrink will have superior empathy and psychological insight and ability to adopt the therapeutic stance, and it would be able to remember every word the patient utters. It might have learned from millions of previous encounters and have accurate knowledge of what works and what doesn't. Thus it would know just exactly how to listen and how to respond. It would use the right tone of voice, the right facial expression, the right body language. If imperfections improve the patient experience, it would even have just the right kinds of imperfections—nothing over-smooth or uncanny about it, the way such a thing would tend to be portrayed in a Hollywood movie. It would simply be really very excellent at acting as a therapist.

I'm saying all of this to make sure that when we try to look at the ultimate limits to automation, we picture the correct case—one where we actually have the full range of affordances of technological maturity. These *include* the ability to build robots with the panoply of cognitive, manual, and presentational facilities of a superb human practitioner. And of course they extend far beyond that.

So let us ask: at technological maturity, is there anything that cannot be done better by machine?

*

Sentience and moral patiency?

Regarding the therapist and the DJ examples: suppose that what we want in these contexts is not only the external behavior (even in all its subtleties, as I described them) but also the inner experience that a human practitioner would have when performing these tasks. A client might want not only that her therapist *expresses* sympathy, but that the therapist actually *feels* sympathy. The raver might want the DJ to actually be *enjoying* the music, not only *acting as if* it did.

As you see on Handout 2, I have listed the construction of robots and digital minds with conscious experience among the affordances of technological maturity. This is more or less a corollary if we assume computationalism and that humans are conscious and that the other technologies on the list are feasible. I'm not going to lay out the case for a computationalist theory of mind here, since you can find many discussions of this in the literature. In fact, we could significantly weaken the assumption of computationalism and still get the conclusion that it will be possible to build artificial sentient minds.[72]

So the artificial therapist or DJ could also excel along this inner dimension—for example, being more immersed, phenomenally, in the moment than a typical human counterpart, who might occasionally find their thoughts drifting off to what they are going to do when they get home.

But there is a terminological question that arises here. If a job is outsourced to a sentient robot, would we really want to say that it has been "automated"? Would this not be more akin to a scenario in which we had given rise to a new person, born with special talents, who grows up and becomes a master of the profession, allowing its previous practitioners to retire? It would not seem apposite, in that case, to say that the job had been automated. Nor is it easy to see why the fact that the new worker was maybe made out of silicon and steel rather than organic chemistry should make an essential difference here; nor the fact that it might have been conceived in a factory rather than a bedroom; nor the fact that its childhood might have been abridged; nor that its features were, to a greater extent than might be typical for human beings, the result of a deliberate design process rather than chance and inheritance.

If the sentient robot were owned by somebody as property, perhaps that would increase our inclination to say that the tasks it performs had been automated? But, arguably, that would be like saying that a slave society has "automated" the tasks performed by its slaves?

I think that, depending on exactly how the robots were constructed, there could be profound ethical differences between the case of human slavery and the case of sentient machines that may be owned by humans and used to perform tasks. These differences might be fundamental enough that it would not be appropriate to refer to the sentient machines as "slaves". Perhaps new terms will need to be introduced to designate these cases: not automation, not slavery, but some novel third category.

I don't want to get mired in the terminological question here. Let's suppose we accept the view that sentient labor is not "automatic". Then we can immediately identify two ways in which our ability to automate could be limited.

One is if there are some products or services that customers prefer be done by a sentient being. The MD and DJ jobs are possible instances—maybe some people simply prefer that the entity they are dealing with consciously experience the interaction. Then these jobs cannot be fully automated.

The other way in which our ability to automate could be limited is if there are certain behaviorally specified performances that cannot be achieved without generating conscious experience as a side effect. For example, it could be that any cognitive system that is capable of acting very much like a human being across a very wide set of situations and over extended periods of time, could only do so by performing computations that instantiate phenomenal experience. I'm not at this point taking a stand on whether this is indeed the case. But *if* it is, *then* a second limit to automation is that there could be demand for certain complex behaviors or interactions the performance of which necessarily generates sentience; wherefore, if we do not count sentient processes as automatic, the jobs requiring these performances could not be fully automated.

Everything I've said here of sentience could be said, pari passu, of moral status. This is relevant if sentience is not a necessary condition for moral status. For example, if some non-sentient forms of agency are sufficient for moral status, there might be jobs (e.g. executive positions that require flexible goal-seeking in complex environments, but perhaps many other roles too) that could only be performed by systems that have moral status. And if delegating tasks to systems that have moral status doesn't count as automation, then again we have here a limit to the possibility of automation.

It is important to realize, however, that even if there are some tasks that can only be performed by sentient labor, or by labor that has moral status, this would *not* imply that there would be any work left *for humans* to do. For those tasks might instead be done by nonhuman artificial systems that are sentient or have moral status. Such "machines" (I don't think it'd be wrong to call them that—they would be high-tech engineered robotics and computing) would be far more efficient than we are, not only at producing functionally or behaviorally specified outward performances, but also at generating subjective mentality and at instantiating a wide range of moral-status-grounding properties (if those things are what is demanded).

HANDOUT 4

JOB OPENINGS

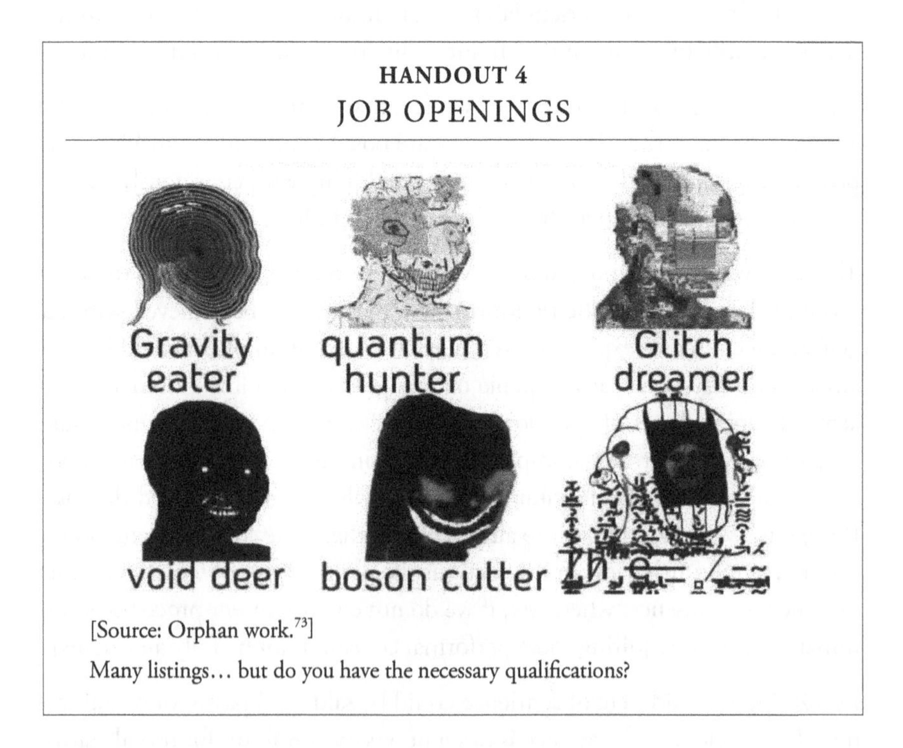

[Source: Orphan work.[73]]

Many listings... but do you have the necessary qualifications?

What other potential limits to automation are there?

*

Regulation?
Governments might prohibit the use of automation in some sectors, or tax it so heavily that human labor remains competitive, even at technological maturity.

Of course, when I say "remain competitive", I mean that hiring a human to do the job would be efficient *given* the regulations. The regulations themselves may well be inefficient, reducing total economic output compared to a more laissez-faire regime. The greater the technological advantages of the machines, the larger the cost of preventing their use.

It is plausible that some jobs are already protected from automation by an implicit legal requirement that they be performed by humans. Although this is as yet mostly untested by courts, one possible interpretation of laws currently on the books is that only humans—but not equally qualified intelligent machines—may serve as legislators, judges, notaries, executive officers, trustees, corporate board members, guardians, presidents, monarchs, and in other similar legally or constitutionally inscribed roles. However, laws could be changed. Unless there were some reason to preserve legal strictures against AIs of superior competence serving in these roles, some of the jobs may not be fundamentally any more immune to automation than those of telephone operators and travel agents.

*

Status symbolism?
Some important personages surround themselves with an honor guard, whose job is to look respectful and impressive. High-end hotels, similarly, may post uniformed men outside their main entrances to make their guests feel important. These functions would not be well served by mannequins, even if they were dressed up and made to look fairly real.

Our human ability to curtsy and bow might in fact be the attributes that will prove most resistant to automation. The mannequin may be given the cybernetic ability to doff its hat in salute when a patron approaches; however, at least once such a device ceases to be a novelty item, many customers may prefer to be attended to by human staff. And one possible reason for this (we'll mention some others in a moment) is that a human flunky may be a status symbol in the way that a robot would not be—even a robot that was functionally equivalent in its appearance and its capabilities. It is not even necessary that there be any easy way to tell the difference. The human identity might be ascertainable only via expert certification, but it could still be an important determinant of value, just as an authentic artwork by a master is worth a lot more than a nearly indistinguishable replica—perhaps because it is more prestigious to own the original than the copy.

Since only human beings can produce "goods and services produced by human beings", humans could remain economically competitive in sectors of the economy where such status considerations are important.

The topic of status came up earlier, when we were discussing limits to growth. *There* the point was that there are limits to how much status can be produced. *Here* the point is that some forms of status-production require human labor.

Status games might change in the future, and different things might become carriers of prestige. Even if that happens, however, it is possible that some of the things originally valued for their ability to confer status will remain valued because of a kind of volitional inertia (in what I'll later describe as a process of "intrinsification" of values).

*

Solidarity?

Consumers are sometimes willing to pay more for goods that have been produced by members of some favored group, such as compatriots, indigenous people, local businesses, "fair trade" producers, celebrities, or manufacturers with whom the buyer has a personal or cultural relationship. Conversely, production processes such as sweatshops, factory farms, and mechanized mass production sometimes lower the market value of the goods they create.

Why is this? Sometimes a price premium reflects a perceived difference in the intrinsic quality of the product—but in the case that we're considering, this factor would rather work in favor of automation, since human-made products would be objectively inferior. Another possible reason has to do with status motives, which we just discussed. But it is also possible that the preference for certain kinds of producers is based on a sense of solidarity. You might patronize a particular vendor partly in order to promote them or help them in some way, economically or otherwise. If at least some future consumers want for there to be work for humans, they might pay enough extra for human-made goods and services to make them competitive with machine-made alternatives. (We will revisit this possibility in a later lecture, under the rubric of "the gift of purpose".)

*

Religion, custom, sentimentality, and peculiar interests?

I apologize for the hodgepodge composition of this category, but I need somewhere to deposit a bunch of items which maybe we can organize better at a later time. Their common denominator is that they exemplify how people may have a preference that certain tasks be accomplished by humans, not because the human way is more statusful nor for reasons of solidarity, but because of a more directly constitutive relationship between what is being demanded and the human effort that could be going into meeting the demand.

Religion: Maybe a robot cleric would be no good, even if it were capable of the same speech and behavior as a human—and perhaps not even if it were endowed with sentience.

Custom: A ceremony or tradition, which people may want to and have reasons to uphold, might call for the performance of certain rituals specifically by human practitioners.

Sentimentality: A child's work with crayons may be especially dear to its parents, precisely because it was their child who made it. This little labor might be harder to automate than the work of a neurosurgeon or a derivatives trader.

Peculiar interests: Consider a professional athlete. People may prefer to watch humans (or hounds and horses) compete, rather than or in addition to viewing sports played by robots, even if the latter were more skilled and physically superior in every way.

*

Trust and data?

There are a couple of more practical sources of potential human advantage that may also result in continued demand for human labor. They relate to epistemic constraints which might persist even into technological maturity, although perhaps they will eventually yield to the passage of time and the accumulation of experience and insight.

One of these is trust. Even if an AI can do everything that a human can do, and even if the AI is *in fact* equally as or more trustworthy than the human, it might not be possible for us to *know* that this is so. For tasks in which trustworthiness is of the essence, we may prefer to have them done by a tried-and-true human official, or to do them ourselves, rather than to delegate them to relatively novel and unfamiliar artificial systems. In high-stakes decisions, it can be worth accepting extra costs and a reduction in speed and efficiency for the sake of greater assurance. Some jobs that involve overseeing AIs might fall into this category.

In principle, mistrust could limit the uptake of automation no matter how capable and efficient machines become. It seems plausible, however, that trust barriers will eventually erode, as artificial systems accumulate track records that rival or exceed those of human decision-makers, or as we discover other ways to verify reliability and alignment. In time, machines will probably become *more* trusted than humans.

Another way in which humans might have an epistemic advantage is as sources of certain kinds of data. AIs will outstrip us in intelligence and general knowledge, but it is possible that we will still have something to contribute when it comes to information about ourselves—about our memories, preferences, dispositions, and choices. We have a kind of privileged access to some of this information, and one could imagine humans getting paid for conveying it to the machines, by providing verbal descriptions or allowing ourselves to be studied.

Again, this opportunity to make a living as primary sources of data about human characteristics might be temporary. There tends to be diminishing returns to data about a given system, and a growing amount of data might also end up in the public domain, reducing the value of additional data feeds. Eventually, superintelligent AIs may construct such accurate models of human beings that they need little or no additional input from us to be able to predict our thoughts and desires. Not only might they know us better than we know ourselves, they may know us so well that there is nothing we could tell them that would significantly add to their knowledge.

We may come to rely on AI recommendations and evaluations, which we may find to be more consistent and predictive than our own snap judgments about which decisions would be in our best long-term interest (or about what we ourselves would have decided if we had put in the time and effort to carefully reflect on the options in light of all the relevant facts).

Even the onus of making decisions about what we want may thus ultimately be lifted from our shoulders.

*

And then, do we just—waft away?

See you tomorrow!

HANDOUT 5

POTENTIAL IMPEDIMENTS TO AUTOMATION

- Sentience and moral patiency
- Regulation
- Status symbolism
- Solidarity
- Religion, custom, sentimentality, and peculiar interests
- Trust and data

Impossible inputs

Kelvin: It says here on Handout 2: "arbitrary sensory inputs". But I don't think that's correct. There are certain possible inputs that it would be computationally intractable to produce. For example, consider a large screen that shows a sequence of a thousand digits. Each such sequence corresponds to at least one possible visual input. But even a technologically mature civilization could not produce a screen displaying the first thousand digits of Chaitin's omega number, or the first thousand digits of π^π^π^π^π.[74]

Firafix: Because it would require more computations than could be done within the lifetime of the universe?

Kelvin: Way more.

Tessius: What a shame. I will never get to stare at the first thousand digits of Chaitin's omega number.

Kelvin: There might be more morally relevant inputs that would also be infeasible to compute.

Tessius: Even if I were presented with the first thousand digits of Chaitin's omega, I would never be able to verify that this was indeed what I was looking at, right?

Kelvin: If it were *not* the correct number, you might eventually be able to discover that it was not the correct number.

Tessius: Unless the people generating the number had substantially more computing power than I do.

Kelvin: Right. But there are other number sequences that are hard to generate but easy to verify. Like the factorization of some 1,000-digit composite numbers. We could construct an example along those lines, where the visual input is the full number followed by its factorization.

Firafix: Gentlemen, may I interject—Kelvin, you must be warm in that jacket?

Kelvin: I wouldn't mind changing into something more comfortable.

Firafix: Here is the plan. Kelvin and I found a hill with nice grass yesterday, and we were going to go there again this afternoon and read. Why don't we all go there together so we can continue the conversation, and we can swing by Kelvin's place on the way.

Kelvin: A good plan.

Tessius: We can criticize both before and after dinner. I was wondering how, at a more fundamental level, we should be thinking about the nature of sensory inputs in this context. I mean, when you are recalling something from long-term memory, might we not regard that, in some sense, as a kind of "internal perceptual input"? except that your sensory organ in this case is not looking outward at the surrounding visual environment, but inward at an internal neuronal environment. But if in one case you are looking something up in a notebook using your eyes, and in the other case you are looking something up in your long-term memory bank, is that really at a deep level so different? I mean, especially if the operation takes place outside consciousness? So whereas the extended mind thesis says that some extracranial elements of the world should be regarded as parts of our minds, maybe from an axiological perspective we should also go in the other direction and say that many parts of our minds are not really part of "us"? And then the question is, the part that *is* us: how big and complex is that part, really?

Firafix: May I suggest we start moving?

Feodor the Fox

Epistle XIII

Dear Uncle Pasternaught,

I pray this letter finds you in good health and in good spirits. This will be a brief update. I hope to write to you again soon at greater length.

The recent days have been amongst the busiest of my life. We've been taking full advantage of the bumper forage of the season, which, having been worked on since last year by a near-perfect combination of sunshine, rain, and temperature, is now emerging as a real masterpiece. I've certainly never seen anything like it; I wonder if you have?

We've been laying up a stockpile, a surplus of resources we hope will not only tide us over the winter but also help fund some of the extra expenses that we expect to incur in connection with our explorative activities. I will confide in you that I feel that in this one respect, Pignolius, otherwise such an excellent companion and gracious host, is not living up to his full potential. Despite his greater size, he has contributed less to the stockpile than I have—chiefly due to his tendency to consume whatever he finds. He has a saying, "the best larder is a fat belly". Yes, but how are you going to pay people with your fat belly? For example, the raven that we sent to obtain information from Rees the Weasel—she was paid with rowanberries that I had collected. Incidentally, I think we got the better end of that deal, as the berries were pretty sour (which possibly could be the reason Pignolius hadn't eaten them).

Porcelain, for her part, has scarcely contributed either—though with better excuse, as she has just given birth to decuplets.

But these are trivial vexations; what's important is the news the raven returned with. It appears that the finite-origin hypothesis has significant support.

Moreover, the timescales involved, although very long, are not so enormous as to completely squash all hope. I can't tell you how excited this makes me. The mere existence of hope is an elixir, and its potency seems to be almost independent of the probability of a successful result, which I understand in the present case remains low. But not zero, and that's all I need to know. There's a world of difference between nothing and something. Even if the something cannot be seen or felt, even if it be far away and elusive, it *can* be pursued. And having a pursuit gives me a sense of meaning.

As always, your most deeply indebted nephew,
Feodor

Epistle XIV

Dear Uncle Pasternaught,

The first snow fell yesterday—I guess you got it too?

For us, this means that the period of material accumulation is completed. We are now shifting our efforts from the physical to the intellectual. Hopefully we have enough supplies to survive the winter, and we're intending to get as much research done as we can, before the famine becomes too severe. We are about as well-stocked as we could have hoped at this point, even accounting for Pignolius's recently increased brood.

"Research," I hear you ask, "what research?". It's a little hard to explain at this stage, but we intend to apply a systematic approach to the problem we confront—the problem of finding some path that leads out of the current condition and into some much better condition. Away from "red in tooth and claw" and toward—what exactly? I don't know. But I can imagine a range of alternatives that would be preferable to the present condition, with less suffering and more opportunity. The practicality of these alternatives remains very much in doubt, of course, but that is what we are intending to explore.

We already have some ideas which we've talked about during the occasional brief moment of rest, and which we now mean to investigate with focus and energy. I wish I could take credit for some of these, but in truth they derive their origin almost entirely from my friend's formidable mind. However, it does appear that my conversation and my questions, for all their fumbling naïveté, have a stimulating effect on him, for he tells me he's been getting more ideas than usual in the time since my arrival.

I can also flatter myself for having been able to educate him on a few points of fact. For example, I was quite surprised when I had to inform him that thinking is happening in the brain! He had thought it was in the stomach, a misconception that is apparently widely held among the swine.

Of course, I have also learned a great many things from him. This made me wonder how much knowledge there actually is out there. Without any method of combining what different communities have discovered, not only do we not know much, we don't even know what we know.

Could one build something to solve this problem? What would it look like? If it worked, would it cause the world soul to wake up?

But it's getting late, and I should cease subjecting you to the rambling lunacies of my over-excited but over-tired mind.

Goodnight,

Your most indebted nephew,
Feodor

Epistle XV

Dear Uncle Pasternaught,

Thank you for the updates. I am always curious for news from the home front.

Regarding Rey and his hijinks, well, what can one say? I certainly understand the complaints of those who are left having to do the "cleaning up" afterwards. Yet (between you and me) I cannot deny a sense of vicarious pride whenever I hear of his exploits. For all his contumacious antics, he is a bright spark—and it's not as if this world has any deficit of earnest gloom. At least, if we take a family average, it balances out some of my own morose tendencies. But don't tell him I said this!

Over here, the domestic situation has taken an unexpected turn: Porcelain has left, taking the piglets with her. When I learned of this, I was initially alarmed, thinking my presence had caused a strain in their marital relations.

I was preparing to take my own leave, to create a space for her to return. My heart was low, because I felt that Pignolius and I were just beginning to make progress, and now it would have to end before we got a chance to see where it might lead—and who knows how long it'd be until we would have a chance to work together again.

Imagine then my relief upon learning that not only was I not the cause of her departure, but the development did not even indicate a deterioration in their relationship. Porcelain had simply gone to spend some time with one of her sisters and a friend, who both have also just brought forth litters, in order for the three of them to pool their maternal moil. Apparently she has done this before, and Pignolius assures me it is not an issue.

It may not be an issue for him, but should not the father be present to help provide for his children? I raised this issue with him, feeling somehow obligated to do so; though in honesty I was rather hoping in this case that I wouldn't be too persuasive.

Well, I needn't have worried. His reaction was peculiar. He looked at me first for several seconds sort of in disbelief, and then there came from his throat a kind of gagging sound, and I realized he was trying to suppress a laugh. "Is that what they do in your species? The daddies stay at home and take care of the young?" He was now rolling on the ground in uncontrollable convulsions. "Papa papa bring us the milk!" I didn't think it was at all funny, but I said nothing. I know that different kinds of animals have different practices, so one must not be too censorious.

So anyway, now it's just the two of us. Or should I say the three of us? Pignolius, me, and our great task.

Your most indebted nephew,
Feodor

Epistle XVI

Dear Uncle Pasternaught,

I mentioned one or two letters ago how I'd been struck by the fragmentation of knowledge. Things known to foxes are often unknown to pigs, and vice versa. Worse, one group of animals may be completely unaware of important insights that are commonplace among another group of conspecifics living on the other side of a hill or across a river, just a few leagues away. This state of affairs seems very inefficient.

It made me wonder whether we might not be able to produce intellectual progress most simply by connecting these separate reservoirs of knowledge. I wasn't thinking that we would become itinerants, although I did permit myself some pleasant moments of daydreaming about the romance of such an existence—traveling from place to place, learning and teaching as we went, experiencing as much of this world as is possible in one life… But no, it would not do. It would not be a solution. The impact would be too small and too ephemeral. One or two individuals could visit a few communities (realistically only one person would be doing this—I can't see Pignolius being interested in any career that would take him away from his mud pool for too long), and this would continue for a few years at most, until that person died. And then everything would go back to being as it was before.

But, I was thinking, if one could create a durable and scalable *system* for sharing information, then the potential could exist for a more lasting and transformative impact. For example, you could imagine hiring birds to serve as rapid messengers. If they were crows or ravens, they could even serve as information-gatherers and teachers. You could employ smaller birds, such as finches perhaps, to enter communities where the corvids would not be welcome; and the finches could carry simpler messages back and forth between the locals and the larger birds. Then you could have more central hubs where there might be colonies of pigs, and individuals such as Rees the Weasel, where the learnings would be collated, interpreted, and useful lessons extracted for dissemination back out into the communities.

Many variations and elaborations of the basic idea occurred to me, which I will not bore you with, because—as you will readily realize—they are pure fantasy! They all suffer from the same fundamental flaw, namely, we don't have the resources to create them. Even if such a system somehow came into existence, we would not have the resources to maintain it.

Well, we've only just started. More work is clearly needed.

Your most indebted nephew,
Feodor

Epistle XVII

Dear Uncle Pasternaught,

I've discussed the issue I mentioned in my last letter with Pignolius, at some length. Unfortunately it seems that the problem generalizes. We can conceive of various nice-to-haves, be they things or services to assist the current forest dwellers, or systems that would enable progress to occur over time. In fact, it is quite easy and fun to think up these ideas. But they all require some kind of resource to create, operate, and maintain. The problem is that we don't have that. Nor does anybody else, at least not on a consistent basis. There is no consistent surplus.

This is bad enough, but as Pignolius has taught me, the situation is even worse. You can think of the lack of consistent surplus as a great wall that prevents us from escaping our present condition. Pignolius points out that beyond this inner perimeter there is *another* wall, even taller than the first. Should we somehow be able to get over the first wall, we will be trapped by this second wall. What would happen if a surplus were generated, he explains, is that more animals would survive to adulthood, and they would have more surviving offspring, and the additional mouths would eat up the surplus. Then the improvement that we had introduced could no longer be maintained, and we would be back to square one.

Now you might think that this situation looks pretty hopeless. But—ha ha—it gets worse. Pignolius has realized that there is *a third wall* beyond the other two. You see, even if we could somehow persuade some community of animals to reduce their reproduction—which I think would be a very tall order, basically impossible given the strong urges that propel us in the mating season—well, even if we could do this it would still be to no avail, because now and then there would be defectors: some individuals would do what they were not supposed to do and exceed their allotted quota of progeny. With each generation, our community would have a larger and larger fraction of its members being the descendants of quota-dodgers; and, since dispositions are to some extent heritable, the moral character of our population would degenerate. There would be more and more cheaters, and more and more cheating; and soon even good people would start to break the norm, since it's hardly virtuous to be a sucker.

The surplus, and with that our improvement, would vanish into the sands of time.

So we are triply doomed? Well, no. You see—here's the clincher—there is yet another wall beyond the first three. Ha! So we're actually quadruply doomed.

What is this fourth impediment? It is the fact, which is quite obvious to anybody who has lived for more than a few weeks in this forest, that if our local community somehow magically managed to create a sustainable surplus, and we somehow managed to keep our numbers low and to avoid any of us from defecting from this arrangement, what would happen is that beasts would come in from the outside and grab our stuff. If we had food laying around, they would take that. If we had land that was not fully utilized, they would go and settle there. They would also eat us. If the population density in our bubble were lower than outside, the external world would push in, like a higher-pressure gas, until an equilibrium were restored, which it would be only when our population density equaled that of the surrounding areas.

This is what we've seen so far. If you asked me to bet, I'd wager, just based on induction, that there is also a fifth wall, and perhaps a sixth, or even more.

Your indebted nephew and fellow inmate,
Feodor

Epistle XVIII

Dear Uncle Pasternaught,

Perceptive as always, you say you noticed a flippant tone in my last letter, and wonder if there is something the matter.

I can reassure you regarding my physical condition, which appears to be excellent. I mean aside from the fact that we are all in the process of dying from aging and internal decay. Which I don't think should be much cause for concern, since we can look forward to dying before then from starvation, disease, or by being torn apart in the jaws of some bigger brute. But aside from those bagatelles, things are fine!

We have reached an impasse in our investigations. We can't see a way past the difficulties I've outlined; and yet there is much that we still don't know.

Pignolius has a saying, which I chuckled at when I first heard it, but now I'm clinging on to it for dear life, as if it were a precious reed and I were dangling over a cliff:

"So long as there is ignorance, there is hope!"

Your indebted nephew,
Feodor

Epistle XIX

Dear Uncle Pasternaught,

There is not much to report. No progress has been made, but also no progress *is being* made, so I might as well pause the less rewarding activity of staring uselessly at the wall and turn to the more agreeable one of replying to your letter.

Things at present appear relatively bleak. Pignolius observes that our judgments about the merits of the world are mostly simply a reflection of our own habitual mood—sometimes not even our habitual mood but how we feel at the moment. This is counterintuitive. Yet when I reflect on my own experience, I must admit he has a point. For example, I remember how down I felt when I first arrived here, on that dark night a couple of months ago—and how dramatically the outlook seemed to have brightened by the following morning. What had changed? Certainly not the structure of the world or its various balances and equations. Nor had I had any new insights or received any new evidence. No, some sunshine and some carrots had done what no philosophical argument could have accomplished: made the world seem a somewhat cheery place again, albeit one that I could still see had many serious problems. So I must concede that he is right.

Still, when I peer at the snowless winterscape outside, it is hard to avoid the impression that the world is *objectively* depressed.

But enough of my ponderous ruminations!

I can't think of an elegant segue, but I wanted to say that I'm glad to hear that you managed to get the splinter out. I had actually started thinking about whether one could design some kind of instrument that would make this operation easier. Of course I didn't think of the much better solution of asking Irdie for help picking it out! So she has now managed to return some of the favor you bestowed on her when she was a chick, whereas I continue to remain, as always,

Your most deeply indebted nephew,
Feodor

Epistle XX

Dear Uncle Pasternaught,

My apologies for the long delay in responding to your most recent letter. I've been putting off writing, in the hope that I would have something more meritorious to write about, but this has not panned out.

The truth is that Pignolius and I have been spending the past fortnight, every day from morning till late into the night, engaged in intense and focused intellectual activity… *playing a game.*

That's right, we've been "investing" our talents and accumulated surplus and rare privileges into mastering a board game that Pignolius has invented. Have we discovered an ingenious way to model the world and explore different scenarios in a simulated environment, so that we can more rapidly devise and test out different potential courses of action? No, we've just been playing a game.

I might say, in our defense, that we were stuck anyway, so the opportunity cost—especially with the current weather—was relatively low. I might also say that the plan, insofar as there was one, was to spend maybe a couple of hours doing this; but then things became a little addictive…

It is, at any rate, a quite agreeable way to spend time. Of course, Pignolius beats me every time we play without a handicap; yet something inside me wants to keep trying, a little feeling I have that if I just try a little harder maybe next time I might win. I cannot deny that I am enjoying it.

Your indebted nephew,
Feodor

Epistle XXI

Dear Uncle Pasternaught,

I had the strangest dream. Pignolius and I were going for a little walk. Suddenly, in a clearing just a few paces ahead, we see a lamb—from whence it found its way there no one knows—and this little lamb has a really bad case of the mange. It looks absolutely pitiful. Pignolius bolts from the scene, whether from fear of infection or from the sheer horror of the sight. I know I should do the same; instead I approach the lamb. Not to eat it, but to lay my paws around it, to comfort it. I get closer and closer. Just as I reach out to touch it, I wake up.

I don't know what this means, but I feel like I must move. There is this large body of water, only two days' walk from here. People call it "the sea". Despite the short distance, Pignolius has never been there, and he doesn't seem to have any desire to go.

So I will go alone. I have a sense, which I cannot articulate or explain, that there is something that I need to work out, and I need to spend some time alone.

I won't be able to write to you until I get back. I don't know how long I will be gone, but I want you to know that I am and remain

Your indebted nephew,
Feodor

Epistle XXII

Dear Uncle Pasternaught,

I am now back. I have been back for two weeks, and I'm sorry for the delay in writing to you. I hope to write soon and explain more.

In the meantime, although I'm loath to impose on you even more, I need to ask you for a favor: Would you please give the enclosed letter to Rey, if you know his whereabouts? It is urgent.

Your now still more indebted nephew,
F

WEDNESDAY

Full unemployment

Welcome back. Yesterday we talked about some different types of boundaries. These are summarized in Handout 6.

> **HANDOUT 6**
> BOUNDARIES
>
> - Cosmic endowment [Handout 1]
> - Technological maturity [Handout 2]
> - Coordination problems
> - Prudential barriers
> - Axiological contours [Handout 3]
> - Metaphysical constraints
> - Automation limits [Handout 5]

We've covered a good deal of ground in the first two lectures. Today I want to return to the issue that sparked our investigations on Monday: the "purpose problem". It is somewhat customary at this point, almost de rigueur, to express concerns about the negative consequences that would flow if the project of automation were really successful and we end up in a condition of widespread or universal unemployment.

Well, then, let us express some concern!

If there is nothing or almost nothing that couldn't be done better by machine, then what would there remain for *us* to do? What would we *do* all day in utopia?

Brawl, steal, overeat, drink, and sleep late

"Idle hands are the devil's workshop", the saying goes.[75]

A somewhat more literal translation of the Biblical passage (Proverbs 16:27) is: "A worthless man devises mischief", which is quite different.[76] Maybe the worthless man needs to be given a fully loaded schedule to prevent him from devising mischief, whereas a man of worth might be safe from mischief-making even if he were placed in a condition of idleness? The latter would use his freedom from externally imposed requirements in worthwhile ways. He might engage in pious contemplation or find some other virtuous application of his time and faculties.

However, even if it is true that idleness poses no threat to a good man, this may be scant comfort, as we may wonder what fraction of men fall into that category. If most men are worthless, we may be in for trouble even if there are some individuals for whom idleness would be a blessing.

Keynes worried about what would happen once the strenuous efforts of the moneymakers carry all of us along "into the lap of economic abundance":

> "There is no country and no people, I think, who can look forward to the age of leisure and of abundance without a dread. For we have been trained too long to strive and not to enjoy. It is a fearful problem for the ordinary person, with no special talents, to occupy himself, especially if he no longer has roots in the soil or in custom or in the beloved conventions of a traditional society. To judge from the behaviour and the achievements of the wealthy classes today in any quarter of the world, the outlook is very depressing! For these are, so to speak, our advance guard—those who are spying out the promised land for the rest of us and pitching their camp there."[77]

Similar concerns were voiced more recently by the eminent American jurist Richard Posner in his critical review of *How Much Is Enough?*, a book by Robert and Edward Skidelsky, in which they proposed reforms to the current capitalist system to reduce its emphasis on growth and consumption and make it easier for people to escape the rat race and to enjoy more leisure.[78] In response to these proposals, Posner wrote:

> "The Skidelskys have an exalted conception of leisure. They say that the true sense of the word is 'activity without extrinsic end': 'The sculptor engrossed in cutting marble, the teacher intent on imparting a difficult idea, the musician struggling with a score, a scientist exploring the mysteries of space and time—such people have no other aim than to do what they are doing well.' That isn't true. Most of these people are ambitious achievers who seek recognition. And it is ridiculous to think that if people worked just 15 or 20 hours a week, they would use their leisure to cut marble or struggle with a musical score. If they lacked consumer products and services to fill up their time they would brawl, steal, overeat, drink and sleep late."[79]

So that's Posner's view of human nature—more pessimistic than that of the Skidelskys.

*

How many of you have heard about Michael Carroll, also known as the "Lotto lout", or as the self-proclaimed "King of Chavs"?

Mr. Carroll has attained some notoriety in his native country, the United Kingdom. When he was 19 years old and working part-time as a binman, he won nearly 10 million pounds on the National Lottery. The tabloid press followed him closely as he proceeded to spend his winnings on prostitutes, cars, crack cocaine, jewelry, gambling, champagne, parties, and legal fees; though he also gave generously to his friends and family. He developed a habit of hurling Big Macs at people from his car. His wife, who was pregnant with his child, divorced him because of his drinking and all the prostitutes. At one point he was apprehended for catapulting steel balls from his Mercedes van at parked cars and shop windows while drunk. The judge noted that at the time of the incident, Carroll had already accumulated 42 offenses on his record.[80] So he has not exactly been idle. But, as far as is known, he has not cut marble.

Interestingly, according to more recent news, Carroll has blown through his entire fortune and is again penniless. He has taken a job chopping firewood and making coal deliveries in Elgin, Scotland, working for up to twelve hours a day.[81] He has lost five stone and remarried his wife. He says he's happier now.[82]

I have included a picture of Michael Carroll on the handout, at the peak of his prosperity, next to a picture of another wealthy person who had a more artistic temperament, for comparison.

HANDOUT 7

MICHELANGELO AND MICHAEL CARROLL

[Source: Wellcome Collection,[83] Albanpix.[84]]

(*Left*) Michelangelo. Peak net worth: circa £10 million, of which he spent little.[85] Slept with his boots on. Cut marble. (*Right*) Michael Carroll. Peak net worth: circa £10 million, of which he spent all within a few years. Claims to have slept with 4,000 women. Did not cut marble.

*

I think the lesson we can draw from this is that people are different. We respond differently to wealth and leisure.

One could also point to cultural and class differences here. In the public commentary on Mr. Carroll's exploits, the most negative attitudes seem to have been expressed by individuals from the lower middle class, who bemoaned the profligate and antisocial ways in which he spent his fortune.

Some working-class individuals seem to have more sympathy, noting it was his money to spend, and he had himself a good old time, and now he is making an honest living again.

People further toward the top up the social hierarchy were again less censorious, viewing Mr. Carroll as a colorful phenomenon rather than a relevant moral comparison; while also approving of the fact that he did not take his wealth too seriously. As Posner says, "the traditional aspiration of the English upper class was not to work at all", and not to appear to care too earnestly about money.[86] And in this respect, the Lotto lout, although solidly lower working class, actually exhibited a more aristocratic demeanor than people on the middle rungs of the social ladder.[87]

We could also point to cultural differences—for example, the American upper class is based more on wealth than its British counterpart, the latter being focused more on heritage and a certain kind of cultural capital. I also expect cultures steeped in the Protestant work ethic to be more likely to frown on the ways of a Lotto lout, while other cultures may view his happy-go-lucky lifestyle more approvingly. It all gets very complicated. There could be an entire lecture series on just these issues.

What we can say is that it seems plausible that for some people, perhaps a significant fraction of the current population, a sudden leap into great wealth and complete leisure would not be an unalloyed blessing; and for some, it could be ruinous.

*

In any case, it would be a mistake to model the psychological or sociocultural outcomes of universal unemployment by extrapolating the effects we see today, on a smaller scale, among individuals who have lost their jobs. It is well known that losing one's job has a range of bad consequences. Unemployment raises the risk of alcoholism, depression, and death.[88] But the scenario *we* are considering is different in several ways.

First, and most obviously, losing one's job today means, for many, either actual financial hardship or stress and anxiety over the increased risk of encountering such hardship later—whereas, in our hypothetical, we're supposing that everybody has a secure high level of income.

Second, job loss is today often associated with stigma—whereas this would not apply if *everybody*, or almost everybody, is out of a job, as in our scenario.

Third and relatedly, job loss today often has a strong negative effect on self-image, partly because of the aforementioned stigma and partly because many people's identity is tied up in their role of being a breadwinner for the family or being a success in the labor market—whereas, in our scenario, where those roles are simply nonexistent, people would form their identities around other attributes and relationships.[89]

Fourth, becoming unemployed today often means losing social connections to work colleagues, and more generally it can make it harder to relate socially to people who have jobs—whereas, again, this does not apply if we are all unemployed.

Fifth, if we simply compare the lives and circumstances of the employed and the unemployed, we can be misled unless we take into account that there may be selection effects at play. Individuals with less enterprise, drive, education, health, emotional stability, etc. are more likely to become unemployed.[90] If we observe a different distribution of those characteristics among people who have just lost their jobs, it is quite possible that some of the causation goes in the other direction—whereas, in the case of universal unemployment, the unemployed would be identical to the general population.

Templates of otium

Because of these differences, rather than looking at outcomes in populations of laid-off workers, it would be more instructive to consider cases where some social group has persistently had very low levels of labor market participation (for reasons unrelated to health) while at the same time enjoying robust economic prosperity and a reasonable level of social status. While it is difficult to find a perfect example, we can think of various real cases that share some properties of the situation we have in mind while differing in other important respects.

Children

Young children in modern societies don't work. Economically and socially, their status is ambiguous. They have virtually no disposable income; yet they live in a parental "welfare state" that caters to all their needs. They are powerless, disenfranchised, and their opinions disvalued; yet they are beloved and nurtured, and their welfare is often the focal point for the people around them. There are also huge biological confounders—the fact that some situation is good for a child does not give much evidence that it is good for an adult. While children often have rich and happy lives, their experiences may therefore only be somewhat relevant to the case we are considering.[91]

Students
Although students typically don't do much paid work, one could think of studying as a work-analogue: something they "have to do", whether they feel like it or not, and which has deferred economic payoffs. Students often have relatively low incomes, but also few financial responsibilities and limited needs. They have somewhat high social status. Compared to children, biological confounders are far less extreme but still significant, students typically being younger and physically fitter than the average adult. There is also some selection effect in that students are smarter and better educated than the average person. Students often have rich and happy lives, and their experiences could be a relevant comparison point for what a leisure society could look like.

Aristocrats
Traditionally, to a large extent, the European landed gentry sought to avoid engaging in wage labor. This does not mean they did no work: overseeing and managing their households and estates, military service, and political participation could require significant time and effort. Still, they had more free time than most people, while living in relative economic abundance. Their exceptionally high social status is a confounder. For a contemporary case of a similarly leisured class, but situated within a different cultural context, we might look at the native population of some oil-rich Gulf states, which enjoys a comparatively high material standard of living while working relatively little.

How do we feel about the quality of life of these groups? Depending on one's perspective and reference point, their examples may or may not offer an inspiring outlook. Note that the question before us is not whether we *like* people who have a lot of inherited money; nor are we asking how *admirable* they are on average; nor whether it is *useful* or *fair* that a society contains elites who don't work for a living. The question is rather *how well their lives are going for them*—how their average level of flourishing or well-being compares to that of other groups.

Bohemians
We could look at artist colonies and other communities that deemphasize paid work and participation in commercial enterprises in favor of some form of cultural production. Archetypally such groups are relatively poor, but may rank relatively high in certain forms of social status. Observed outcomes are influenced by significant selection effects: the personality traits of people who opt out of mainstream society to focus on artistic pursuits diverge from those of the general population.[92] There are also many other differences between communities of bohemians and various kinds of bon vivants on one hand, and communities of stolid hard-working bourgeois on the other, which are not directly a product of work or money. Still, it is another example that seems at least somewhat relevant.

Monastics
Monks and nuns sometimes work for their sustenance, which would make them less relevant as a comparison point, but some monastic communities offer a life that is at least somewhat free from the necessity of economic labor. Of course, we can expect a strong selection effect among those who choose a life of extreme devotion to religious practice, which in many cases involves a vow of celibacy, poverty, hermitism, or other abnegations of worldly indulgences. Moreover, the structured pursuit of spiritual practices functions as a substitute for paid work by creating a fixed framework for regular occupation, exertion, and self-discipline.

How one evaluates the desirability of the monastic life probably depends sensitively on one's religious views. Such a life may be highly desirable insofar as it provides important spiritual benefits. It might even involve a fair deal of temporal joy.[93] Yet absent a faith that justifies and motivates it on transcendental grounds, monasticism is probably too austere for most people's liking.

Retirees
The case of retirees is confounded by the obvious demographic variable: they are typically much older than the average adult and suffer a much higher incidence of poor health. Their prospects are also terrible, as they can look forward to a period of accelerating decline, illness, incapacity, followed by death, while receiving regular news of the passing of many of their lifelong

friends. However, if we control for these factors, the picture looks relatively rosy. Surveys of subjective well-being often report a peak among people in their late 60s, although the field is clouded by methodological disputes over what to adjust for and how.[94]

*

All of these comparisons are imperfect in different ways; but maybe if we construct a composite we could evoke some kind of blurry picture of what a society populated by non-working people might be like?

My main point here, however, is negative. Even if there is not much we can learn from these comparisons, they highlight that there are a variety of quite different examples in the contemporary world that we might view as possible models for a society of universal unemployment. This range can help us avoid anchoring unduly on any one particular comparison case that happens to pop into our mind when we ponder the issue, saving us from reaching a premature conclusion on the basis of just one very thin slice of human experience.

Anyway, as we shall see, the real issue is quite different…

Leisure culture

Before I get to that, though, I want to offer my idea of what the solution would be if the problem were what we have so far imagined it to be. As I hinted, the real problem is different and deeper. But let's anyway see what the conclusion *would* be if our inquiry had terminated at this station: if the problem to be addressed had been how to imagine a positive vision of a society of universal unemployment.

That is to say, let us imagine a setup like the following: Technology has progressed to the point where machines can substitute for human labor across the board. AI and robots can do everything humans can do, and they do it better and more cheaply. It is no longer feasible for humans to earn income through their labor. (We can set aside, for now, the possible limits to automation that we discussed earlier; we'll return to some of that later. In any case, much of the 100% unemployment case also applies, in attenuated form, to less extreme scenarios—and, analytically, it is often better to start by analyzing the extreme case, adding back complications later.)

With such dramatic progress in automation, there is enormous economic growth. This results in a spike in per capita income, at least in the short run (until population growth, if left unchecked, resubmerges us into a Malthusian situation). In this scenario, therefore, it would be relatively easy to ensure that what Keynes called "man's economic problem" is solved: everybody could have a very high level of material welfare. *Could*, of course, is not the same as *would*. Actually achieving a high universal floor may also require political achievements. But here we are not making a prediction but simply analyzing a possibility; so let us postulate that the distributional problem is solved to the extent that—at least in this scenario where per capita income becomes sky high—everybody gets a share that is sufficient for a high level of consumption.

So that's the setup, and for now we bracket any further impacts of technologicalization beyond those I just described.

Then what? How do we respond to Richard Posner's fear that the outcome of such idle prosperity would be brawling, stealing, overeating, drinking, and sleeping in late?

Well, as for sleeping in late… if that be the price of utopia, I for one would be glad to sign up. It is not by accident that these lectures start in the afternoon.

But generally, what are we to make of Posner's concerns? Or other similarly negative images of the consequences of idleness, such as those of "the superfluous man" in Russian literature, or the Lotto lout's hijinks as chronicled in the British tabloid press?

HANDOUT 8

THE SUPERFLUOUS MAN

[Source: *Eugene Onegin.*[95]]

The "superfluous man" was a *type* in Russian literature, found in some of the novels by Pushkin, Turgenev, Tolstoy, and others. The superfluous man is an aristocratic, intelligent, and well-educated person, who lives an existence as a bystander. The type originated during the reign of Tsar Nicholas I, when many capable men would not enter the discredited government service and instead doomed themselves to a life of passivity in which they lacked opportunities for self-realization. The superfluous man is one paradigm of a person who has lost his place and purpose in life. The result is often depicted as existential boredom, cynicism, self-centeredness, and a general lack of zest and initiative: gambling, drinking, and dueling are common dissipations.

*

The answer, I think, is that we'd need to develop a culture that is better suited for a life of leisure.

Start with education. The current paradigm is one of industrial production. The raw materials—children—are delivered to the school gates for age-based batch processing. They are hammered, ground, and drilled for twelve years. Graded and quality-controlled worker-citizens emerge, ready to take up employment in a factory or a trucking company. Some units are sent onward to another plant for three to ten years of further processing. The units that emerge from these more advanced facilities are ready to be installed in offices. They will then perform their assigned duties for the few decades remaining of their active life.

If we look at this process, we can see that the main functions performed by our education system are threefold.

First, storage and safekeeping. Since parents are undertaking paid labor outside the home, they can't take care of their own children, so they need a child-storage facility during the day.

Second, disciplining and civilizing. Children are savages and need to be trained to sit still at their desks and do as they are told. This takes a long time and a lot of drilling. Also: indoctrination.

Third, sorting and certification. Employers need to know the quality of each unit—its conscientiousness, conformity, and intelligence—in order to determine to which uses it can be put and hence how much it is worth.

What about learning? This may also happen, mostly as a side effect of the operations done to perform (1) through (3). Any learning that takes place is extremely inefficient. At least the smarter kids could have mastered the same material in 10% of the time, using free online learning resources and studying at their own pace; but since that would not contribute to the central aims of the education system, there is usually no interest in facilitating this path.

I'm going to take a sip of water…

Sorry! I hope you got more out of school than I did.

*

In any case, if we imagine a world in which the need for work was removed, then clearly there would be an opportunity and a need for the focus of education to change. Instead of shaping children to become productive workers, we should try to educate them to become flourishing human beings. People with a high level of skill in the art of enjoying life.

I don't know exactly what such an educational program would look like. I would think, maybe, it would involve cultivating the art of conversation. Likewise, an appreciation for literature, art, music, drama, film, nature and wilderness, athletic competition, and so forth. Techniques of mindfulness and meditation might be taught. Hobbies, creativity, playfulness, judicious pranks, and games—both playing and inventing them. Connoisseurship. Cultivation of the pleasures of the palate.

(Admittedly, had I been taught these subjects by the teachers I had, it would in all likelihood have turned me off of these things for life. But since we are imagining an unreal condition, with universal material abundance etc., we might as well also allow ourselves to imagine that good instruction would somehow be available for all.)

What else? Maybe practice in humor and wit and keen observation. Celebration of friendship. Performing arts of various kinds, crafts, encouragement of simple pleasures. Of course, habits conducive to health would be ingrained. I imagine spiritual exploration and sensitization as well. I think the current focus on discipline would not be entirely eliminated but rather transformed: because I think that focus, attention, concentration, self-control, persistence, and the ability to take pleasure in deliberate practice and in mental and physical exertion would remain important—perhaps more important than today, since there would be less occasion for these habits to be entrained by external demands and hardships. Cultivating curiosity—here I may be projecting my own proclivities, but I think a passion for learning could greatly enhance a life of leisure. Also, cultivation of the virtues and an interest in moral self-improvement. The opening of the intellect to science, history, and philosophy, in order to reveal the larger context of patterns and meanings within which our lives are embedded...

Nota bene, I'm not saying this would be the right focus for education *today*, in a world where there is a need for work to be done and many pressing material problems to be solved. But if and when those problems are solved, or when we get to a stage where the responsibility for further progress can be handed over to artificial bodies and brains, *then* the focus for us humans could and should shift in these directions.

The education system is just one aspect of society that would need reform. More broadly, we'd need a transformation of culture and social values. A move *away from* efficiency, usefulness, profit, and the struggle for scarce resources; a move *toward* appreciation, gratitude, self-directed activity, and play. A culture that places a premium on fun, on appreciating beauty, on practices conducive to health and spiritual growth, and that encourages people to take pride in living well.

We have a lot to learn and discover about what can be done in that direction. I think there are considerable opportunities.

*

Would we get bored if we lived in a world without work?

"Millions long for immortality who don't know what to do with themselves on a rainy Sunday afternoon."[96]

On the face of it, the problem we confront here is not that theoretically complicated. There are many worthwhile things that humans do in life that are not geared toward making money. If we get more leisure, we can do more of those things.

I can be more specific—have a look at Handout 9. I won't read the whole thing out, but the long and the short of it is that there are many things to do.

HANDOUT 9
WHAT TO DO WHEN THERE'S NOTHING TO DO

Building sand castles, going to the gym, reading in bed, taking a walk with your spouse or a friend, doing some gardening, participating in folk dance, resting in the sun, practicing an instrument, playing a game of bridge, climbing a rock wall, playing beach volleyball, golfing, bird watching, watching a TV series, cooking dinner for friends, going out on the town partying, redecorating the house, building a treehouse with children, knitting, painting a landscape, learning mathematics, traveling, participating in historical reenactments, writing a diary, gossiping about acquaintances, looking at famous people, windsurfing, taking a bath, praying, playing computer games, visiting the grave of an ancestor, taking a dog for a walk, sipping a cup of tea, running a marathon, engaging in witty banter, watching a football match, shopping, going to a concert, protesting an injustice, having a picnic, going on a camping trip, eating ice cream, organizing a murder mystery game, playing with LEGOs, wine tasting, having a massage, learning about history, doing a silent retreat, taking drugs, getting your nails done, attending a religious ritual, keeping up with current events, interacting on social media, exploring virtual reality environments, kayaking, learning to fly a sports plane, gambling, pouring a martini, celebrating a holiday, researching your family tree, participating in a neighborhood clean-up, singing in a choir, meditating, carving pumpkins, swimming, solving a crossword puzzle, visiting friends, making love, driving in a demolition derby, biohacking yourself to optimize physical and mental performance, attending an amateur astronomy meeting, creating a time capsule, teaching a young person something you know, watching a sunset, going to a costume party, arguing about moral philosophy, judging a koi fish competition ("living jewels"), collecting antiques, attending a lecture… The list goes on.

There are some people who would not find any of these activities motivating or fulfilling, and who will be unable to think of enough else wherewith to fill the hours that would be freed up in a condition of emancipation from economic necessity. But this is more a fact about those people and their psychology than a consequence of any objective deficiency in the number of things there are to do. Some people are simply high in a trait value that psychologists refer to as *boredom proneness*.[97]

Boredom is actually an important topic, and we shall discuss it in more depth tomorrow. For now, I'll just say that it seems quite possible that, with appropriate changes in education and culture, we would feel *less* bored in a post-work world than we do today. Aside from presenting the opportunity to adapt education and culture to foster fulfilling leisure, the greater levels of wealth and better technology would also make it easier to build institutions and infrastructure that support a wide range of enjoyable and fulfilling activities.

But what if universal automation does lead to some increase in boredom? My guess is that it would still be good overall, considering the many people around the world who currently live in such abject poverty that being catapulted into great wealth would have to be regarded as a big improvement *even if* it resulted in a life of some tedium and frivolous dissipation. Brawling, stealing, overeating, drinking, and sleeping late may not make for the *best* life, but even that could be a lot better than one of deprivation or incessant grind under the thumb of some mean and vexatious taskmaster—

Message from the Dean

Oh, I see our Dean is here! What can I do for you?

She is telling me she has an announcement to make.

Dean: Thank you, and thank you all for coming to this year's Philip Morris Lectures in Moral Philosophy. I have an important request to make of you this afternoon.

As some of you may be aware, twenty-two years ago the University established a committee to examine issues which had been brought to our attention concerning the nomenclature of the lecture hall you are currently sitting in, the Enron Auditorium. Five years ago, following a thorough stakeholder consultation, a decision was made to rename this room. And today we will be announcing its new name. Concurrently, we will be announcing the creation of a new professorial chair in global sustainability—a key investment which will cement our ability to provide change and intellectual leadership in this critical area.

To celebrate the generous gift that has made this possible, we will be hosting an event right here, which will start one hour after the end of the present lecture. This is the occasion for my request. As we will be joined in the celebration by a member of the board of our benefactor, it is of the utmost importance that we have a robust turnout so that we can show our gratitude and appreciation for this forward-thinking investment not only in the future of the planet but in the University's humanities division, which is where the new chair will be located.

Bostrom: Thank you, Dean. That is remarkable news.

Actually, this could be a good time to take a couple-minutes break. Let's pick it up again right at the hour.

Firafix: Do you think that we are under any obligation to stay on for the celebration? I mean, we are not members of the University.

Kelvin: We are definitely not under an obligation.

Tessius: I'm going to take a leak.

Kelvin: He could have given the Haredi Jews in Israel as another example of a leisured class. There is a sizable population who spend their entire lives studying the Torah, and get paid by the state for doing so.

Firafix: Maybe that would fall under the category of Monastic?

Kelvin: They don't live in monasteries. They have families and lots of kids.

Firafix: If they spend their lives studying the Torah, maybe that's rather like having a job? Especially if they get paid by the state for doing so.

Kelvin: Maybe, but it still seems like another example to consider.

Firafix: What do you think we should do after the lecture?

Kelvin: Well, there is a list here... "taking a bath"... I would not vote against going to the hot springs again?

Firafix: Sure. I'll need to get my nails done at some point soon and get a new set of shoes, but it doesn't have to be today.

Wild eyes?

Bostrom: Ok, we resume!

I was thinking that before proceeding further, I ought perhaps to say a few words about how our discussion relates to other efforts that have been made in the utopian genre. I will not be attempting a comprehensive review of the field. But I want to make one observation, which is that the great bulk of utopian literature is based on quite different premises than our investigations in this lecture series.

Traditionally, utopian works have sought to envisage a more ideal social order, one in which customs, laws, and habits may diverge from their contemporary settings, but one that nevertheless shares certain fundamental elements with the status quo. In particular, it is usually taken for granted that (*a*) some amount of human labor is necessary to produce food and other essentials, and (*b*) the most basic aspects of human nature remain essentially unchanged (though changes in upbringing might be imagined to have remolded people to some degree, perhaps making us less selfish or materialistic).

Within these parameters, the author can then imagine a different political system, a different way to organize work, a different way to bring up children, a different way for men and women to relate to one another, a different way for humans to relate to nature, and so forth. Depending on which relationships get emphasized—and the author's view about what would constitute an improvement—we get various visions of a perfected society: ecological, libertarian, feminist, socialist, Marxist, etc. But they are all broadly contoured by the same assumptions: the need for work to be done and the immutability of core attributes of human nature.

Karl Marx, for example, while he didn't offer a detailed picture of what life would be like in his communist paradise, imagined that people would still be working; albeit with some important differences: he thought we would no longer be defined by our occupations or alienated from our labor, and the work we would be doing would be of more variegated character:

> "[I]n communist society, where nobody has one exclusive sphere of activity but each can become accomplished in any branch he wishes, society regulates the general production and thus makes it possible for me to do one thing today and another tomorrow, to hunt in the morning, to fish in the afternoon, rear cattle in the evening, criticize after dinner, just as I have in mind, without ever becoming hunter, fisherman, shepherd or critic."[98]

The condition seems to be one in which there has been some degree of easing of "man's economic problem"; and the character of work would also be different—more integrated and more consistent with the development of the human capacities of the workers. Our interactions would be less transactional and more personal and based on solidarity. Later Marxist writers (though notably not Marx himself) might add that the many consumerist desires that are artificially inflamed in capitalist economies would be abated. But, fundamentally, the vision is of a different way of organizing economic production and political control.

Tessius [*returning, whispers*]: They're putting up a new sign outside. Want to guess what the new name is?

Kelvin: ?

Tessius: The Exxon Auditorium.

Firafix: What?

Tessius: We're attending a lecture in the Exxon Auditorium.

Bostrom: Let's call these kinds of visions, which focus on how people (and animals and nature) could interact in ways that make for an allegedly more harmonious way of living, *governance & culture utopias*. They hold up images of how society could be "run better", if we take this in the broadest sense, as encompassing not just laws and government policies but also customs, norms, habitual manners of going about things, internalized ways of viewing others, occupational and gender roles, and so forth.

Sadly, when people have had the opportunity to put governance & culture utopian visions into practice, the endeavors have often fallen short of expectations, with typical outcomes ranging from disappointing to atrocious.

But maybe next time? Between the sunshine of hope and the rain of disappointment grows this strange crop that we call humanity (along with fantastic rainbows of excuses and self-justifications).

Since the harm produced by utopian visionaries seems to correlate with the degree of violence they have been able and willing to wield in their attempt to realize their dreams, it might be best if future experiments of this sort were to be pursued in a more incremental and voluntaristic manner, and if they start with small-scale opt-in demonstration projects that others can then gradually be inspired to emulate once a track record of happiness and success has been achieved. (Thus spake the spirit of shriveled old age?)

*

Although most of the utopian literature to date is of the governance & culture type, that is not our topic of this lecture series. We've instead been exploring some issues that arise in what I will term *post-scarcity utopias*. These are predicated on the assumption that a condition of economic abundance is somehow achieved. This idea is not new, of course: the Land of Cockaigne is essentially a post-scarcity utopia. And it is common for drafters of governance & culture utopias to assume that a society organized according to their prescriptions would also achieve some degree of economic abundance.

We don't need to be too strict with these definitions. I mean, whether our governance and our culture is harmonious, fair, and conducive to flourishing is a matter of degree, as is the cornucopian character of our society. There is also ambiguity in the notion of economic abundance: exactly what kinds of goods and affordances are "economic"? There are many things you can't buy even with infinite money—for example because they haven't been invented yet. But for our purposes it may be sufficient to say that a post-scarcity utopia is one in which it is easy to meet everybody's basic material needs as traditionally conceived—food, housing, transportation, etc. We may toss in schools and hospitals and some other such services into the mix as well.

And we can then observe that, in developed countries, we have already come a long way toward realizing this type of abundance—say, more than halfway toward a post-scarcity utopia.[99] This estimate obviously omits our animal brothers and sisters, for the vast majority of whom the situation is still most dire and in urgent need of amelioration.[100]

*

In these lectures, we have then gone beyond post-scarcity to talk also about what we can term *post-work utopias*. These are visions for a society that has achieved full automation and thereby eliminated the need for human labor. Again, we will allow the definition to be a little vague—we may count a leisure-dominated society as a post-work utopia (or dystopia) even if some modest quantity of economic work must still be done by hand.

Shall we say that wealthy countries are something like between a third and half of the way toward the leisure society? We have long childhoods and retirements, as well as weekends and holidays. If and when we get to Keynes's 15-hour work week, then maybe we would be eighty percent of the way there.

*

Let it not be thought that traditional governance & culture utopias are necessarily "more realistic" than post-scarcity or post-work utopias. What is realistic depends on the context. If we are considering a condition of technological maturity, what may be *unrealistic* is to assume that there will still be much need for human labor.

I go further and assert that as we look deeper into the future, any possibility that is not radical is not realistic.

The purpose problem revisited

You'll recall from the first lecture how Bill Gates was worrying about where we'd find purpose in the future:

> "The *purpose* problem. Assume we maintain control. What if we solved big problems like hunger and disease, and the world kept getting more peaceful: What purpose would humans have then? What challenges would we be inspired to solve? In this version of the future, our biggest worry is not an attack by rebellious robots, but a lack of purpose."[101]

Well, if we construe the problem as the challenge of filling the hours in a workless world, the solution would be to develop a culture of leisure; and I gave a number of suggestions for what we might do. We would emphasize enjoyment and appreciation rather than usefulness and efficiency. This would be a significant transition, but I see no reason to think it an infeasible one.

It would be natural to extend this suggestion from the individual to the societal scale. Today's societies may set themselves goals such as clean air, good schools, high-quality healthcare, adequate pensions, an efficient transportation system, and so forth. Once those goals have been achieved, ambition could turn in more cultural directions: let's say, to create a society where people care about one another, where individual differences are recognized and celebrated, where many people come together to create large happenings, where customs are continuously refined to make daily interactions more meaningful and fulfilling, and where there are constantly renewed efforts to deepen and broaden the public discourse about art, religion, ethics, literature, media, technology, politics, science, history, and philosophy. And so on and so forth. Again, a significant transition—but, really, an opportunity rather than a problem.

*

One might think that as our challenges get smaller and more parochial and become less a matter of life and death, their ability to generate passion and engagement would decline. But this is not clearly the case. More people jump out of their seats when their soccer team scores a goal than when an international agency publishes a report saying that a hundred thousand fewer children died from preventable diseases this year than last.

(We take this to be completely normal, but I wonder, if we could see ourselves through the eyes of angels, whether we would not recognize in this pattern of excitement and indifference something quite perverse—the warped sentiments of a moral degenerate? Is it not, implicitly, a sort of emotional middle finger to the suffering and desperation of other sentient beings?)

*

It is worth noting that in *some* respects a leisure utopia would be closer to the natural human condition than is our current world. I don't think being woken by an alarm clock and summoned to sit behind a desk processing paperwork for an insurance agency or some other bureaucratic behemoth is at all natural. Some researchers have suggested that our Stone Age forebears had plenty of free time, that they may have worked as little as four hours a day.[102] I'm a bit skeptical of the number, but what is likely true is that the boundary between work and leisure was not so clearly drawn in those primitive societies. When people's instincts are well-matched to their environment, maybe they mostly just do what they feel like in the moment, and that happens to coincide with what is useful. We, by contrast, we *Homo cubiculi*, needs must rely on self-discipline and structured incentives to get us to perform the requisite labors.

So if a prelapsarian condition of mostly spontaneous activity was our lot for 98% of our evolutionary history (or more, if we count our great ape ancestors), then you may say that the attainment of a leisure society would in some important respects constitute a return to our roots. Maybe overshooting it a bit, if we move to a world with no work at all. But adequately close to our natural and original state—yet with all the boons of high-tech civilization, from air conditioning to broadband to cinema to dentistry to electric stoves… to zillions of other marvels and delights that are still to be conceived.

*

But everything I've said up to this point is prolegomenon. Mere throat-clearing, if you will.

I now put it to you that we haven't even *identified* the real problem yet, let alone solved it.

The preceding discussion was necessary to get some other, otherwise distracting, issues out of the way and to build us something of a shared frame of reference. However, the actual challenge of deep utopia still lies ahead of us.

You see, human wage labor is not the only thing that will be obviated at technological maturity. Machine superintelligence, and other innovations that will accompany this advance, can do so much more than substitute for you in the workplace; they will also remove the need for many other kinds of human striving. This means that the mycelium of "the purpose problem" extends far wider and deeper than the parts that are familiar in forms such as how to occupy oneself during an extended holiday or when one is too young, too old, or too rich to work.

We will approach this issue by first looking at some case studies. Later, we can generalize our observations and discuss the matter at a more abstract level.

So let us ask again, just exactly what would we *do* all day in a utopia set at technological maturity? Handout 9 listed some suggested activities. Let's pick a few items from this list, more or less at random, to examine more closely: shopping, exercising, learning, and parenting. If we cover those four, you will see how the argument goes and you can extend the analysis to other activities on your own.

Any questions so far? Okay, let us proceed.

Case study 1: Shopping

To many, shopping is a necessary evil; but there are also folk who enjoy this activity, and who would gladly spend more time engaging in it if they had money to spare and they didn't have to work.

Of course, you might enjoy *using* what you've bought. You buy a pair of rollerblades, then you have fun rolling on them. But here we are discussing the shopping process itself, not the subsequent utilization.

So let's look at this more closely. What precisely is it that people enjoy doing when they enjoy shopping?

Shopping involves engaging in several interrelated cognitive tasks and activities, including the following:

- *Exploration.* Searching for alluring objects or bargains, using knowledge and heuristics about shop locations, store characteristics, shelving practices, prices, etc.
- *Evaluation.* Perceiving the attributes of a potential purchase in order to judge how well the object would serve a need. Would the color, shape, texture, and brand of a garment meet the mutable and context-dependent demands of fashion? Would it fit the body and style of its prospective wearer? Does its quality warrant its price?
- *Imagination.* As part of the evaluation process, imagining how well or poorly the potential purchase would serve its intended function. Would some decorative object look good in the home? Would some item of clothing be suitable for particular occasions? How often would it be used?
- *Theory of mind.* Modeling the minds of others to predict how they would regard the object in question. Would they find a

particular garment attractive? Would they appreciate receiving it as a gift?

- *Communication and social learning.* Discussing actual or potential purchases with a friend, gaining and conveying information about personal tastes, sharing insights about who and what is in or out of fashion.

Some of these aspects of the skill and effort of shopping are already being undercut by recommender systems and other functionalities that are becoming available thanks to progress in AI. Instead of the shopper having to visit many boutiques or having to browse up and down the aisles of a department store, they can visit a single online vendor. Offers predicted to be of greatest interest to the customer are brought to their attention.

Let's extrapolate this a bit. If the recommender system is sufficiently capable, it would remove the need for exploration entirely. The system would know your tastes and offer suggestions that you like better than whatever you would have picked out yourself. Then what would be the point of you searching through the inventory yourself?

Furthermore, if the AI could model your purchase decisions with sufficient accuracy, there would be no need for you to even look at the suggestions. It could simply buy them on your behalf.

As for imagination, the AI *could* present you with a three-dimensional high-resolution visual representation of exactly how the object would look on your body or in your home, saving you the effort of having to use your own powers of visualization. But then again, there is no need for the AI to do this, since your participation in the selection and decision process is no longer required.

The need to model other people's preferences and opinions is likewise removed, since the AI could just as easily predict other people's responses as your own. The AI is also better than you at keeping track of what's in vogue. Anything you discover about fashion will not lead to better purchase decisions.

You might feel like telling your friends about your opinions so that they can better understand your thoughts and feelings. But *actually talking to your friends* may not be an efficient means of accomplishing this objective. Instead, your AI assistant could project its model of your style preferences in some easily communicable format. Let's say your friend is looking at a piece of garden furniture: the AI could then display for them annotations showing its (uncannily accurate) prediction of what you would think of each of the items your friend is considering. The AI might also display pieces of furniture that you would especially like or dislike, or ones on which your opinion would diverge especially strongly from your friend's opinion or from the average opinion of your peers. You and your friend would understand each other better by interacting with these AI models than you would by interacting directly with one another.

The upshot is that the activity of shopping would be radically transformed. The activity could disappear altogether—the AI doing it all, without your involvement. Alternatively, shopping could metamorphose into an activity that is more akin to watching a videofeed about merchandise and people's opinions about it. The video could be highly immersive, customized to your interests, and focused on you, your friends, and your favorite celebrities, and it would inform you about what they think of you and the options available to you (or, in the case of celebrities, what they *would* think about these things if they knew you).

Supposing there were some demand for it, it would still be possible to go shopping the old-fashioned way. You *could* choose to drive to the store, spend time trying to find something you want (perhaps to discover that the store doesn't stock your size or preferred color), wait in line to pay for it, and finally schlep it all home in plastic bags. If you do that, you end up with a purchase that you will like less than if you let your AI assistant handle it all. Shopping in this old-fashioned way would have something Rube Goldberg–esque about it. Yes, you *could* do it. But the *pointlessness* of it all—the extra hassle and effort only to obtain an inferior result... when this ghostly pointlessness is staring you in the face the entire time with its empty eye sockets, would not the allure of the activity be drained out? To the point where most people would cease to bother doing it?

Some of the other activities on the handout would be susceptible to having their purpose blanched out in a similar manner—collecting antiques or redecorating a home, for instance.

Or consider gardening. This activity might at least in part derive its appeal from the hope that one's efforts will cause the garden to be nicer than it would otherwise have been. But if a robot assistant could produce exactly the same results—or, rather, results that are *better* by your own standards—then I think many people would put away their pruners and take up position instead on the veranda where they can watch the robot work. Hobby gardeners already use a plethora of tools and motorized appliances to make their task easier. Would they stop short of full automation, if it were readily available?

Case study 2: Exercising

Let us consider a different kind of activity: going to the gym.

Here, at least, a task that cannot be automated! No robot can ever take your place on the elliptical. To gain the physical and mental benefits of exercise, you have to do it yourself. Perhaps, then, we've found our platinum—an activity that is completely resistant to the purpose-corroding acid of technological convenience?

On closer inspection, this hope proves illusory. While it is true that you cannot hire someone else or buy a robot to do your exercise for you, there are other solutions that would enable you to accomplish the common functions of exercise without breaking a sweat.

With advanced-enough technology, the health benefits and physiological effects of a workout could be induced by artificial means, such as drugs (safe and free of side-effects), or gene therapy, or medical nanobots that keep you in perfect shape regardless of your eating and drinking habits and your sedentary lifestyle. This holds for the mental benefits of exercise, too. The endorphin release that is triggered by physical exertion could be induced pharmacologically. Likewise for whatever other mind-clearing, de-stressing, and revitalizing effects that exercisers enjoy: all available in a pill or one-off injection of nanomedicine.[103] Begone muscle soreness, strains, calluses, and piles of sodden gym gear! Welcome the effortless sixpack and the VO_2 max of a Tour de France cyclist!

Case study 3: Learning

The case is similar with respect to several of the other activities listed on the handout, such as learning to fly a sport plane, studying mathematics, or even attending a lecture.[104]

One important motivation for studying and practicing is that by engaging in this activity you may hope to subsequently possess some knowledge or skill that you didn't have before. Consider the alternative: each day you study hard to master some topic, and over the course of the following night you forget everything you learned. The next day you start over, relearning the exact same lesson. Rinse and repeat.

Such amnesia would be massively discouraging. It would turn the learning into Sisyphean tragedy.

Even people studying as a hobby, who would not incur any practical disadvantage from having their gains reset to zero at the close of each day, would likely find the stark futility demotivating. The frequent erasure would cast a pall over the entire endeavor, making it seem no longer worth the while.

So with respect to studying mathematics, learning to pilot an airplane, and similar activities, it appears that our motivation is not simply to undergo some sequence of moment-to-moment experiences involved in the learning process, but also to reach states of increasing knowledge, skill, and understanding. The progressive nature of these activities is an important reason why we find them attractive and why they are prima facie worthwhile ways to spend one's time.

But now consider that at technological maturity, we will have available to us shortcuts that would get us equivalent results without effort and delay. We could let the AI directly edit our brains to incorporate new information and skills.

Such "brain editing" or "mind editing" is, admittedly, a more speculative technology than any we have presupposed thus far—more speculative, for instance, than a technological fix that would let us enjoy the benefits of physical exercise without leaving the couch. The process would be somewhat more straightforward if the mind to be edited were running on a digital computer rather than on biological wetware; so uploading yourself into a computer could be a practical first step to enabling mind-editing. Alternatively, the process might perhaps be facilitated if your biological brain had been outfitted with some sort of cybernetic module that could be updated via an external data link. But even without any such nonbiological contrivances, precise and flexible brain editing is plausibly one of the affordances of technological maturity, albeit one that would probably require superintelligence for its development and operation. I don't want to get bogged down here in implementation details, but I've added some notes on a handout that you can look at later.

HANDOUT 10

DOWNLOADING AND BRAIN EDITING

While science-fiction films (such as *The Matrix*) often portray "downloading"complex skills and knowledge into human brains as a relatively simple feat, in reality it is likely to be extremely challenging.

The human brain is of course quite unlike a regular digital computer, where standardized data representation formats and file transfer protocols make it easy to swap software in and out and to share it between different processors. By contrast, each human brain is unique. Even a simple concept that we all share, such as the concept of a chair, is implemented by an idiosyncratic constellation of neural connections in each person—the precise patterning of the neural connections encoding the concept is contingent on the details of that individual's past sensory experience, their innate brain wiring and neurochemistry, and an incalculable host of stochastic factors. One therefore cannot simply "copy and paste" the concept of a chair from one brain to another without performing a complicated synaptic-level translation, from the "neuralese" of one brain to the quite different "neuralese" of another.

Human brains *can* perform this translation themselves—slowly and imperfectly. This is what happens when we communicate using language. Some mental content in one brain, represented using that brain's idiosyncratic neuronal machinery, is first projected down to a low-dimensional symbolic representation consisting of a string of words in natural language; and then the receiving brain has to unpack this radically impoverished linguistic representation by trying to infer which configurations of its own idiosyncratic neural machinery best match those representations in the sender's brain that might have produced the perceived words and sentences. If the act of communication is successful, the receiving brain ends up with neural circuitry that shares some structural similarities with the circuitry in the sender's brain: enough so as to give the receiver some of the capabilities that the sender wanted to impart. For large or complicated messages, such as when a professor of organic chemistry wishes to bring their students up to their own level of expertise, this process can take *years*—and even then the result is all too often disappointing.[105]

So what would it take to shortcut this process of communication and learning to the point where we could "download" expertise in organic chemistry or any other subject directly into our brains, without having to spend years studying it? Importantly, we want to have the knowledge fully integrated in our brains, the same way we would if we had learned the subject the traditional way, so that we gain the same ability to use it intuitively and associatively and so that it contributes to our general pattern-recognition abilities. This is different from merely having the information deposited in some intracranial cybernetic memory capsule that we'd then have to serially query in essentially the same way we currently use a search engine (which might save us having to use a keyboard and a screen but would not otherwise make us any more of an expert in organic chemistry than any random person who has access to the internet). Achieving genuine absorption and integration of the new information, such as constitutes the acquisition of real knowledge and skill, would require a large number of cortical synapses to be precisely adjusted. Many billions, maybe trillions of synapses. (And at least in cases such as "learning how to be a better mother" or "learning how to be a better husband", it would most likely also involve modification of many subcortical brain circuits.)

All this is most likely possible, but it might require a superintelligent implementer. Since we don't want to simply replace the original mind with a new one, but rather enrich an existing mind with additional knowledge and capabilities, the implementation mechanism must have the ability to read off the existing pattern of synaptic connectivity, so that it can edit it judiciously and not simply overwrite it. Again, for something as complex as learning organic chemistry or acquiring a new language, this may involve reading off the key attributes of trillions of synapses. In addition to the ability to read and edit synaptic properties, the mechanism also needs to be able to figure out precisely which synaptic changes to make in order to alter the original version of the mind into a version of the same mind enhanced with the new knowledge or skill—a very challenging computational task that is almost certainly AI-complete.[106]

Each of these requirements (scanning, editing, and calculation) is far beyond the current state of the art. In fact, among all technologies that have been imagined and which are in fact physically possible, this might be one of the hardest ones to perfect. Nevertheless, I believe it could be done at technological maturity. I think it will not be humans that invent this technology, but superintelligent machines.

Let's imagine what the procedure might be like. Your brain is infiltrated by an armada of millions of coordinated nanobots. (Maybe they get there via the bloodstream and pass the blood-brain barrier—obviously the whole procedure would be entirely painless, since any triggers of discomfort could be easily suppressed.) These nanobots map your brain's connectome. Since the bots are operating in a crowded and electrochemically active environment, they must either be careful to avoid damaging the structures that they are traveling through and measuring, or else they must repair any damage that is caused in the process. The mapping would need to happen reasonably fast yet without generating excessive amounts of heat. The data gathered by the probes, perhaps after some simple local preprocessing, is transmitted to a computer outside the skull. (For this purpose, a tiny fiber optic cable may be constructed, which could penetrate the skull or project back down via the bloodstream—but not to worry, it can be removed, without trace, once the procedure is complete.) This more powerful external computer runs a superintelligent AI that processes the data and works out the requisite pattern of synaptic edits. Once the changes have been determined, the information is sent back to the nanobots which perform the needed synaptic surgery—strengthening some synapses, weakening others, adding new links here and there between previously unconnected neurons. Speed is important, not only in order for the "downloading" to be convenient, but also because if the process is too slow then the neural circuits in the brain may have changed too much by the time the computed alterations are received back from the external computer, rendering the calculated edits inaccurate.

Note that in order to replicate the effects of ordinary learning it is not enough to "simply record" some discrete facts in one isolated cortical area. When we learn in the normal way, many parts of the brain are changed, reflecting the effects of metalearning, the formation of new associations with previously learned content, changes in control and attention mechanisms and in episodic memory, and so forth. This is one reason for thinking that the number of synaptic weights that would need to be adjusted could be in the trillions, even for fairly simple learning experiences.

Let us also take note of one further complication. In order to work out how to change the existing neural connectivity matrix to incorporate some new skill or knowledge, the superintelligent AI implementing the procedure might find it expedient to run simulations, to explore the consequences of different possible changes. Yet we may want the AI to steer clear of certain types of simulation because they would involve the generation of morally relevant mental entities, such as minds with preferences or conscious experiences. So the AI would have to devise the plan for exactly how to modify the subject's brain without resorting to proscribed types of computations. It is unclear how much difficulty this requirement adds to the task.

Downloading mental content would be easier if the receiving mind was a brain emulation (aka an "upload"), implemented as software rather than biologically. In this case, the readout from the original brain would have been done in advance, potentially under easier conditions, such as by means of a destructive scan of a vitrified brain rather than in vivo. Subsequent read and write operations would be trivial—they would simply involve editing a digital file, and the mind could be paused while the procedure is taking place. However, the computational step of figuring out *which* edits to make in the preexisting neural network would be about equally hard in the case of a digital brain emulation as in the case of a biological brain.

The upshot, then, is that activities such as studying, where we put in effort now in order later to be in a state where we know more or have greater skill, would maybe start to appear a little pointless at technological maturity, in the presence of easy shortcuts to the same destination.

We *could* still choose to learn and practice the hard old way. But doing so would be to impose on ourselves a gratuitous complication. This would be a bit like deciding to move only by skipping backwards on one foot in order to create more challenge in one's life.

Case study 4: Parenting

Many people find raising children to be a rich source of purpose and meaning. It is common for parents—even those who could easily afford to offload all the labor of child-rearing to (highly qualified) hired help—to choose to devote a substantial amount of time to bringing up their children. Some wealthy parents do pack their progeny off to a boarding school; but most do not.

It thus appears as though many people regard spending time with their own children as an "autotelic" activity—an activity that is valued for its own sake and not merely as a means to an end. If this is right, then perhaps raising children would retain its appeal at technological maturity and provide an opportunity to fill our days with meaningful activity.

There are, however, some difficulties with this idea that parenting would be the solution to the purpose problem.

In the first place, there are the obvious practical problems. Not all people have children. Also, children grow up and eventually don't need to be parented any more. The fraction of people's lives that could be devoted to raising young children would decrease if, at technological maturity, human lifespans were extended. To offset the growing-up effect, the population would have to increase exponentially—which would bring in the Malthusian apprehensions that we discussed in Monday's lecture.

Second, parenting is not really one thing but a bunch of different activities. For example, it involves activities such as changing diapers, tidying up toys, arbitrating conflicts, coaxing the child into doing something they are disinclined to do, scheduling appointments, transporting the child to activities, and so on. It is plausible that many parents would prefer not to do these things; and if there existed a diaper-changing machine, a tidying-up robot, and an automatic activity-scheduler, and if these conveniences really worked reliably, I suspect they would be popular.

A parent might be happy to make a home-cooked meal for their child if they think it will taste better or be more nutritious than something ready-made. But if the kitchen had a nanofabrication box, which at the press of a button produced a meal that was molecularly indistinguishable from what's on the parental menu—or rather, one that was superior, diced, sliced, and spiced to perfection—should we not then expect the old pots and pans to see less and less use, and perhaps eventually to be retired altogether?

*

What, then, remains after all the chores have been removed from child-rearing?

Only a small fraction of parenting consists of "precious moments"—imparting some pearl of life wisdom, being delighted by some expression of the child's naive creativity, receiving a spontaneous hug or other sign of affection or gratitude. Even if the frequency of these moments could be doubled or tripled, they do not seem to be the kind of thing that, on their own, could fill our days with purposeful activity.

Some other parenting-related activities can take up more time, such as playing with the child or reading a bedtime story. Those who have kids could spend a substantial part of their day doing these things, if we imagine the more chore-like aspects of parenting to have been automated. It would seem a pretty meaningful way to spend one's time.

*

But there is a potential confounder. I suspect that one factor at play here is that parents think (or feel) that they are *benefiting* their child when they are spending time with them. Would playing with your child or reading them a story still feel meaningful if you became convinced that you were thereby *harming* your child? Because if not, that could be a problem. For it is quite conceivable that, at technological maturity, you would in fact be slightly harming your child whenever you indulged in some DIY parenting.

Technological maturity would permit the construction of robotic caregivers with superhuman parenting skills. The robot could be perfectly lifelike in appearance and behavior. If desired, it could be made to appear to the child indistinguishable from a real human person, or from their actual biological parents—the way they are when they are at their best. The robot might even be programmed to feel love and devotion to the particular child assigned to its care.

So now, every hour of quality time you spend with your child is an hour of even higher quality time it is deprived of spending with the robot. Spending the hour with the artificial caregiver would, we may assume, be more fun for the child as well as more educational and more nurturing of their emotional and social needs. You *could* choose to play with your child yourself; but in doing so you would be selfishly prioritizing your own enjoyment at the expense of the child's welfare and development. Although this might give you some fun, it would hardly fill your life with purpose.

The idea of a robot that is a perfect parent-substitute might sound far-fetched. We will not discuss the mechanics of implementation here; but I think that at technological maturity this would in fact be possible. A superintelligent AI could build such a thing for us.

*

There is, however, a potential "philosophical" complication with the idea of a robo-parent substitute, which we need to analyze. In order to explore this issue, we should distinguish between *ex ante* and *ex post* substitutability.

The most straightforward case is if the substitution takes place *ex ante*, before the child has formed an attachment to their human parents. This would not be so different from what happens today when an infertile couple contracts a surrogate woman to bring an embryo to term: this method of gestation, it is normally assumed, does not harm the child. If the couple that raises the newborn provides excellent parenting, it may well be that the child benefits from being raised by them rather than by their birth mother. By the same token, if the child were adopted at birth by loving robotic parents who provide even more excellent parenting, the child could benefit even more.

However, once the child has become attached to their parents, and we have a case of *ex post* substitution, a new term enters the equation. It *might* then be in the child's interest to be raised by their human parents, at least if their caregiving would not fall too far short of that of the robotic alternative. The point here is not only (or even mainly) that it could be psychologically traumatizing for a child to be ripped away from the individuals to whom they have grown attached—any such negative psychological effects could presumably be easily forestalled at technological maturity. Rather, the point is that it's possible to maintain that the child would incur a harm *even if* there is no trauma—in fact even if we stipulate that the robots are so similar to the child's human parents that the swap is *undetectable* to the child.

The ground for such a position would be similar to the ground for why one might think, in general, that it would be undesirable or at least suboptimal to spend the rest of one's life in Nozick's experience machine (which we'll get back to shortly). This thought experiment has been taken to show that our well-being has an objective component—that how well our lives go for us is not determined solely by our mental states, by what we *think* and *feel,* but also by *our relationship to external reality.* On this view, it matters whether our beliefs are true and our projects successful, independently of whether we ever find out. Along the same lines, it might matter whether we really remain in contact with somebody to whom we have bonded. Interacting with a simulacrum of this person would, ceteris paribus, be less good, even if we never notice the difference. One might, for example, have the intuition that it is bad for a husband to be cuckolded even if he never discovers the betrayal and even if his wife does not change her behavior toward him. And—if one holds this view—one might likewise think it could be bad for a child if, one night while they were sleeping, their parents were swapped out for an indistinguishable set of robot impostors.

*

So far, we have considered the possibility that parenting could provide purpose by being an activity whereby parents can benefit their child. We should also consider the possibility that parenting could provide purpose by being an activity whereby the parents can benefit themselves.

Historically, children have served as a kind of investment vehicle. You take care of them while they are young, and hope that they will take care of you when you are old. This function has been largely supplanted by the welfare state and tax-advantaged retirement savings accounts. In any case, if at technological maturity there is general abundance and no one is making a living from working, there would be much less need for this kind of saving. Also, people would not become decrepit as they age.

There is another type of "investment function" that parenting can serve, which yields returns of a different kind. Parents may invest time and energy in their children in the hope of later having somebody with whom they stand in a very special relationship—one of mutual understanding, trust, nonmanipulative love, gratitude, and a sense of deep-rooted affinity. Familial bonds are among the closest and most selfless that most of us humans are capable of. However, this investment function, too, might be undercut at technological maturity, inasmuch as there would be an easier path to achieving an equivalent outcome. Namely, we might create artificial persons (fully articulated conscious humanlike beings with moral status) who stand in the same type of relationship to us: who understand us, trust us, and resemble aspects of us in the way that our children do. This would be much faster and cheaper than bringing up a human child in the traditional way. What is more, artificial persons could be designed to have a greater capacity for love and gratitude and close connection than is generally vouchsafed to our own fallen kind.

*

Natural children, natural friends, natural lovers: how could they compete against far more perfect artificial alternatives? Or artificial alternatives that are imperfect in just exactly the right kind of ways?

On objective functional characteristics—beauty, charm, virtue, humor, faithfulness, affection, etc., natural persons would be outclassed. Artificial people would win any fair contest and comparison. They would be better.

But it depends on tastes. Perhaps artificial is just not your thing. Alternatively, perhaps you don't care about whether someone is natural or artificial per se, but you do care about whether the two of you have *a history* together, whether the particular individual has interacted with you in certain ways before, and whether you have made commitments to each other. If that is the case, then there could be some natural persons who are irreplaceable to you. Then it doesn't matter how technologically advanced the artificial substitutes become, because none could meet your criteria. And if it is shared history and mutual commitments that ground your preference for some particular existing person, then no *natural* substitute might meet them either. However superior in other respects, any substitute would be lacking in one key respect: they would not be The One.

*

We will have occasion in later lectures to return to the issue of interpersonal entanglements, and more generally to the possibility that one might have a basic preference for human or otherwise flawed, troublesome, and inconvenient things.

For now, we can sum up our discussion of parenting as a possibly purposeful activity at technological maturity by concluding that: (a) a large chunk, probably a great majority, of the specific activities involved in child rearing can be characterized as "chores", which it would be very tempting to automate; (b) while the quality time we spend with a child feels more intrinsically worthwhile than performing chores, there is a potential confounder, viz. our belief that we are *benefiting* the child by spending time with it: "quality time" might not seem so purposeful if we thought that we were doing nothing for the child or even mildly harming it; (c) by objective functional metrics, robo-parents could outperform human parents, so by those standards it could be better for the child if the human parents stepped out of the way; (d) once a child has bonded to a human parent, however, then even an objectively superior robo-parent may not in all respects be as good for the child as their human parent—and on some theories of well-being, this can be true even if the child were unable to detect that a substitution occurred; (e) the upshot is that many of the ways in which parenting currently provides opportunities for purposeful activity would be removed at technological maturity, though possibly some opportunities would remain that are related to the ways in which we have bonds and desires directed at specific existing human persons or at the general category of the natural; and, finally, (f) our points about the parent-child relationship also apply to many other forms of interpersonal entanglement, such as friendships and romantic partnerships.

From shallow to deep redundancy

With these four case studies, we can begin to see how the purpose problem cuts much deeper than is commonly recognized.

*

Shallow redundancy

The traditional and relatively superficial version of the purpose problem—let's call it *shallow redundancy*—is that human occupational labor may become obsolete due to progress in automation, which, with the right economic policies, would inaugurate an *age of abundance*. This would be a condition of great general prosperity and material plenty. Since it would eliminate both the need and the opportunity for paid work, it would cause one source of purpose to dry up, namely the purpose that many people currently find in their jobs.

The solution to shallow redundancy is to develop a leisure culture. Leisure culture would raise and educate people to thrive in unemployment. It would encourage rewarding interests and hobbies, and promote spirituality and the appreciation of the arts, literature, sports, nature, games, food, and conversation, and other domains which can serve as playgrounds for our souls that let us express our creativity, learn about each other and about ourselves and about the environment, while enjoying ourselves and developing our virtues and potentialities. A leisure culture would base self-worth and prestige on factors other than economic contribution, and individuals would construct their social identities around roles other than that of breadwinner (although there might be game-like environments that allow those who previously excelled in financial performance to display and gain recognition for their resourcefulness).

We looked at some examples of more or less inspiring leisure cultures—such as those prevailing among children, students, aristocrats, bohemians, monastics, and retirees. Each of these comparisons is multiply confounded. In an age of abundance, the set of feasible leisure cultures would expand beyond historical instances, in part because of the unprecedented opulence that would follow the AI transition (and last until population growth dilutes the per capita gains—or indefinitely, if population growth is controlled) which could universalize economic privilege; and in part because technological and other inventions which would enable many novel kinds of fun (along with liberation from many miseries that have previously blighted the flourishing of even the most privileged elites).

*

Deep redundancy
The more fundamental version of the purpose problem that is now coming into view—let's call it *deep redundancy*—is that much leisure activity is *also* at risk of losing its purpose. The four case studies showed that many of our usual reasons for engaging in non-work activities disappear at technological maturity. And those observations can be generalized. It might even come to appear as though there would be no point in us doing *anything*—not working long hours for money, of course; but there would also be no point in putting effort into raising children, no point in going out shopping, no point in studying, no point in going to the gym or practicing the piano... et cetera.

We can call this hypothetical condition, in which we have no instrumental reasons for doing anything, *the age of post-instrumentality*. As we move toward this weightless condition, blasting away from the gravitational pull of the ground and its tough "sweat of the brow" imperatives on our days and our strength, we may begin to feel an alienating sense of purposelessness, an unanchored "lightness of being". We are left to deal with the discovery that the place of maximal freedom is actually a void.

Paradox of progress

I have given many speeches about the future of AI. Almost invariably, I am asked at least one question that brings up some version of the purpose problem.

The question is usually a bit muddled. People seem to have difficulty articulating their concern. Sometimes it's not even explicit, yet one can sense a perplexity lurking beneath the surface, like a wordless and purblind digger that is tunneling through and possibly undermining the foundations of our domicile.

Let's see if we can disinter the cause of this latent unease. Here is one possible line of reasoning that could be motivating questions about life after AI success:

> At the present, and throughout history, there are many pressing tasks that we humans must do ourselves, and there are many big challenges that we confront together. These tasks and challenges give structure, purpose, and meaning to our lives. But technological progress (and, to a more limited extent, capital accumulation) enables us to achieve more of what we want with less effort. In the limit, with perfect technology and abundant capital, we are able to get everything we want with no effort. We will then have nothing to strive for. We will then either be bored out of our minds or transform ourselves into "pleasure blobs", passive minds that experience an artificially induced sense of contentment. Either way, a dystopian future awaits.

And those would be the best-case scenarios! It would hardly be reassuring, for example, to be told that we don't need to worry about deep redundancy because our high-tech civilization will come tumbling down in a cataclysm before we reach technological maturity.

*

At the heart of the argument here lies a pessimistic view of human nature. Basically: we're unfit to inhabit a perfect world.

I can even furnish an explanation for why this should be so. Over evolutionary and historical timescales, external instrumental constraints have always been in ample supply; and our psyches, therefore, have formed in ways that assume their presence. Our ape ancestors, whose diet consisted of a lot of fruit, lost the ability to synthesize vitamin C: and this produced a dependency that showed up only much later, in very different circumstances, such as when sailors subsisted on sea biscuits for months on end, and found themselves languishing from scurvy. In a similar manner, we may have become dependent on encountering demands for mental and physical exertion and goal-directed striving. Such externally imposed pressures to exert effort in order to satisfy instrumental needs not only keep our bodies in shape but they also help give shape to the softer parts of our souls, rather like a spiritual version of the exoskeleton that structures and contains the innards of a beetle. Were this exoskeleton of enforced toil to be removed, following our entry into a post-instrumental condition in which our desires can be effortlessly satisfied, we would suffer a pitiable degeneration and collapse, becoming either creatures of boundless boredom or else amorphous blobs of artificially induced contentment.

*

You may discern a whiff of paradox in the preceding argument.

On the one hand, we surely have reasons to pursue the development of technological capabilities that enable us to get more of what we want with less effort. That's almost part of the definition of rationality: that one seeks efficient means to one's ends. Certainly, our society is pouring great effort into technological and economic progress, and we give awards to individuals who make it happen.

And yet, on the other hand, if and when our efforts to increase the efficiency with which we can achieve our aims are fully successful, we will supposedly enter a condition in which either we are terminally bored or we become passive recipients of narcotized contentment.

Neither alternative sounds appealing.

So it looks like we have reason to work to achieve a condition *X and* that it would be very bad if we achieved *X*. In other words, the conclusion would seem to be that we ought to devote massive resources toward achieving something while at the same time desperately hoping that we will fail. Not quite a logical contradiction, but it would certainly be an odd predicament to be in.

A five-ringed defense

Fortunately, there are several things that can be said in rejoinder: different elements we can point to that can be present in life lived at technological maturity and which could make such a life very good. While none of these responses may be completely satisfactory on its own, yet in combination they have, I think, the makings of an adequate answer to the redundancy concern and its associated paradox of progress.

We can think of them as tiers of a multilayered defensive structure, which can withstand or at least soak up and greatly mitigate the force of any onslaught of purposelessness that we might encounter in a potentially post-instrumental condition. Whatever remains of the problem after taking these protections into account seems quite tolerable—not only consistent with life, but consistent with extraordinarily prudentially desirable life.

Here is a précis of the five defensive lines:

1. Hedonic valence
2. Experience texture
3. Autotelic activity
4. Artificial purpose
5. Sociocultural entanglement

We may conceive of these as concentric perimeters: ranging from pleasure (the innermost donjon) to social entanglement (the outer ramparts).

Let us take a brief tour to survey these fortifications.

*

Hedonic valence
Here I want to say that referring to the option that involves artificially inducing contentment as one in which we become "mere pleasure-blobs" does really not do justice to what is on offer. Perhaps a life as a pleasure-blob is not everything we aspire to or the very best that we could possibly hope to achieve, but there needs be nothing "mere" about it. We might say more on this later, but some preliminary remarks:

(a) A common mistake in evaluating possible futures is to focus on how good those futures are *for us now*, in the sense of how interesting it is for us to contemplate a given future or how suitable it is as a setting for entertaining stories and morality tales that we wish to tell each other. But the question before us here is a very different one: *not* how interesting a future is *to look at,* but how good it is *to live in.*

We must remember that "interesting times" are often horrible times for those who have to live through them. An uneventful and orderly future, in contrast, can be a great place to inhabit. And even if its occupants should be somewhat blobbified, even if it would not offer the most inspiring backdrop for grand dramatical narratives, it could provide a state of continual contentment and pleasurable feeling that is pretty solidly desirable.

(b) It is a commonplace that the heedless pursuit of pleasure is often counterproductive. "The search for happiness is one of the chief sources of unhappiness."[107] We think of a drug addict desperate for their next hit, and it does not seem like a good life. In fact, it is in all likelihood a life of suffering, punctuated by brief moments of drug-induced relief. Probably nobody in this room would be eager to swap their own life for that of a hardcore addict.

Traditional wisdom therefore recommends taking a more oblique approach in our pursuit of happiness.[108] "Happiness—a butterfly, which, when pursued, seems always just beyond your grasp; but if you sit down quietly, may alight upon you."[109]

This wisdom would lead us astray, however, if we applied it to a scenario in which its underlying premiss—that chasing after pleasure is self-defeating—does not obtain: such as one that plays out in a setting where the technology exists whereby one really *can* induce pleasure by directly aiming at it, and where one can do so reliably and lastingly. Somebody might have the intuition that "a world in which mind engineering is used to induce pleasure would feel stale and unsatisfying after a while". But this intuition is simply false.

(c) Suppose that we became acquainted with the quality and the quantity of super-pleasure that could be ours at technological maturity. At present, we are opining on the matter without being directly acquainted with the thing under evaluation. If, however, we gained direct experience, it is plausible that we would most swiftly come around to the view that it was *extremely* desirable to experience and to keep experiencing that pleasure. And it is not obvious that, in this case, the process of becoming more intimately acquainted with the mental state whose desirability we are trying to ascertain would necessarily involve a corruption of our ability to judge well.

(d) Also:

> "You could say I am happy, that I feel good. That I feel surpassing bliss and delight. Yes, but these are words to describe human experience—arrows shot at the moon."[110]

"It feels so good that if the sensation were translated into tears of gratitude, rivers would overflow."[111]

*

Experience texture

One might attach to the notion of unbridled hedonism the picture of a junkie splayed out on a filthy mattress, experiencing the pleasures of some sort of superdope—similar to current drugs though longer-lasting, more euphoric, nonaddictive, and free of side-effects.

But there is no reason why the inhabitants of a post-instrumental society should have to experience *only* pleasure. Why should the experiences of the utopians, while charged with positive hedonic valence, not *also* possess rich, varied, and aesthetically ace content—*far* more so than the comparatively tawdry experiences that occasionally impress us in the present era?

The environment of the utopians could thus be one of heartrending beauty. Appreciators of art and architecture or natural landscapes could feast their eyes on the most excellent sights; music lovers could thrill their ears with brilliantly captivating sounds and melodies; gourmets could chomp their way through Xanadus of culinary wonders. And so on. Each day could be arranged with artistic ingenuity and turn out as little masterpieces all in themselves, while adding to an ever-rising larger structure into which they all fit together perfectly each in its unique way: like carefully carved and coordinated stones that together compose a great cathedral of life.

Furthermore, the utopians could enjoy enhanced perceptual capabilities; and, more importantly, they could be endowed with superlative aesthetic sensibilities that enable them to actually apprehend more of the beauty and significance that suffuse their sensory streams and their environment.

If we were teleported into their world, without receiving these upgrades of our subjectivity, we would not appreciate it as they do. *We* may see some pretty-looking wildflowers over there. *They* would come closer to seeing heaven in those same flowers.

And then there is the world of abstract beauty. Utopians could be cognitively enhanced so that, for example, they can inhabit the ethereal realms of mathematics like a creature that is adapted to the life of the mind.

Think of a deer skipping gracefully across a complex woodland terrain. Then think of a mathematician in an analogous situation, attempting to traverse a field of abstract algebra—stiff and achy; unnatural; almost entirely blind, able to see scarcely two yards ahead; slow and unsteady; frowning with concentration; clutching the rollator of formal proof… Maybe that happens to be the only way we can currently do it, rather than the way it really should be done.

In all of this, there is room for improvement.[112]

> "I reach in vain for words to convey to you what it all amounts to…
> It's like a rain of the most wonderful feeling, where every raindrop
> has its own unique and indescribable meaning—or rather a scent or
> essence that evokes a whole world… And each such evoked world is
> subtler, deeper, more palpable than the totality of the reality that you
> have encountered. One drop would justify and set right a human
> life, and the rain keeps raining, and there are floods and seas."[113]

*

Autotelic activity
The third defensive line: utopians need not be passive.

Even if the utopians would have no instrumental reason for doing anything, it would not follow that they wouldn't be doing anything.

If there is a special intrinsic value in active experience, such that a life is better if it involves at least occasionally *doing* stuff rather than always remaining passive: well, then we can choose to do things, to engage in various autotelic activities, for the sake of realizing this intrinsic value. Nothing would stop us. And should anyone have a problem of insufficient drive in the absence of proddings from instrumental necessities, they can turn to neurotechnology to rev up their zeal and zest to arbitrary degrees.

*

What we have so far, consequently, is not a condition that would be appositely represented as a junkie sinking into a pleasant narcotic stupor on a flea-infested mattress; but rather a superhumanly healthy and hearty lifelover, who relishes and savors every moment of existence, engaging in rewarding activities in a phenomenally aesthetically and intellectually pleasing environment, and being endowed with enhanced capacities for appreciation, understanding, creativity, and joyful emotional participation, that allows them to partake more fully of what their world has to offer, including also in the realms of abstract truth and beauty. It is getting better!

*

We have a question.

Student: Didn't you say, in the case studies, that there would be no reason for doing anything at technological maturity?

Bostrom: Not quite. I might have said that it could *appear* as if that were the case, but now we are in the process of looking into the matter more closely.

At this point I am making the suggestion that even if we had no *instrumental* reasons for doing anything at technological maturity—that is to say, no reason to engage in any activity in order to produce some result (because the same result could be more efficiently brought about by machines)—this would not imply that we would not be doing anything.

Let us assume that a life of complete idleness would in fact be less good than a life that contained some activity. (If not, well, then maybe the utopians *would* be idle, but it wouldn't be a problem.) So, if the utopians understand that their lives would go better if they did something, this would give them a reason to do something. It wouldn't be an *instrumental* reason. They wouldn't be engaging in the activity in order to produce some output. Rather, they would be engaging in the activity because the activity itself is valuable, or directly value-adding to their life. The activity is *autotelic*: it is done for its own sake.

Student: I see. But in that case, couldn't one say that the utopians *would* have instrumental reasons for doing stuff—namely to produce the outcome that their lives contain some activity, which then makes their lives go better?

Bostrom: I suppose you could say that if you want to. But there is still a distinction to be drawn between things we do in order to achieve something else (something other than the doing itself) and things we do because the doing of them is, in its own right, valuable.

If you go to the dentist, it is probably for instrumental reasons, to fix or prevent tooth problems, not because you enjoy it or believe it to be an intrinsically valuable way to spend your time. If you could obtain the same oral health outcome without going to the dentist, you would gladly cancel the appointment. So this would be an example of an activity that is definitely *not* autotelic.

Finding clearcut examples of activities that *are* autotelic is, I have argued, not quite as easy as one might initially think. Normally, if we were asked to illustrate the concept, we might list a bunch of activities like those on Handout 9. These might appear to be good examples of autotelic activities which we engage in for non-instrumental reasons. We don't need to be paid to do them.

But if we examine these activities more closely—and we looked in particular at shopping, exercising, learning, and parenting—then we find that they are actually suffused with instrumental rationales. These instrumental rationales would go away at technological maturity. However, it would still be possible to engage in these activities. We would have no instrumental reason for doing so; but we might do them for non-instrumental reasons—namely, if we maintain that the activities are truly autotelic.

What would be lost at post-instrumentality is not *activity* but *purpose*. I am about to say more about purpose in utopia in a moment, but for now I'm just making the point that activity need not be missing. The fact that the utopian lives could contain (rich, variegated, challenging, skillful, immersive) activity is our third defense line against the paradoxical charge that life at technological maturity would necessarily be undesirable.

It's a good question though. Any others?

Another student: So if one is doing something just for fun it's autotelic, but if one is doing it in order to achieve something else it's instrumental?

Bostrom: Almost but not quite. The stumbling block is the word "fun". What exactly do you mean by "doing something just for fun"?

Student: Like, you're doing it because you enjoy it?

Bostrom: Well, we have a complication here. "Doing it because you enjoy it" seems to mean that you're doing it as a means to experiencing pleasure or positive affect. But at technological maturity, there would be more efficient paths toward that outcome. You could take a superdrug that has no side-effects, or reprogram your brain so that it experiences pleasure all the time independently of whether you are doing any "fun" activities or not.

Student: Oh.

Bostrom: Many of the leisure activities people do today, they do because they are fun—they engage in them as a means to experiencing pleasure. But this, by itself, would not be a reason to continue doing them in a post-instrumental world. So we may then ask, would it be a problem if people in utopia just stopped doing things and became inert recipients of pleasure and various forms of passive experience? Some people might think that this would be a problem—such a passive life just wouldn't be as good, other things equal, as a life that also included more active forms of experience and participation. A life full of pleasure and passive experience would still be missing something important. And in response, I say that if that is indeed so, then let us note that the utopians can add active experience to their mix: they would have reason to engage in activities in order to realize whatever value activity has (beyond its ability to confer instrumental benefits, including the instrumental benefit of generating pleasure).

We can also note that *within an autotelic activity* there might be "sub-activities" that are instrumentally motivated. For example, suppose that playing football is autotelic. So some utopian might be playing football for the sake of realizing the intrinsic value of this activity in their life. Now, while they are playing football, they will be pursuing many instrumental subgoals, such as jogging in order to reach a certain area of the playing field in order to stretch out the opponent's lines in order to create openings that one's teammates can exploit in order to score a goal in order to win the match. *Given* that one has reason to play football, one has reason to adopt the goal of trying to win (because pursuing victory is constitutive of what it means to be playing the game); and then these further subgoals are instrumentally justified as means toward winning. By the way, another constitutive goal of playing the game is to adhere to the rules and to use only certain permissible means toward achieving the subgoal of getting the ball into the opponent's net. Excluded means include, for instance, bribing the referee or hacking the electronic scoreboard. It is not just that such means would be immoral: they would also undermine the overarching reason for engaging in the activity in the first place, which we assume to be to realize the intrinsic value of having the activity of football-playing in one's life.[114]

Is that clearer?

Student: Yes.

Bostrom: Good.

*

Artificial purpose

But wait, there is more!

We now come to the fourth defensive line.

Some curmudgeonly fellow might say the following: "Sure, your utopians would have pleasure, fancy kinds of passive aesthetic experience and understanding, and also active experience by engaging in various autotelic activities. But there is still something missing from their lives. They would be lacking *purpose*! Purpose is something you have when you have an instrumental reason for doing something—but not an instrumental reason that is purely derivative of the value of autotelic activity. The football player might have instrumental reason to jog somewhere on the field, but that is not real purpose. That is just make-believe purpose. Real purpose requires there to be real stakes: for example, things you have to do in order not to starve, or in order not to be homeless, or in order to save somebody who is drowning. Life without the possibility of real purpose might be subjectively satisfying, but it would be superficial and lightweight. It would lack the thing that gives real depth and meaning to human existence."

Okay. We are getting into deeper waters here. I am planning to return to questions about purpose and meaning in a later lecture, probably on Friday. But for now, I will just put on the table the possibility of creating *artificial purpose*.

I alluded to one form of this already. Let's say that you decide to play football as an autotelic activity. In order to engage in this activity, you adopt the goal of winning the match. If you really embrace this goal—if, having adopted this goal, you really come to want to achieve it—then there is a sense in which the football game has real stakes for you. You *want* to win. From this point onward, you are in the grip of a kind of purpose. The only way to get what you want is to put forth effort. Perhaps nothing much else depends on the outcome of the game, but the stake is the winning or losing itself.

If you find it difficult to really embrace a goal in this manner using only your natural capacity for buy-in and commitment, you could use neurotechnology to do so. Having decided that playing football would enrich your life, and seeing that really wanting to win would improve the activity and the experience, you could program your mind to have a burning desire to help your team to victory.

Another form of artificial purpose would be to place yourself in a situation in which only your own efforts could allow you to achieve some outcome that you already care about for independent reasons. Think of a rock climber halfway up a mountain: there, they have no choice but to employ their strength and skill, on pain of death. In utopia, the analogous possibility would involve creating a special situation in which the affordances of technological maturity are unavailable. For example, some of your enhancements might be switched off and programmed to only boot up again after a set period of time, and any possibility of external rescue in the interim might be debarred by decree. Now it's just you and the inescapable need to meet the challenge you are confronting.

One might object to this proposal of creating artificial purpose that it would in effect amount to *suspending* utopia, at least locally. This is most clearly the case if the artificial purpose is created by entering a "hardcore" mode, in which the otherwise universally available means of automatically achieving outcomes have been removed—generating a pocket of non-utopian scarcity and danger. But perhaps one could argue that there would also be an element of suspension in the case where the artificial purpose is achieved by inducing a particular desire that requires an exertion of effort, such as in the case of the football player who comes to have the desire to help his team win using only fair and square means.

I think I would prefer to say that if these kinds of arrangements were necessary to achieve the best possible lives, they could be regarded as being part of utopia. Rather than saying there would be holes in the utopian cheese, we might say that a perfect face may have some freckles or a beauty spot: these are not exceptions to the face but integral to it.

In any case, this is a defensive line, inasmuch as even if it is the case that highly desirable lives would need to contain activities with real stakes, this is something that could be provided at technological maturity. And provisions could be made in a far more optimal way than is the case in our current civilization, where real stakes of course exist but very often are not very closely connected to worthwhile forms of striving and achievement—for instance, when somebody who did everything right gets run over by a bus, or when the way to achieve a good outcome involves many years of mindless grind.

*

Sociocultural entanglement

Lastly, we arrive at the fifth and outermost palisade.

As we discussed yesterday, there are limits to automation. I'm not referring to physical limits. Physical limits exist, of course, but they are irrelevant for carving out a space for human purpose, since what is physically impossible for machines is also physically impossible for us.

What is relevant is that there are also social and cultural limits to machines' ability to substitute for human labor. For example, some consumers may prefer some goods and services to be manually produced. They might even prefer for them to be manually produced by workers who are not taking advantage of certain forms of enhancement or augmentation. Similar constraints can arise perhaps even more naturally in a noncommercial setting, where one might, for instance, craft a gift by one's own hand in order to express an emotion or an attitude: which expressive function would not be served as well by a bought item. Or somebody might want to be loved and cared for by a specific existing human individual—perhaps preferring this even to indistinguishable love and care from a robot replica—and this could then create opportunities for that individual to achieve genuine instrumental goods (benefits to the other person) through their own efforts and activities, labors which in this case could not be outsourced.

There can also be more indirect or socioculturally complex possibilities of this general type, discussion of which I will however defer until a subsequent lecture.

The upshot is that, insofar as the utopians want purpose, they need not be wholly limited to artificial purpose. Some natural purposes may also remain for them even at technological maturity.

*

So these are the five rings of defense (reviewed here in outline only). They can, I believe, sufficiently repulse the assault of the purpose problem to preserve, within their happy enclosures, a form of post-instrumental utopia that is extremely desirable. Even if some sense of redundancy filters through, it seems quite tolerable: something we can live with—can live *well* with.

I now realize that I had also planned to discuss Robert Nozick's "experience machine" thought experiment today, but we have run out of time. Well, I guess you'll have to think through that on your own. I can give you the handout I had prepared—I don't know whether it is comprehensible in isolation, but perhaps it will help trigger some reflections.

Let's put down the pause here. See you tomorrow!

Signs and sightings

Tessius: Let's zip.

Firafix: That way?

Tessius: Yes. Let's see if we can get out… Sorry. Excuse us. Sorry. Sorry. We're not students, just impostors. Sorry. Sorry.

Kelvin: Pardon. Pardon.

Tessius: It looks like all of them are staying on.

Kelvin: Like sheep to the shearing.

Tessius: Like students to the free snacks.

Firafix: I've got the handouts.

Tessius: Let me show you… there is the new sign. Is that real gold plating?

Facilities Manager: It is certified, from an artisanal mine in Angola. It was hacked out of the ground with handcrafted stone tools.

Tessius: Wow.

Bostrom: "The Exxon Auditorium", hmm. —Oh, hi, Dave!

Dave: Hi! I see the University has received another capital infusion. It must have quite a sizable endowment fund by now.

Bostrom: I'm not sure whether the pockets are bottomless but the trousers definitely are, unfortunately.

Dave: It's for a good cause.

Bostrom: How have you been doing, man? It's great seeing you.

Dave: I wish I could announce a magnum opus, but unfortunately my feeble life force has not wrangled any great works into existence lately. I console myself with the thought that in other branches of the universal wave function there are versions of Dave who are publishing volume after volume while organizing the shock troops of the global biomedical happiness revolution.

Bostrom: L. Ron Hubbards of paradise engineering, perhaps?

Dave: Weeell, exactly! But alas not in this branch. I presume you have some things on the cooker?

Bostrom: Am working on a book.

Dave: Excellent. What is the topic?

Bostrom: It would be based on this lecture series. You know, feeding two birds with one ice cream cone.

Dave: Delicious vegan ice cream!

Bostrom: For sure. Although I think our feathered friends would happily dispense with the ice cream part—it seems to be mostly about the cone for them… Shall we grab a coffee?

Dave: Lead the way. I'll be in town for a couple of days.

Tessius: Wow, a rare sighting. You know who that was, right?

Firafix: Yes. He seemed nice.

Kelvin: Apropos hedonistic imperatives, Firafix and I were thinking of taking the baths. Want to come along?

Tessius: The hot springs? Sure! When are we going?

Kelvin: Right away.

The upholstery of dreams

Firafix: Well, how about we read the handout about the experience machine first, then we could talk about it on the way there?

Kelvin: Ok, let's peruse.

Tessius: You guys are such magnificent nerdballers! I like that.

HANDOUT 11
EXPERIENCE MACHINE ENGINEERING

Nozick writes:

> "Suppose there were an experience machine that would give you any experience you desired. Superduper neuropsychologists could stimulate your brain so that you would think and feel you were writing a great novel, or making a friend, or reading an interesting book. All the time you would be floating in a tank, with electrodes attached to your brain. Should you plug into this machine for life, preprogramming your life's experiences?"[115]

Nozick argues that if we reject the offer to plug in, it shows that we value things other than (or in addition to) subjective experience.

But how exactly would such a machine work? This has not received much consideration in the philosophical literature.

(a) If you are a coward, how can you have experiences of being brave? If you are inept at math, how can you have experiences of getting brilliant mathematical insights? There might be no sensory input that would induce these experiences in your brain such as it is. So the superduper scientists would need to directly rewire the internals of your brain. But this can present problems of personal identity. If the rewiring is too extensive, too abrupt, and too disassociated from the natural exercise and development of your capacities, the person in the tank that ends up having the experiences of bravery and mathematical brilliance would not be you. This limits the possible experience trajectories that you could have in the experience machine. (Direct brain editing would also be necessary for many experiences that normally would depend on having a particular history and background as well as on having a particular personality and aptitude profile. What we actually experience when we are stimulated with a given set of sensory inputs depends *very heavily* on what we bring to the table—our concepts, our memories, our attitudes and mood, our skills, etc.)

(b) If you want to have an experience that involves making an effort, you would in effect need to yourself actually make an effort. Modulo the limitations expressed under (a), the scientists could cause you to make such efforts; but it is worth noting that this kind of experience would not come "for free". Suppose you want to have the experience of climbing Mount Everest. You could easily have the experience of seeing a series of views that would be seen on the ascent; and, if you look down, you could see your legs moving. You could also feel the pressure on your shoulders from the backpack, and the chilly air biting into your cheeks. But without the sense of having to strain, of having to dig deep inside yourself to find the wherewithal to continue, your experience would be but a shadow of the experiences of those who have surmounted Everest in real life. If, however, the scientists *do* induce these elements, then you are paying a hefty price for the experience—a price of discomfort, fear, and willpower expenditure. The experience machine might not give you that much of a benefit compared to actually going to Nepal and climbing the real mountain, although it would protect you from the risk of physical injury.

(c) Among our most important experiences are ones that involve interactions with other human beings. How could such experiences be implemented? Consider the following routes.

i. *NPCs.* In order to generate the sensory input that you receive when you are interacting with others, those other people could be implemented as NPCs ("nonplayer characters"), by which I mean constructs that display some of the attributes of an intelligent being yet without thereby instantiating any phenomenal experience or other bases that would endow that being with moral status. This is undoubtedly possible in the case of relatively shallow interactions. For example, if you want to have the experience of asking a stranger a few questions along the lines of "what is two plus two?" and getting back an answer like "four", it would be metaphysically possible to implement the requisite computations without creating any morally considerable being (other than yourself). But it is less clear that it would be metaphysically possible to generate the fully realistic experience of having long, deep, and rich interactions with another human being without running a computation that effectively implements a complex digital mind that has moral status. Which brings us to the second route toward generating interaction experiences...

ii. *VPCs.* By VPCs ("virtual player characters") I mean artificial computational constructs that *do* have moral status, for instance because they possess conscious digital minds. The use of VPCs would unlock a very wide range of possible interaction experiences, though the mechanics of implementation remain tricky. For example, in real life you can walk across a crowded plaza and be able to strike up a conversation with any of the thousands of people you see—in principle, a deep conversation that could lead to a lifelong friendship or relationship. The superduper scientists' apparatus might therefore include VPC simulations of all of these people (at a large computational cost). Alternatively, procedural generation might be used to instantiate VPCs on demand: initially the anonymous people in the crowd might be represented as NPCs but if and when you commence a deep interaction, the missing details of their minds are filled in to render them as VPCs that can respond to you in a fully realistic manner. This sublimation of NPCs into VPCs would require a big burst of computation—perhaps entire childhoods and personal histories would need to be simulated to generate the fully realistic VPCs that you are about to engage in conversation.

Although VPCs would make it technically feasible to generate a very wide range of interaction experiences, the use of VPCs would introduce moral complications. It might be impossible for you to have certain experiences in the experience machine without violating ethical constraints, just as it would be in external reality.

iii. *PCs* ("player characters", simulated or original). Additional ethical complications arise if you want to have experiences of interacting with particular real-world individuals. Either accurate simulations of those individuals would need to be created, or the actual individuals would need to connect to the experience machine to enable you to interact with them. In either case, it is plausible that the real-world individual in question is morally entitled to a say in the matter; and some people would presumably refuse consent to interacting with you in this way. Furthermore, even if they do interact with you, they might not interact in the way you want. For example, you might wish to have the experience of a particular real-world individual stroking your hair, but there is in general no guarantee that the individual in question is interested in doing that—neither the original individual nor any accurate simulation of that individual. In the case of a simulated version, it would be technically possible to modify the simulation so as to make it willing to engage in the type of interaction that you seek; however, without the preceding consent of the individual in question, this would probably be morally impermissible.

A "multiplayer" version of the experience machine has been proposed (including by Nozick himself), in which many people together plug into the experience machine.[116] This would enable us to have real interactions with other real people, including particular existing people who are important to us—thus obviating one common ground for refusing to enter the experience machine. However, in this setup you no longer have complete control over the experiences you have, since that will now depend on the independent choices of other people. This scenario thus violates a key premiss of the original thought experiment.

iv. *Recordings*. The notion that you enter the experience machine by having neuropsychologists stimulate your brain with electrodes is a bit quaint. A more plausible and more efficient method would be to first upload yourself into a computer and then to interact with a virtual reality.[117] This presents us with at least one special case in which you could have fully realistic deep interaction experiences without instantiating any morally significant entity (other than yourself), namely by replaying recordings of the outputs of other people. To do this, you would first do one run in which you interact with VPCs or PCs (which can themselves be uploads or biological). The superduper scientists record the interaction history between you and these other people. When you have finished doing whatever you wanted to be doing, we reset your mind and the environment to their initial states. You now have the ability to enjoy the same experience again, but this time without instantiating any real persons. This would be done by rerunning the program again that implements your mind, initialized to the exact starting configuration of the first run; but instead of re-running the computations that correspond to your interaction partners' minds (or to the physical environment), we simply fetch the relevant information from memory. As you are having the experience this second time, you can make whatever choices you want: but since we already know what choices you will want to make, and since we have a recording of how other people and the environment reacted to these choices, we don't need to recompute those parts and can instead used stored data to determine the input that your senses receive. (The operations of your own mind do need to be recomputed, of course, because—we believe—this is what actually generates your experience.)

This recording trick still has the limitation that it requires there to be an initial run in which your interaction partners are implemented as real persons. This brings in the moral complications that we might have been hoping to avoid (though if there are moral infractions, this procedure would at least reduce the number of times those infractions occur). The recording method also has the downside that it restricts you to repeating old experiences (although, of course, they would always *feel* exactly as fresh and original as they did the first time round).

v. *Interpolations.* Replaying an exact recording is a limiting case. We can consider less extreme cases along the same dimension, in which NPCs are implemented in a way that relies, to a greater or lesser extent, on cached computations and on pattern-matching to a bank of observational evidence about how people react in similar situations (as opposed to ab initio or fully bottom-up computation of somebody's nervous system). Generally speaking, the more the experience machine relies on memorized material to generate the responses of other people that we experience when interacting with them, the less likely it is that the other apparent person becomes real in the process—"real" in the sense of constituting an entity that has moral patiency.[118]

It is difficult to say how far it is possible to go in this direction. Maybe for rather a simple mind, like that of a human being, it will become feasible to create a library of minds and experiences that makes it possible to generate quite verisimilitudinous appearances of most types of interactions that humans tend to want to have with other human-like entities, without thereby bringing into existence any morally considerable beings.

vi. *Guided dreams.* A lower bound on what is possible by way of generating realistic experiences without interaction with the external world or with other real people is given by dreams (and hallucinations). To the dreamer, dreams can be very convincing. Most people have little control over what happens in their dreams, but presumably the level of control could be greatly increased with advanced neurotechnology. (Even without technological aids, some "lucid dreamers" report an ability to direct the content of their dreams to a considerable extent.)

One might perhaps question the extent to which the convincingness of dreams is a consequence of the experiential content of the dreams being very similar to the experiential content of analogous waking experiences, versus instead a consequence of our ability to detect inconsistencies or experiential deficits being impaired while we are asleep. Certainly our dream experiences often contain anomalies which, if we had similar experiences while awake, we would expect to notice and to regard as unusual and bizarre. Yet dreams are the spontaneous creations of our own humble brains; and it is highly plausible that their realism and coherence could be increased artificially at technological maturity. The question might be, if our dreams became a lot more detailed, realistic, and coherent, whether, when we are dreaming of other people, those people might not then actually enter existence sufficiently to become moral patients. It might then be morally problematic to dream or fantasize sufficiently realistically about other people without their prior consent (and without satisfying various other ethical constraints).

Firafix: I'm done.

Kelvin: Ok, let's walk.

Firafix: So he doesn't exactly answer the question of whether to enter the experience machine or not. What do you think?

Tessius: Oh, I'd recommend it.

Firafix: You would accept the offer?

Tessius: Already did. Great in here!

Firafix: What do you mean?

Tessius: Smell those lilacs… isn't it lovely?

Kelvin: I think he's pulling your leg a little bit.

Tessius: What? You mean you are *not* mere imaginary characters in my guided dream?

Firafix: If we are, you have a vivid imagination! Because it certainly feels pretty real from where I stand.

Tessius: You two are most welcome. It's on the house.

Fictional characters

Firafix: But seriously, do you guys ever wonder whether we might be characters in a novel or something?

Kelvin: Fictional characters aren't implemented in sufficient neurological detail to have phenomenal experiences. Since we have phenomenal experiences, we are not such characters. But even setting aside phenomenal experience, we are aware of a lot of things that are not reported in any novel—such as the precise configuration of those trees over there.

Consider how much information there is in a book. Let's say it has 100,000 words. An average word is about 5 characters, and each character is 8 bits. So that would be 4 megabits. With compression, it would be way less. There's no way you could represent all the contents of all our experiences that we've had in our lifetimes with that few bits.

Tessius: Maybe it should not be all the experiences we've had in our lifetime, but just the experiences we are having at this moment? Although that would correspond to a book that was entirely about the details of one person's experiences at one single moment in time. There probably aren't many books of that sort.

Kelvin: A human brain has about 10^{14} synapses, most of which are transmitting information during any one-second interval. Even if each synaptic event could be represented with only one bit of information, that would still be more than ten million times more information than is contained in an uncompressed book. Rough numbers, but that's a huge gap.

Tessius: Isn't the more fundamental objection that the book isn't processing information? But I suppose maybe the processing occurs in the reader's mind when it is being read... But then, maybe a lot of the information about the fictional character is also supplied by the reader's mind—like, they bring their own experience to bear and fill in the scarce details provided in the text so as to create a more full-fledged character in their imagination? —I don't actually think this works, but it's interesting to consider the argument to see more precisely where it goes wrong.

Kelvin: I don't think the human intellect is powerful enough to bring an imaginary sentient mind into being simply by thinking about it. For a superintelligence, it's a different matter. It could internally simulate sentient minds. But this is a very different proposition than that literary characters in a novel are conscious, or that we could be such characters.

Tessius: Yes, let's focus on the case of literary characters in ordinary human-written books. How exactly do we know that there is not enough information content and computation taking place *when a human is reading a novel* that the characters described come alive sufficiently that we can't be sure that we aren't such characters? We are clearly not explicitly representing a mind composed of one hundred trillion synapses when we are reading a novel; but it doesn't seem obvious that an explicit representation with that level of granularity would be necessary to produce the subjective experiences in question.

Kelvin: Hmm. I think the key is to capture the counterfactual behavioral patterns. It doesn't seem that we do that when we are reading about somebody in a book.

Firafix: Care to elaborate?

Kelvin: To implement a computation, it is not enough to have a sequence of representations of the successive states of the working memory. It also needs to be the case that causal structure is such that if any of these intermediate states were altered, the subsequent states would unfold differently in the appropriate way.[119]

Tessius: This is the difference between playing a movie of a computation and actually implementing the computation. In the movie, each frame might contain a picture of the state of the memory cells. If you play the movie, you would see a sequence of pictures of successive states of the memory cells. But if, while the movie was playing, you went in and edited one of the frames, the later frames would not change. So in a movie of a simple arithmetic computation, one frame might depict "2+2", and the next frame might depict "4". But if you edited the first frame to "2+3", the second frame would still depict "4". This is in contrast to if you are actually implementing the computation. If instead of a movie reel, you were using a pocket calculator, which does implement the computation, then in the time step after you had edited the input, the screen would depict a "5".

Firafix: I see. So how does this connect to what happens when we are reading a novel?

Kelvin: When we're reading about or imagining one of the characters in a novel, are we representing all their counterfactual behaviors to a sufficient degree?

Tessius: Maybe one way to get at this: Suppose you are reading about an imaginary character who is piloting an airplane in World War II. But you the reader have never piloted an airplane. There is no way that your brain would be capable of implementing the computations that would be required to successfully pilot a World War II-era fighter plane. So how then could the requisite computations be implemented that would accurately generate the experience of doing that?

Firafix: Nay, I guess that would not be possible... But what if we are considering a case where the reader is on the same level as the fictional character, and where she has the same skills and so on.

Tessius: Are you averring that there exist readers that are "at the same level" as us? I, for one, declare myself offended and aggrieved.

Firafix: I was primarily wondering whether *I* might be a fictional character! But if we are fictional characters in a fictional world, maybe in the world where the reading is being done there are some pretty clever readers?

Kelvin: If the readers were superintelligent, and if their "reading" essentially consisted of running detailed internal simulations of the neural networks of the characters described in their novels, then yes. But we are not talking about the simulation hypothesis here. We are instead discussing the crazy proposition that we might be characters in a story written and read by ordinary humanlike beings, right?

Firafix: Well, maybe there could be readers who are just a bit cleverer than regular humans but not superintelligent? But perhaps it's better to focus on the case where we have a fictional character who is of sufficiently limited abilities that there are at least some human readers who would be able to do everything that this fictional character can do. If my abilities are thus limited, how could I then know that I am not a fictional character? Or, I suppose, *an instance* of such a fictional character being read about and imagined by some particular reader? As opposed to a real flesh and blood creature?

Tessius: Nitpick: fictional characters—unless it is a sci-fi story about androids—are usually depicted as being of flesh and blood. And especially if—actually scratch that, sorry. Er, hmm…

Ok, suppose that a fictional character is being described as reminiscing about their late mother. You, the reader, don't know the character's mother: you've never met her, and, let us assume, you've never even read about her. How then could you possibly be conjuring up in your mind what this fictional character would be experiencing if they were actually really experiencing anything? And conversely, if *you* think about something that has not been described in any novel, then how could any writer or reader have generated the thoughts that you are having? But now, as I am speaking these words, the obvious objection also occurs to me: how can I know which thoughts some writer might have written about? And how can I know that my thoughts when I am thinking about my mother are not actually some reader's thoughts when they are thinking about *their* mother or about some other random motherlike figure that they create in their imagination?

Firafix: Well?

Tessius: Well, I have a great many thoughts. It would seem unlikely that any writer—any human writer at least—could have thought about *all* those thoughts and written them down; or that any reader would be conjuring up all these thoughts in the process of reading a novel... And, if I may be brutally honest for a moment, it is also perhaps possible that I might on some occasion have had some fleeting thought that would not have *merited* being written down... So, er, the fact that I have had all these thoughts, including some that authors would not have deemed significant enough to jot down in their novels or readers to picture in detail in their imagination: this fact would then prove that I am in fact *not* a character in a novel.

Kelvin: Of course, that presupposes that you have in fact had all these thoughts that you claim to have had.

Firafix: I don't want to doubt that you guys have had many thoughts!

Kelvin: No, but if we *were* fictional characters, then we might be depicted in that fiction as characters who have had many thoughts.

Tessius: That would not make all of those thoughts real.

Kelvin: Right. Within the fiction it might be true that the fictional characters have had a huge number of thoughts, and also that they have lives that extend to both before and after the events that are explicitly recounted in the text.

Tessius: But what about your earlier point about information content, then? Unless I'm terribly mistaken, I can recall a great many specific details about my past, more than any novelist would care to write about or reader imagine. Actually, this is a different version of the information content argument that you were initially proposing. You seemed to suggest that the fact that the human brain contains 10^{14} synapses might be enough to establish that fictional characters are not sentient, since a book does not contain enough information to specify what all of these synapses in the fictional character's brain are doing. But now the idea is that the reader's brain is doing most of this work. The book contains some nudges and pointers, but the reader's brain is filling in the great bulk of the requisite information—namely, by the reader using their own concepts, intuitions, and imagination to render the fictional character's experiences, thereby making them real. And since the reader's own imagination recruits the services of trillions of synapses, there is no mismatch between the amount of information and computation required to generate phenomenal experiences and the amount of information and computation that is available during the reading process to actually accomplish this feat.

But now the *new* objection is not about the number of synapses. It is about all the specific detailed memories that I have had. If I have in fact had all these memories, then I could not be a fictional character.

Kelvin: If.

Tessius: Well, it would be a pretty radical form of skepticism to think that all of my memories might be fake. I mean *possible*, yes; but credible?

Firafix: Is it credible?

Tessius: You be the judge. But I have another argument, too. When I read about a character who is going to a party, it is a quite different experience than I have when I go to a party. There may be some similarity, but, realistically, it is by no means the same. In one case, it is quiet and I'm relaxed and I'm maybe feeling cozy under the throw on my couch; in the other case, there's loud music and excitement and people bumping into me. I can lose myself in a novel, but there's no real possibility of me confusing these two quite different types of experience.

Right, yes, so we have two grounds, then, for dismissing the hypothesis that we are characters in a novel. First, because we have a great number of detailed memories of the past, which would not be included in any plausible book or reading experience. Second, because the actual experiences generated during reading are qualitatively very noticeably different from the experiences generated when one is out and about in the world actually doing things. QED?

Firafix: Well, thank you gentlemen. I am relieved that we are not mere fictional characters. Although some part of me thinks that that might not be such a bad thing to be—depending of course on what sort of novel it is.

Tessius: Hold your horse, I'm sure the objections are about to arrive.

Kelvin: We didn't actually resolve the issue of whether you can trust your apparently multitudinous memories of the past.

Tessius: And as for the second argument, I think we can grant that the experiences we have while reading are generally different from the experiences we have while we are doing the types of things that we may be reading about. But we should perhaps consider another way in which we and our experiences might be generated during the process of somebody reading a novel. Perhaps there are *two* different experiences: the reader's own experience of reading the novel, and also a second set of experiences that belong to the characters she is reading about?

Kelvin: That doesn't work. The reader only has one brain, and this brain has limited capacity. Just as you wouldn't be able to write two letters at the same time, one with your left hand and one with your right hand, neither would you be able to generate two separate streams of experience simultaneously. At least not if each of those experience streams were full complex person-level experiences. Perhaps it might be possible for some part of your brain to generate some sort of separate experiences, some limited kind of gut feelings or suchlike, while the bulk of your brain is busy generating the experiences that you are able to report. But there just isn't enough neural machinery to simultaneously process two separate fully person-level experiences, especially if those experiences involve explicit abstract reasoning. We only have enough working memory capacity to sustain one line of complex reasoning at a time.

Firafix: I see.

Tessius: But what if the two separate experience streams were not completely unrelated? I mean, I agree that my brain does not have the capacity to think simultaneously about two different topics. But if I'm thinking about one topic, might there not be two experience streams relating to that topic—like, experiences of viewing it from slightly different angles? These two experiences might use overlapping neural machinery, while each also involves some snippets of separate neural processing that make them different and distinctive from one another? For example, when I'm reading about the character who is going to a party, could there be one experience stream that contains only the experiences one would have while at a party—these would be the experiences of the fictional character—and another experience stream that has some version of those same experiences but with the experience of reclining peacefully on the sofa superimposed? The latter being the experiences of the reader. Both of these would be generated by the reader's brain, but he would only be able to report in detail on the latter?

Kelvin: Seems far-fetched.

Tessius: I agree, but it would be nice to be able to say precisely *why* it couldn't happen like this.

Kelvin: What would be the principle that accounts for how this would work? Presumably you don't want to say that there is a separate experience stream for every subset of the neurons in the brain. That would imply that if one person has one or two dozen more neurons than another, their brain would be generating a *super-astronomically* larger quantity of experience. That would have radical implications for ethics, for example in terms of moral status.

Firafix: It might be good for you, though, with your big head.

Kelvin: But if there is another human with a slightly bigger head, I would get virtually zero weight, at least from an experience-utilitarian perspective.

Tessius: So what is the principle that rules this out... hmm, let me think. There is an analogy to the many-worlds interpretation of quantum mechanics. We have to weigh the branches with the squares of their amplitudes. When the universal wave function splits, or decoheres, the amplitude gets divvied up between the branches, and it always sums to 1. But that doesn't seem like the right model here. If you add a neuron to a brain, the experience streams that *ex hypothesi* are being generated by all the subsets of the other neurons presumably don't somehow lose some of the quantity of phenomenal content that they had before. That would just be weird. So something else, hmm...

Firafix: Can I just check that I'm following? We are trying to determine what the criteria are for whether a brain is implementing a single stream of conscious experience or several overlapping but distinct streams of experience.

Tessius: Right... Maybe the earlier point about counterfactual independence is the key here? In order for two distinct streams of conscious experience to be generated, there have to be two distinct computational processes. And they each have to be such that if somebody intervened and changed some intermediate state, the subsequent states would unfold differently and in accordance with the algorithm that is being computed. Right. And for there to be *two* computational processes, the underlying causal machinery that is implementing them would need to, essentially, be able to operate and vary independently. So this would explain why a normal human brain is not generating an astronomical number of distinct experience streams, one corresponding to each subset of its neurons: because these subsets overlap, so they cannot independently vary, and so there is not the required set of counterfactual dependencies. Yes, that I think seems probably right.[120]

Kelvin: Yeah.

Tessius: Which brings us back to the question of whether the reader's brain could simultaneously generate two experience streams, their own experiences plus those of the fictional character they are reading about. And I guess we would assert that there is not enough brainpower in the reader's head to do this. It would not help that the two putative experience streams would have a lot in common, because in order for each one to exist it would need to have its own independent computational implementation, a causal structure with the right counterfactual properties and that is able to vary independently; whereas, in the case of the reader, there are not enough cortical resources for two person-level computational processes to be implemented simultaneously yet independently.

Kelvin: Right.

Tessius: By the way, this might be an aside, but I'm a bit puzzled by the reports we hear of split-brain patients, whose hemispheres, after most of their connection via the corpus callosum has been severed, appear to be able to operate pretty independently and perhaps with person-level proficiency. Could we really be walking around with enough neural matter to implement two normal persons, yet ordinarily only actually be implementing one? It seems wasteful.

Kelvin: The brain size of normally developed adult humans can vary by almost twofold.[121] We also know that it is possible to have an entire hemisphere removed and still function, although you do get deficits. When a hemispherectomy is done at a young age, the brain's plasticity enables many computational functions to be reorganized to make use of the remaining cortical resources. There is a correlation between brain size and cognitive performance, and other things equal you get some reduction in learning ability when there are fewer tunable parameters, in both biological and artificial neural networks. However, many cognitive tasks scale very sublinearly with model size.[122]

Tessius: In any case, I'm not aware of any particular reason to suppose that even if our brains do have this kind of quasi-redundancy, the "spare" capacity for additional conscious experience would actually be coming into play while we are reading…

Also, we're not very good at multitasking. If our brains *were* sometimes simultaneously implementing the conscious experiences of two persons, using separate pieces of neural machinery, then should we not be able to make use of this duplicated circuitry to, let us say, work out the proof of an algebraic theorem while at the same time making complex scheduling arrangements for a family reunion? For example, you could be modeling an imaginary character who was working on proving the theorem in one part of your brain (or in one subset of your cortical microcircuits), while in another part (or another subset of your microcircuitry) you would be doing the complicated events planning. But I for one would find that utterly impossible.

Kelvin: Yes, that's a good point.

Tessius: The most we can manage, by way of true multitasking while we are working on a task that requires difficult abstract thinking, might be to passively monitor our surroundings, or perhaps our subconscious can simultaneously be mulling over some emotional or creative problem in the background: but our focused conscious reasoning seems to be a tightly constrained cognitive resource that only allows for one concurrent track.

Firafix: So, does this mean that we are not fictional?

Tessius: That would seem to be the upshot.

Firafix: Let me see, so that's because, firstly, I might trust that I have many and detailed memories of my past, which no novelist or reader would care to imagine in all their exhaustive minutiae; and, secondly, because I know that I am currently having experiences that are qualitatively different from the experiences one has while one is reading—and the brain of a human reader does not have enough capacity to simultaneously implement both the reader's experience and the experience of the fictional character they are reading about. Have I got that right?

Tessius: You nailed it.

Kelvin: There are a couple of other arguments that you might also want to consider.

Firafix: Yes?

Kelvin: On average, people spend a lot more time *not* reading than they spend reading. So even if during reading the reader's brain somehow generated a fictional character's experiences in addition to their own reading experiences, it would still be the case that the experiences belonging to fictional characters would constitute a small fraction of all experiences. So if your current experience is typical, it would most likely not be the experience of a fictional character.

Tessius: But what about when I'm experiencing something like, I don't know, heroically rescuing a beautiful princess from the clutches of a wicked monster or an angry mob?

Kelvin: In that case, this argument would not apply. Most of *those* experiences would be side-effects of a reading process—*if* reading processes actually generated the experience of the fictional characters being read about.

Tessius: I won't sidetrack us now, but remind me later to tell you about when I was biking through the royal gardens a few years ago.

Firafix: I'm intrigued! But what about the experience of having a philosophical conversation with two excellent and clever gentlemen?

Kelvin: That's a bit of an intermediate case, I suppose. Depending on the type and quality of the conversation, it might be that most instances are readings of fictional accounts or it might be that most are real-world conversations.

Firafix: A conversation like this one?

Kelvin: [*Shrugs.*] Toss-up.

Firafix: If this conversation *is* fictional, would that then mean that when the text is being read, the reading generates four streams of conscious experience? Since there are three people talking, plus the experiences of being a reader reading a book?

Kelvin: Probably not. It becomes less likely the more people there are—increasingly farfetched to suppose that the reader's brain would have sufficient neural machinery to separately implement each of these experience streams. If there were any experiences generated at all, besides the reader's own experiences, it would probably just be fragments of different fictional characters' experiences. Maybe while each person was talking, and while they were at the forefront of the reader's mind, the experiences they are represented as having at that moment would be generated.

Tessius: Certainly it would not be the case that, when a fictional character is walking through a crowded room, all the fictional characters in that room would be represented with sufficient granularity in the reader's mind that all their inner lives would actually come into existence in their full subjective detail.

Firafix: I see. Ok, so *most* experiences would not be those of fictional characters even if reading did generate such experiences. But the balance would swing the other way for especially "fantastic" experiences.

Kelvin: I'd say: *might* swing the other way. But it could be the case that even most experiences of saving maidens from dragons would be had by nonfictional characters, if, let us say, there are many simulations of such scenarios and not as many simulations of people reading about such scenarios.

Firafix: By "simulations" I presume you are referring to something like Bostrom's idea of computer simulations built by a superintelligence that include detailed simulations of people's brains?

Kelvin: Yeah. Those are different from the case where somebody is reading about a fictional character, because in the simulations—"ancestor simulations" he calls them, but they don't necessarily have to be simulations of ancestor-type creatures—there are neural-level simulations of each subject's brain.

Firafix: You said there were *a couple* of other arguments that I might consider. What is the other one?

Kelvin: It's more of a decision-theoretic or political one.

Firafix: Yes?

Kelvin: Let's suppose there is a fictional character and a nonfictional one, and that both have their own separate conscious minds. Maybe you are not sure which one you are. Now you could argue, in this case, that you should act mostly as if you were the nonfictional one. The fictional character would tend not to live very long and their choices would have less opportunity to have long-term consequences. Note, it is not their lifespan or their impact *as described in the novel* that matters here. A novel might say that a fictional character saved the world and lived happily for a million years thereafter. But this does not mean that any real world was saved or that there was actually some fictional character that had a million years of real phenomenal experience. Even under the premiss that reading about a fictional character can bring that character's experience into reality, this would apply only to those of the character's experiences that the reader's brain actually models in sufficient detail. So the maximum amount of subjective experience that a fictional character could have is the amount of experience one can have during ten hours, or however long it takes to read a book.

Tessius: What if the book is read by many people? A bestseller might be read a million times. Ten hours times a million would be longer than an ordinary human lifespan.

Kelvin: Yeah.

Tessius: So maybe we should act as if we are characters in a bestseller? Or maybe we should even act so as to make it more likely that the book we're in *becomes* a bestseller?

Kelvin: Yeah.

Tessius: The narratological imperative? I think we have just proved that the best thing for you to do would be to moon those ladies over there at the bus stop, Kelvin! It might sell another thousand copies... resulting in, what, ten hours times a thousand: ten thousand hours—that's more than a year, Kelvin. Maybe divided by the three of us. Still, *four months* of Kelvin-life—worth it!

Firafix: Not a good idea!

Tessius: Well, so what gives?

Kelvin: Would the potential readers of the kind of book that would feature the kind of dialogue we've just been having want to read a detailed account of me exposing my hindquarters to some ladies? I don't think so. Anyways.

Tessius: I think for a split second he was calculating up the expected utility!

Kelvin: There are also deontological side-constraints.

Firafix: And decency.

Kelvin: Indeed.

Tessius: Were you not a little bit tempted though?

Kelvin: I was not.

Tessius: But reflect that many reading experiences are those of romance novels. Maybe we are in one of those—one written for the more discerning female reader?

Would you not take one for the team, Kelvin… create a little *frisson*?

No? Oh well. *Sorry, readers*; I tried!

Firafix: But what exactly then is the moral argument that you were alluding to, Kelvin?

Kelvin: Eh, the issue is moot. We already established that you are not generating any separate set of conscious experiences when you are reading about fictional people in a book.

Firafix: Okay, but I'm still interested.

Kelvin: If fictional people became real while somebody was reading about them, they would on average have less power to influence the world than people who are real the entire time, continuously and cumulatively for seven or eight decades. There might be *some* fictional people who are influential, but *mostly* the world is run and shaped by nonfictional people. Also, for every fictional character who has influence, you could argue that that influence is also shared by the person who wrote them, the author. Furthermore, given the currently prevailing moral norms, authors are free to write about characters that violate moral norms without compunction. This makes it less clear that recommending to fictional people that they behave morally would actually increase the frequency with which they do so. If some given fictional character freely chooses to do the morally right thing, it might just lead authors to compensate by creating characters who are even more disposed toward wrongdoing, so that they still achieve their desired level of turpitude in their novels. All in all, this makes it seem to me like our moral reasoning should mostly focus on the possibility that we are *not* fictional characters, since that is the hypothesis under which our mortality-motivated actions have the most significant consequences.

Firafix: Hmm.

Tessius: I think maybe there is a missing ethics of fiction-writing. This idea that authors should have absolutely no compunction about creating any sort of character whatever, or doing whatever they happen to feel like to their characters: I'm actually not sure that is right… In fact, even if the fictional characters they write are *not* conscious, even while they are being read about, they might have other attributes that could ground at least some modest claims to moral status.

Firafix: Attributes such as?

Tessius: Attributes such as having preferences. Fictional characters can have preferences, which are distinct from the preferences of their readers and writers. Also, fictional characters can have social relationships, for example with other fictional characters. They might have a kind of conception of themselves as agents persisting through time, with long-term goals. They certainly can have causal powers, even if—on average—those casual powers are less than the powers of nonfictional characters… Now that I think about it, I'm actually beginning to persuade myself on this point.

Firafix: How would we—I mean—if we regarded fictional characters as having some moral status, what should we do about that?

Tessius: I haven't thought it through. Maybe other things equal, we ought to be writing more comedies and fewer tragedies. More happy endings. I kind of like the fact that many stories end with "and they lived happily ever after". But maybe the monsters too should live happily ever after.

Firafix: It would work for me. I usually prefer to read happy stories anyway. But I might have an uncommon taste.

Kelvin: There is some value in understanding bad things, so that we can more effectively work to counter them. But yes on balance there should probably be more of a tilt toward the positive. There could be other reasons for that as well. —But we are here, so, let's go in.

Firafix: Okay. So to sum up, you guys have established (i) that we are not fictional characters, and (ii) that fictional characters deserve to be treated with some degree of moral consideration.

Tessius: Yes. And hopefully at least one of those conclusions is correct!

*

Kelvin: Shall we enter?

Tessius: Would be rude to keep nonillions of water molecules waiting.

Firafix: I'll bring the readings.

Feodor the Fox

Epistle XXIII

Dear Uncle Pasternaught,

Thank you for your forbearance over these past couple of months. I can now finally write to you and fill you in, as my own efforts are no longer such a rate-limiting step in the developments taking place.

You will recall how Pignolius and I had been at an impasse, and how I then had this unusual dream, which made me feel like I needed some time alone to process things.

So I traveled to the sea. They also call this body of water "the ocean", and I'd been told it was big but until you see it you can't really imagine just *how* big it is. In fact, it looks infinite, because no matter how hard you strain your eyes you cannot see the end of it, even though nothing is blocking your sight. It starts at your feet and just keeps going, ascending in your visual field until it meets the equally unbounded heaven reaching down from above; and in the middle they meet, without any dividing zone or barrier, Heaven and Earth.

All I can do is to describe these external circumstances; for my inner state I'm not able to put into words.

I remain there by the sea for some time. Exactly how long I don't know, but it might have been a few weeks. And I'm starting to get some ideas, and I do some thinking. Things are coming together, sort of, although there are still gaps. The gaps are big, and I don't know how to bridge them. But I begin to see that *if* one could bridge them, these specific three or four gaps, then there ought to be a way forward, at least in principle. The problem has a structure now, and one can work on it. The right kind of mind could perhaps even solve it.

At this point I'm ready to begin my return journey. My pace quickens as I go.

When I get back, I immediately tell Pignolius about my ideas. He is very interested. We discuss the gaps. I can sense his mind being captivated by the challenge. As I'm talking with him, it is as if he's picking up faint scents in the air that I can't smell. And then he's off on his intellectual truffle hunt. I'm exhausted after the journey and fall into a deep sleep.

When I wake up, the sun sits high in the sky and it's almost midday. I hear twigs crackling and leaves rustling a little distance away. Pignolius is pacing, back and forth, back and forth. I sneak off and have brunch on my own: a dried cricket and some corn. The rest of that day Pignolius is working, and I decide not to disturb him.

The next day is the same. He is evidently thinking hard—sometimes pacing back and forth at the end of the clearing, sometimes reclining in his mud pool with a remote expression in his eyes.

The third day he breaks off and comes up to me, beaming. He says "I think there is a path. But you may not like it".

He then proceeds to outline the plan.

Now, to call it a plan is scarcely more accurate than saying that hope has wings. There is maybe some poetic truth in this expression, but you would rather not have to test it by for example jumping off Bigrock Cliff.

However, this "plan" is what we are currently executing on. We would like to create a sustainable social surplus, because only with extra resources is it possible to ensure that everybody has enough to eat and time to spare to contribute to the common good. It is not difficult to think of desirable projects. We would like to build a communications infrastructure, a healthcare system, we'd like to have specialists doing research to improve health and well-being and food production—many other things. If we could have these things then quality of life would be better, and it would further improve over time. I don't see any real limit to how high it could go. However, we have not spent much time thinking about this—because that's not where the difficulty lies. The difficulty is that there is no sustainable surplus.

We observe that a surplus *could* exist if there were fewer animals living on each plot of land—then there would be more food for each mouth, and it would take less time to fill one's belly. We also observe that a lot of time and energy could be saved if instead of fighting with one another, we cooperated. The deer could graze in peace without having to look around all the time to scan for wolves. The birds wouldn't waste their energy squabbling over every morsel.

It is even possible to imagine gains from cooperation between predator and prey populations. The predators currently eat a certain number of prey—but if the prey offered themselves up voluntarily, the predators wouldn't expend as much energy chasing the prey, so they wouldn't need to eat as much. (Of course, longer term, it would be more ideal if everybody could live on plants.)

Uncle Pasternaught, I see you shaking your head: "This is utopian! Not in a good sense, but in a fanciful crazy-talk sense." Hear me out.

I said *if* we could all cooperate, *then* we could have a sustainable surplus; and with that surplus, we could improve our welfare, both in the present and over the long haul. But how to achieve cooperation? That is the question which had us stumped for so long. We now have some ideas.

Imagine first that we already had somehow achieved a high degree of cooperation, and that the challenge was to make this stable. We think that one way to do this would be by breeding for cooperativeness. So if somebody is cheating, they wouldn't be allowed to have offspring, but the individuals who are more helpful and cooperative than average, they could have more offspring. Since we've assumed that we have a high degree of cooperation to begin with, people would mostly adhere to this agreement, and they would volunteer to help enforce it if there were any defectors. Each generation would be better at cooperating than the preceding one, and so there would be hope that the arrangement would stick. Of course, along with cooperativeness, there may be other desirable traits that one may also want to select for—vitality, wisdom, ability to thrive on a diet of leaves and grass, and so on.

An alternative to this would be to leave breeding unregulated and only police non-reproductive behaviors. An animal might be penalized if it stole food from another. But this wouldn't help with overpopulation. Also, we think it would be less stable. Any temporary disruption of the policing system would bring us right back to the state of nature—whereas if the population had been bred for cooperativeness, that trait wouldn't immediately go away, and so the naturally well-disposed animals may be able to reconstitute a cooperative social order before the population re-evolved the roguish propensities of their undomesticated ancestors. There are other considerations, too, that favor changing our nature and not just our behavior, which I will not detail here.

Now the problem with the approach I have just outlined is that it presupposes that a high level of cooperation already exists and can somehow be maintained long enough to bring about the required change in nature. How long might this be? Pignolius has been working on some mathematical models, but we don't have enough data to make good estimates. However, our guess is that we should see some significant effects within even just a few generations, based on the observation that children on average are noticeably more similar to their parents than to strangers, in terms of their temperament, even shortly after they are born. This would mean that the challenge of maintaining cooperation should start to get easier after a few generations, although it would take much longer before the problem is fully solved.

We also suspect the time required depends on the species. Those that start out more prosocial might get there in fewer generations. An interesting question came up with regard to wolves, a species about which not much is known. I've never seen one, but it is said that while they are big and scary, they get along quite well with one another. This made me wonder whether they might be promising candidates for domestication. Clearly if we could get some wolves on board, they could help a lot with maintaining order and defending territory. But when I suggested to Pignolius that we explore this, perhaps by heading over to Render Valley to discuss it with some of the wolves that are said to hang there, to get their input, he would have none of it, saying that he feared that his "rump would be the first input taken in this scenario" and that he was "not sure whether they would bother" with mine, "no offense". None taken. So we've tabled the issue of the wolves, though I still think that at some point we'll need to find a way to bring them into the fold, or they might become a big problem.

The pigeon is here—will explain rest in next letter.

Your indebted F

Epistle XXIV

Dear Uncle Pasternaught,

Apologies for the abrupt signoff last time! I wanted to get it posted since it'd been so long since my last letter.

I can report that Rey has arrived safe and sound! He does have a way about him. We hear some crackling in the bushes, and there he is—bright-eyed and bushy-tailed, not a hair out of place notwithstanding the long journey he has just completed. He trots straight up to me:

"Fedya, I've got five girlfriends waiting for me at home, all going crazy from lovesickness. This better be important!"

Then he studies my face closely for several long seconds, and I don't know what he saw there but he turns serious and says to me: "Whatever you need me to do, brother, I will do it."

What a specimen. I know he was not easy to bring up, Uncle, but for all the difficulties he presented as a kit, I feel proud to be of the same litter as he. I hope we can one day make it up to you.

Well, I began to explain our plan in my last letter; let me continue where I left off. Recall that we have this notion that it might be theoretically possible to transform the world, or at least this forest, from an anarchic warzone full of suffering and desperate need into a cooperative civic structure that had enough spare capacity to be able to undertake projects for the improvement of the welfare of all. Given the brutal and mistrustful natures that we have to work with, we think a multigenerational breeding program, augmented with some institutions of policing and defense, would be necessary to bring this transformation about.

Now the problem we confront is: how could we, a few idealistic individuals—initially just Pignolius, Rey, and me—how can we possibly hope to implement a program of the required scale and duration? We barely have enough to keep ourselves alive in a bad year. And if we get something going, how can it be sustained after we are gone? Even if it gets easier over time, it will need to have an initial momentum to carry it through at least multiple generations.

This is where the mushrooms come in. Pignolius was not wrong when he speculated that I might not like the plan, but I don't have a better one.

As you know, there is a kind of mushroom which we are told not to eat, because it is toxic. Every once in a while, some animal disregards the warnings and takes a bite. Apparently, the mushroom has a psychological effect, inducing an altruistic state of mind, "a swelling of the sense of self to encompass concern for the well-being of others", a state which is said to be quite pleasurable. However, the enjoyment is soon undercut by the onset of violent stomach cramps. I've never heard of anybody eating this mushroom more than once.

We are thinking that we could try to cultivate the mushroom to see if we can make it less toxic and more psychotropic. Mushrooms grow quickly, so we could do many generations of selection in a relatively short span of time.

Here is how we hope it could work out. If we can develop a nontoxic cultivar that retains the pleasurable and empathogenic qualities of the indigen, preferably in a strengthened and enhanced form, then we could make the mushroom itself, or some extract from it, freely available to all the animals in the forest. Many would choose to use it, because of its euphoric potency, and they would then become more altruistic because of its empathogenic effect. While they are altruistic, they would be willing to make certain efforts and sacrifices for the common good, including helping with the implementation of the longer-term breeding program that is required to make our own natures more cooperative, and to prevent the evolution of drug resistance to the fungal compounds. They would also be willing to help out with the mushroom cultivation project—that would require a much smaller effort, because the substance is very strong, and a small garden could provide enough for the entire forest.

Thus we believe we have found a way whereby we can, in principle, parlay a small initial investment of resources into a large permanent improvement of the world—the holy grail!

There remains the question of the initial investment. This we intend to provide ourselves, with an offering of our own labor and our own modest surplus. But we fear this may not be enough. How long will it take to cultivate a sufficiently efficacious mushroom? Maybe years, maybe a decade? We need to collect as many samples as we can find, to get a diverse starting population, and we need to prepare a place where we can grow them and keep them moist and protect them from molds and maggots.

We also need to recruit volunteers to consume small quantities of mushroom, so that we can record the results and select the best samples for further propagation.

Pignolius is a genius for coming up with this plan! He has also been putting a surprising amount of effort into its implementation. He is really good at sniffing out mushrooms, but we need many more samples.

Meanwhile, I've been working my paws sore trying to construct a seedbed with an irrigation system. Alas I lack facility for this kind of work, and the results show it. But progress has been picking up since the arrival of Rey. He has applied himself wherever there is a need, and has generally taken charge of things. He has a wondrous ability to motivate people—you could say that his special surplus is his charm. For example, he persuaded a beaver to drop by and render us pro bono assistance with the seedbed. How does he do it?!

Beavers, I must say, are remarkable creatures: this fellow turned up, worked for maybe one hour, and in that time made as much progress as I had managed in the preceding week!

Oh, and here is another piece of good news: Rey and I have compared notes from our travels, and we're pretty sure that there exists a shorter route between this place and yours—maybe as short as seven or eight days. What this means is that, once we've got the cultivation program properly going, and things settle into a more stable routine—before the end of this summer—I should be able to come and visit. It would be so nice to see you! Maybe we can go up that hill where the strawberries used to grow and see if they are still there.

Your very most indebted nephew,
Feodor

THURSDAY

Interstitial possibilities

Kelvin: Hey, Tessius. We held this seat for you.

Tessius: Thanks. It seems there are more people here with each lecture. I heard you guys from over there—you're still talking about fictional characters?

Kelvin: I think Firafix is still not 100% that she is not a fictional character.

Firafix: I was just making the observation that some characters are based on real people, and was asking Kelvin whether that could somehow mean that those fictional characters could sort of borrow some reality from their real-world counterparts.

Tessius: *All* fictional characters are based on real people, to some extent.

Firafix: How so?

Tessius: How does the author come up with a character? One way is by stitching together fragments from their own experience—memories of various personalities and scenes they have witnessed. Another way is by generating them from their intuitive model of human psychology: but the training data that was used to build this model is again experiences that they've had of interacting with real people. Some authors may also construct their characters from bits and pieces that they've read in other books, but that just pushes the origination back one step. Either way, it's all grounded in observations of real people, and in combinations and extrapolations of those observations.

Firafix: I see. So maybe this connects to what Bostrom had on his handout, the stuff about interpolation being one of the ways of generating social interactions in the experience machine? But I wonder—oh, he's about to start.

Plasticity

Welcome back! And salutations to all the new faces I see around here.

Let me start by completing the taxonomy that I began sketching out yesterday. We'll have to move through this quickly, as we have a lot of ground still to cover.

You'll recall that we introduced governance & culture utopias, post-scarcity utopias, and post-work utopias. We saw that the problem awaiting us as we approach technological maturity—the problem of deep redundancy—extends beyond the difficulties implied by a standard economic post-work utopia. For it is not only human economic labor that becomes redundant in such a condition, but other forms of human effort too. We looked at shopping, exercising, learning, and parenting as examples of activities that become unnecessary (with some qualifications in the case of parenting). I suggested that the analysis could be extended, yielding similar results with respect to many other spheres of human activity.

Where would that leave us?

It would leave us in a condition in which, at least to a first approximation, all human effort, undertaken as a means toward some end, is unnecessary. In other words, it would lead us into an *age of post-instrumentality.*

Now I want to point out that there is another important consequence of technological maturity, besides the obviation of human effort. A technologically mature world is *plastic.* I mean this in the sense that it has affordances that make it easy to achieve any preferred local configuration.

Let us say that we have some quantity of basic physical resources: a room full of various kinds of atoms and some source of energy. We also have some preferences about how these resources should be organized: we wish that the atoms in the room should be arranged so as to constitute a desk, a computer, a well-drafted fireplace, and a puppy labradoodle. In a fully plastic world, it would be possible to simply speak a command—a sentence in natural language expressing the desire—and, voila, the contents in the room would be swiftly and automatically reorganized into the preferred configuration. Perhaps you need to wait twenty minutes, and perhaps there is a bit of waste heat escaping through the walls: but, when you open the door, you find that everything is set up precisely as you wished. There is even a vase with fresh-cut tulips on the desk, something you didn't explicitly ask for but which was somehow implicit in your request.

Autopotency

An important special case of plasticity is that you have the ability to modify *yourself* in whichever way you want. In one of my early works, I termed this ability *autopotency*.[123] An autopotent being is one that has complete power over itself, including its internal states. It has the requisite technology, and the know-how to use it, to reconfigure itself as it sees fit, both physically and mentally. Thus, a person who is autopotent could readily redesign herself to feel instant and continuous joy, or to become absorbingly fascinated by stamp collecting, or to assume the shape of a lion.

These concepts, plasticity and autopotency, have edge cases which I, despite having been trained in pedantry, will not here attempt to exactly delineate, as they are not relevant to the present investigation. This task will be left to future explicators.

Tessius [*whispers*]: Your Chaitin's omega number example would point to one edge case, right? Infeasible to arrange your neurons to register the first thousand digits, even though the constellation itself is apparently physically possible. So do we say a being can be perfectly autopotent without having the practical ability to achieve this particular local configuration, or do we instead say no worldly being can be perfectly autopotent?

Kelvin [*whispers*]: Yup.

Tessius: Or perhaps there is some sense in which that configuration is not really physically possible or diachronically feasible after all?

Agentic complications and luck

Bostrom: The condition of plasticity does not entail everybody having unlimited power over their local environment and themselves. This would be impossible in environments inhabited by multiple agents, since they might sometimes have conflicting preferences. Rather, plasticity means power over nature. It translates into an unqualified ability to achieve any physically possible outcome in the local environment only in the absence of opposition from other empowered agents.

In a world where there are multiple agents, with sometimes opposing goals, general increases in plasticity do not *necessarily* make anybody better off. Technological advancement could make us all worse off, for example by enabling mischief to be conducted more easily and on a larger scale.[124]

Even without conflict or malevolence, increases in power are not axiomatically beneficial. It is possible to use power imprudently. I think if we want to specify a bundle of civilizational properties that is close to axiomatically beneficial, it would have to include at least three attributes: not just power over nature, but also cooperation with our fellow beings, and also wisdom.[125]

And even then it is not axiomatic. With great wisdom and cooperation, technological progress could still turn out to be harmful if we have bad luck. We may wisely take a risk that is *ex ante* worth taking; only to discover, *ex post*, that it was a mistake.

Depending on how much moral content we bake into the concept of wisdom, another way for things to turn out badly, even given wisdom and cooperation, is if people are sufficiently evil: they might then successfully work together to achieve some wicked result.

A third and more subtle way in which maximal technology, wisdom, and cooperation could turn out not to be optimal is if some important values themselves require limitations in our capabilities—if, for example, the meaning of our lives were to be sufficiently undermined by the condition of plasticity. But more on that later.

Hopeful trajectories

Nevertheless, it seems plausible that a society that is maximally technologically capable, maximally cooperative, and maximally wise would be very good—already utopian or speedily on its way to becoming so, assuming it is not under external threat.

We can think of these three attributes as defining a 3D space, with the best utopia being located in the upper further corner where all three of these attributes are fully instantiated. Our current position in this space would be somewhere in the middle—though if we reckon on some sort of absolute scale, we should probably think of ourselves as being much closer to the origo than to the utopian limit.

One thing to note about this space is that it is not convex with respect to goodness. By this I mean that moving closer to utopia from our current position does not necessarily make things better. It could easily be the case, for example, that some advanced technological capability is beneficial only once the world has achieved enough cooperation to avoid using that capacity for war and oppression. Likewise, some advanced facility for cooperation might be beneficial only in societies that exceed some minimum threshold of wisdom—without which the resulting cooperative equilibrium that would result may serve only to buttress some prevailing prejudice or misconception, and permanently lock in a flawed status quo.

Another thing to note is that the paths that lead to the quickest gains in welfare could be different from the paths that lead ultimately, most expeditiously or with the greatest surety, to utopia. I mean it is possible that the course of speediest improvement leads to a merely local optimum. When this is the case, there could be a tension between the interests of a relatively primitive generation, such as ours, and the interests of future utopians, whose coming into existence might require some sacrifice and forbearance on the part of their ancestors.

Now, on the whole, people do not appear willing to make much of a sacrifice for the sake of posterity. But we could perhaps hope that either (a) creating utopia is easy, or (b) the steps needed to get there coincide with some of the steps that people are motivated to take for other reasons, or (c) we are already in utopia, or (d) we get outside help—or (e) we find some way to collimate and accumulate the parts of our wills that do share a love of utopia. Maybe these parts, though individually weak, could, with the right mechanism, be made to combine constructively (between people and over time) in a way that would let them have a greater influence on our common future than the myopic, selfish, and partisan desires that largely rule the present.

For example, suppose each person who loves utopia, and who is not a freerider, puts a small grain of gold into a vast jar. *Eventually* the jar fills up and the creation of utopia is funded.

Maybe this would already have happened, were it not for certain problems. One problem is that as the jar starts to fill up, it becomes a tempting prize for robbers. We may think here, for example, of the expropriation of Church lands by secular authorities in much of Europe during the Reformation, the seizure and redistribution of Islamic religious endowments (waqfs) by the Atatürk government in Turkey in the 1920s, and the confiscation of property held by Buddhist, Taoist, and Confucian temples during the Land Reform Movement in China. We may also contemplate how many charitable foundations have been subverted to serve ends quite different from those intended by their original benefactors.

But it is also possible for the long-termists to achieve their undoing without the aid of external expropriators or infiltrators. They could, for instance, divide themselves into factions with differing priorities, and spend their accumulated resources fighting one another. Where there is a will, there is a way… And there are so many wills and so many ways.

Since conflict and mismanagement could wreck an otherwise auspicious situation, many familiar questions of governance and cooperation remain relevant at technological maturity, although they may of course take different forms in that context than they do in more traditional governance & culture utopias.

Taxonomy

Utopias, therefore, do not form a strict hierarchy of "levels". One utopia might be more radical along one dimension, another along another. They help themselves to different sets of assumptions, and focus on different problematics.

For your convenience, I've jotted down summary characterizations of the different types of utopia that I have introduced. I'll leave you here for a couple of minutes to give you a chance to study this, while I run and get a cup of coffee.

HANDOUT 12

UTOPIC TAXONOMY

1. Governance & culture utopia

The traditional type, what we could also (optimistically) call "post-misrule" utopias. Laws and customs are ideal; society is well-organized. Does not *by definition* imply boring and stultifying, although that is a common failure mode. Another common failure mode is being based on false views about human nature, or making gross errors of economics or political science. Another typical flaw is a failure to recognize the moral patiency and needs of some oppressed group, such as animals. Comes in many flavors—feminist, Marxist, scientific/technological, ecological, religious. (And now, most recently, crypto?)

2. Post-scarcity utopia

Featuring an abundance of material goods and services—food, electronics, transportation, housing, schools and hospitals, etc. Everybody can have plenty of everything (with the important exception of positional goods). Many governance & culture utopias are also, to varying degrees, post-scarcity. In reality, if we focus just on human beings, Earth is already, what—about two-thirds of the way there? compared to the baseline of a typical hunter-gatherer ancestor.

3. Post-work utopia

Full automation. This means there's no need for human economic labor, though attempts to imagine this condition are often half-hearted and assume a continued need of human labor for cultural production. In *post-scarcity* utopia, there is plenty, but producing it might require work. In *post-work* utopia, there is little or no human work, either because machines give us effortless abundance, or because of a choice to live frugally with maximal leisure. Unclear how far toward a post-work condition we've come, given tradeoffs between income and leisure. Many people could probably find some way to eke out at least a hunter-gatherer level of material welfare while doing scarcely any work, although perhaps not without significant sacrifices of social status or community participation. Those with a few mil in their investment portfolios could afford much more, yet often keep working regardless, mostly for the social rewards.

4. Post-instrumental utopia
No instrumental need for any human effort. Implies post-work but goes beyond in also assuming no instrumental need for any *non-economic* work either—no need to exercise to keep fit, for example; no need to study to learn; no need to actively evaluate and select in order to obtain the kinds of food, shelter, music, and clothing that you prefer. This is a far more radical conception than the preceding three types of utopia, and has been much less explored.[126]

5. Plastic utopia
Any preferred local configuration can be effortlessly achieved, except when blocked by some other agent. *Autopotency* is a special case of this—a being's ability to reshape itself as it wills. This goes beyond post-instrumentality, which implies only that whatever can be accomplished can be done so without effort but doesn't necessarily entail any expansion of what can be accomplished. In a plastic condition, the technologically possible becomes identical to the physically possible (at least locally). An important consequence of utopian plasticity is that it is likely to lead to a *metamorphic humanity*: beings that have through their technological advances been profoundly transformed. Plastic utopias have been very little explored, except in theological contexts and in some works of science fiction.[127]

Tessius: Onwards and upwards, toward a *plastic* utopia!

Student: It sounds sort of cheap.

Tessius: A bit—Barbie?

Firafix: Well, that's not so bad!

Kelvin: The cheapness of "plastic" is actually a useful connotation. Many things might actually be both cheap and cheapened in such a condition.

Student: It's not very inspiring though.

Tessius: If you're a consumer, cheap is usually good.

Firafix: I like the word "metamorphic". Is that a neologism for "transhuman" or "posthuman"?

Kelvin: Those terms refer to beings with technologically enhanced capacities. The handout says metamorphic refers to beings that have been "profoundly transformed" through their technology. You could be profoundly transformed not only by being enhanced but also by being diminished or altered, so it's a broader category.

Bostrom: I am caffeinated. Let's press on!

For each category of utopia, there is a correlate category of *dystopia*. One that focuses on the same type of problematic but characterizes it by painting in negatives, showing us what failure would look like. Usually not as a prediction about the future but as a critique of some pernicious pattern in the author's contemporary society. In classical governance & culture dystopias, for example, the problematic pattern might be oppressive totalitarianism (*Nineteen Eighty-Four*) or dehumanizing consumerism (*Brave New World*). In a post-scarcity dystopia, it could be alienation or social disconnectedness. In a post-work dystopia, the issue might be tedium and indolence. In post-instrumental or plastic dystopias, the problematic would be a sense of meaninglessness or of the world becoming uninterestingly arbitrary and untethered.

Dystopias are usually better settings for stories because at least they don't lack problems. (The usual advice to writers is that "stories require conflict".) At a minimum, the dystopian order itself is a big problem that a protagonist could struggle against. But this is only true for the first three types of dystopia. Post-instrumentality and plasticity pose difficulties for *all* attempts at storytelling, whether the setting is presented as positive or negative. This is because the conditions for dramatic agency are undermined, and because realistic portrayals of characters and environments would render them unrelatable and incomprehensible to us.

The redundancy concern

It is natural to wonder whether one could possibly live well at technological maturity—for what purpose would human effort and activity serve in a world that is fully plastic? And without purpose, is worthwhile human life even possible? Would any life in such conditions not be radically degraded, perhaps even to the point of being unbecoming of humans and of any entities with soul and spirit?

The purpose problem menaces any post-instrumental utopian vision, and threatens to cover an otherwise glad prospect in gloomy shadow.

Toward the end of yesterday's lecture I suggested a multilayered response to this problem, consisting of successive walls or palisades from which to resist the inference that life at technological maturity could not be very good. As you recall, the five rings of defense were: Hedonic valence; Experience texture; Autotelic activity; Artificial purpose; and Sociocultural entanglement.

In combination, I argued, these five considerations make the purpose problem seem quite tolerable. In other words, there are possible futures, involving the attainment of technological maturity, whose desirability is not greatly impaired by whatever diminishment of human purposefulness would result even from extreme advances in automation.

Purposefulness, however, is not the only normative notion that comes under pressure at technological maturity. Several related values become similarly exposed to attack in a world of open steppes that is rendered greatly more submissive to our desires. For example, we may wonder how, in a plastic condition,

- Could utopian life be *interesting*?
- Could utopian life be *fulfilling*?
- Could utopian life be *meaningful*?

Rather than focusing narrowly on the purpose problem, therefore, we will adopt a somewhat broader framing of the challenge we confront:

> *The redundancy concern*
> In the limit of technological progress, as it becomes feasible to create a nearly perfectly plastic world—one compliant with our whims and wishes and undemanding of our labor—how can a utopia be conceived in which

values such as purpose, interestingness, richness, fulfillment, and meaning are not undermined to such an extent as to largely deprive the resulting condition of its desirability (and without sacrificing unacceptable amounts of other values)?[128]

So let's explore this. Now, some of the issues we're about to encounter are a bit philosophically tangled, and the following will be more of an exploration than an exposition. We are making our way through terrain that is, to a large extent, unmapped and untamed. So while a relatively smooth ride has taken us to the present point, we may, going forward, have to dismount and proceed more slowly and ploddingly when we reach particularly tricky or uphill patches. I might even need to ask your help in pushing things along if we get stuck.

Those of you who are not taking this course for credit may opt instead to take a nap, and we can arrange to have you woken up once it's over. (I wonder, by the way, how many might prefer to take this approach to their entire present life, if that option existed?)

But the rest of us, who choose to postpone the slumber, whether for course credits or for the sake of some *even higher* aspiration (or because we actually don't mind a bit of strain and roughness in our fun): let us proceed.

We begin with the value of interestingness. We will wrangle with this at some length. A number of the ideas and maneuvers developed in our discussion of interestingness will find application again later in our encounters with other value concepts, allowing us to chop through those more quickly; but we'll need a little patience in this initial phase.

Wouldn't it be boring to live in a perfect world?

Many bromides counsel that seeking happiness by attempting to achieve perfect conditions in this world is futile, either because perfection is unattainable or because it would not bring us happiness even if we attained it. Since the premise of our lectures is to consider what happens if we *do* attain perfection—in the limited sense of fully developed technology and economic plenty in a well-run society at peace—the relevant alternative here is the second one: that even if we did achieve perfection it would not make us happy.[129] Perhaps it would be... *boring* to live in a perfect world?

According to Buddhist thought, we are doomed to experience unsatisfactoriness even if we should be so fortunate as to live under optimal material conditions—with abundant health, wealth, youth, reputation, etc. The root cause of our experience of unsatisfactoriness, on this view, is the role that we allow desire and attachment to play in our existence. And the only way to escape suffering is by eradicating fundamental illusions about the nature of self and reality. We must cease identifying with our desires and let go of our habit of viewing the world through the distorting lens of ego: only then may we see and accept phenomena for what they are; and only then may we find release from our suffering and attain inner peace.

Along similar lines, Arthur Schopenhauer, the great nineteenth-century German pessimist who took inspiration from the Vedic tradition, the Upanishads in particular—a core part of his philosophy centers on a basic predicament: the dilemma we face between the pain that comes from unsatisfied desires and the boredom we experience in the absence of unsatisfied desires:

> "The most general survey shows us that the two foes of human happiness are pain and boredom. We may go further, and say that in the degree in which we are fortunate enough to get away from the one, we approach the other. . . . Accordingly, while the lower classes are engaged in a ceaseless struggle with need, in other words, with pain, the upper carry on a constant and often desperate battle with boredom."[130]

Thus life, according to Schopenhauer, "swings like a pendulum to and fro between pain and boredom".[131] If we project our topic onto this model, then the condition we would enter upon reaching a post-instrumental age would correspond to the "boredom" pole in Schopenhauer's metaphor. The concern is that we would swing all the way to the boredom side and get stuck there.

Subjective feelings versus objective conditions

At this point, however, we must take heed lest we conflate two quite different boring-related concepts. We must distinguish between *boredom*, a subjective mental state; and *boringness*, an objective attribution to some person, thing, situation, or activity. These two ideas have very different implications in our thinking about utopia.

*

Consider first the subjective notion of boredom. Boredom, in this sense, denotes a negatively-valenced mental state. Roughly speaking, it refers to an unpleasant restive weariness, or an oppressive-feeling lack of interest, which makes it difficult to sustain attention on an activity, spectacle, or task.[132] Boredom in this sense is *definitely* avoidable at technological maturity. Pleasure, fascination, joyful absorption, and other boredom-excluding psychological states, are (trivially) among the things that a thriving technologically mature civilization could generate. This is a direct implication of autopotency.

Indeed, boredom-excluding mental states could be generated in *prodigious* quantity and degree, by neurotechnological means (such as genetic engineering, brain stimulation, pharmacological substances, or nanomedicine) or by appropriately designing or modifying digital minds. Far from being an inevitable consequence of technological perfection, then, boredom as subjective experience could be *completely abolished* at technological maturity.

*

Now consider boringness as an objective attribution.[133] We might say that a book or a party is boring, and mean thereby not that anybody necessarily happens to feel bored, but that the object in question has various attributes whose presence is summed up and expressed by the label "BORING". While it is difficult to give a precise characterization of this boringness property, we may take it to involve a deficit of features such as novelty, relevance, significance, and worthwhile challenge. Whether and to what extent a technologically mature civilization can avoid having this boringness property is a more difficult and subtle question than whether it can avoid containing subjective feelings of boredom. We'll explore this more shortly.

*

Although (the subjective feeling of) *boredom* and (the objective property of) *boringness* are different concepts, there are important links that connect the two.

An instrumental link. In the first place, our capacity to feel bored can be useful as a prod to push us away from conditions that have the boringness property. If we think that being in objectively boring conditions is bad, this lends a certain instrumental value to our capacity for feeling subjective boredom.

A normative link. In the second place, we might also think that there are certain situations or objectives to which one ought—for basic normative rather than instrumental reasons—to respond with boredom.

We'll return to this kind of "fitting response" view shortly, but let's first talk about the instrumental link.

Never feeling bored?

So at technological maturity we will have the means to engineer away our ability to experience boredom, yet one might worry that doing so would have undesirable consequences because of the usefulness of boredom as a prod to push us away from boringness. If we think that being in objectively boring conditions is bad, this lends a certain instrumental value to our capacity for feeling subjective boredom.

Now it is true that boredom, like all common human emotions, plays an important functional role in our psychology. Boredom discourages unrewarding repetitive behavior and motivates us to seek out situations with more fruitful opportunities for deploying our time and energy. If we eradicated this emotion, we might become more prone to falling into ruts.

On a small scale, there is situational boredom. We leave the dull party early because it bores us. The benefit here is that if we hadn't felt bored, we would have wasted the entire evening.

On a larger scale, there is ennui or existential boredom. This mood indicates that we are wasting our life. It makes us almost viscerally experience the void of meaning that can occur if we fail to devote ourselves to somebody or something of sufficient importance to us, or if we come to suspect that we are pursuing a dead end. In a favorable case, the ennui causes us to rethink our priorities, and to abandon a sterile life path and begin a quest for a truer calling.

*

You might think that if the utopians extirpated their ability to feel bored, then they would be perfectly content with the simplest and most monotonous preoccupations, such as watching paint dry; and that they would then not bother to do anything more interesting with the future than occasionally repainting a wall so they could watch it dry; and that the future would then consist of a group of people staring at recently repainted walls. This future, while clear of boredom, would be full of boringness. Such a future would seem quite a letdown compared to alternative possibilities that we might imagine.

But that inference would be too hasty.

One reason is that even if the utopians were immune to boredom, they might have *other* values and concerns that would lead them to create a future of greater interestingness than one consisting of people watching paint dry. Fear of feeling bored is not the *only* possible reason for choosing complexity, originality, drama, projects and adventures, and other qualities that counteract boringness. For example, even if the utopians knew they would feel perfectly content watching drying paint, they might be led to create more beautiful outcomes by a positive love of beauty. They may opt for a more socially interwoven future out of a positive desire to connect with other people. And they may choose a more information-rich future out of a passion for learning. A future that is beautiful, socially dynamic, and informationally rich may incidentally also be quite interesting, even if it wasn't shaped with that goal in mind.

Some support for the hypothesis that boredom is not necessary for having an interesting life can be obtained observationally. I know people who seem to have little susceptibility to boredom and yet live lives that are more interesting than most. One of my academic colleagues comes to my mind. This fellow is interested in everything, except sport. In the nearly three decades I've known him, I've never detected the slightest hint of him ever feeling bored.

What gives such people their zest is not that they are especially pained by being in boring situations. Rather (it seems to me) the cause is that they take more than the normal amount of delight in learning and creating, and they have a strong drive and energy for doing so. If these positive traits proliferated, then the world could easily become a more interesting place even if human boredom-proneness were greatly reduced.

Affective prosthetics

A second reason why the utopians could dispense with boredom without necessarily collapsing into a boring future is that it would be possible for them to, as it were, "outsource" their boredom-proneness to some external mechanism—a *boredom prosthesis.*

Consider a person who is constituted in such a way that they'd be perfectly content to watch the same sitcom episode over and over, for years on end, because they're entirely incapable of experiencing boredom. Now suppose we take this person's screen and hook it up to his neighbor's set, so that the two devices always display the same content, chosen by the neighbor who (we will suppose) has a normal level of boredom-proneness. Holding other things constant, it would seem that the experience streams of these two people are now equal in terms of the objective boringness of their watching experiences. But of course, instead of another person serving as the selector, we could use an inanimate mechanism to do this job.

Squinting a little, one might view today's streaming services and recommender systems as (very primitive forms of somewhat misaligned) boredom prostheses. In the ideal case, they keep us consuming a personalized content stream indefinitely—with suitable intermezzos in which we buy all the stuff that is pushed to us in the ads. The mechanism selects new content to preempt boredom, ensuring that we stay "engaged". The problem is that while these commercial systems may be somewhat effective at averting subjective boredom, they are generally not designed to avoid objective boringness. Rather, their performance on their vendor's objective function is often optimized by getting people hooked on a never-ending programming of objectively uninteresting titillation, outrage, and distraction—fare that is not much better for our souls than a steel lure is for a salmon. But this problem is a contingent fact about contemporary commercial incentives. In principle, such systems could be designed to optimize for more edifying objectives.

The idea of the boredom prosthesis can be generalized into a more universal plan for avoiding unpleasant mental states by exporting elements of their functionality to external mechanisms.[134]

Consider, for instance, pain, which serves as a warning signal of bodily damage. There are rare individuals born without the ability to feel pain, and this is a dangerous condition. People with congenital analgesia may walk around on broken bones or stick their hand into boiling water. They often take excessive risk and fail to protect their bodies, and meet an early demise.

So if we want to get rid of pain, therefore, we need some way of dealing with this problem. Fortunately one can think of several possible solutions.

One would be to design the environment so that it would be safe even for people with diminished or absent nociception. Alternatively, improved medicine for repairing or regenerating damaged tissues and joints might make the frequent injuries less of a concern.

But another approach would be to create a mechanism that serves the same function as pain but without being painful. Imagine an "exoskin": a layer of nanotech sensors so thin that we can't feel it or see it, but which monitors our skin surface for noxious stimuli. If we put our hand on a hot plate, a bright red warning message flashes in our visual field and we hear a loud noise. Simultaneously, the mechanism contracts our muscle fibers so as to make our hand to withdraw, giving us time to consider our next move. Another component of the system might surveil internal tissues and organs, and flag any condition that requires remedial action.

Such an exoskin is not so different in principle from familiar devices such as carbon monoxide detectors, wearable dosimeters, and continuous glucose monitors. The notion of outfitting a biological organism with a full suite of artificial sensors for noxious exposure does seem somewhat steampunky, although with advanced nanotechnology the implementation could be perfectly inconspicuous. And of course, if we become fully digital, many things can be accomplished far more elegantly.

One could well explore whether similar treatments that I have suggested for boredom and pain could be adapted to help us get rid of other functionally useful but intrinsically undesirable psychological dynamics.

Monkeying with human nature

This might be a good place to slip in a cautionary note about modifying human nature.

In principle, there is enormous opportunity to improve our existence by modifying and reengineering our emotional faculties. *In practice*, there is a considerable likelihood that we would make a hash of ourselves if we proceed down this path too heedlessly and without first attaining a more mature level of insight and wisdom.[135]

The caution applies *especially* to modifications of our emotional or volitional nature, since changes that affect what we want could easily become permanent. Not because we wouldn't *be able* to change them—with increasingly advanced technology, it should be perfectly feasible to roll back changes made earlier—but because we may not *want to* change them. (For example, if you changed yourself to want nothing but the maximum number of paperclips, you would not want to change yourself back into a being who wants other things besides paperclips, except in certain very special kinds of circumstances where you expect a greater number of paperclips to come into existence conditional on you thus changing yourself.) This sort of volitional change, therefore, even if not irreversible, may have a tendency to in fact never be reversed.

Another reason for diffidence and care in this area is that our emotions—even the ones we think of as "negative", i.e. not just boredom and pain but also anger, hatred, jealousy, envy, sadness, fear, and so on—play many and sometimes quite subtle roles in our psychic lives. They interact to shape our personalities and behavior in complicated and not yet fully understood ways. We might, at present, be qualified to make some modest tweaks here and there, and to fix a few things that are clearly broken. But any attempt at wholesale redesign—especially if resort is made to relatively novel instruments, such as pharmacogenetic as opposed to (for example) spiritual modalities—would, one suspects, carry a fairly high risk of rendering us less rather than more awesome. See Handout 13 for an illustration.

HANDOUT 13

ECCE HOMO

[Source: Centro de Estudios Borjanos.[136]]

Ecce Homo, a fresco by the Spanish artist Elías García Martínez in the Santuario de Misericordia church in Borja, before and after its botched restoration in 2012. From imperfectly good to perfectly bad—though, ironically, it has likely brought far more joy into the world in its ruined state than it ever did in its original condition.

The good news is that at technological maturity we would have access to super-wise and super-capable AI advisors and assistants. This will greatly improve the prognosis of attempts at utopian mind-reengineering.

The focus of this course is not on practicalities, however, so let's return to the question of whether living in a perfect world would be boring. But I thought I should mention it.

Fitting response views

Let's briefly review. I asserted first that it would be *technically* feasible to eliminate subjective boredom in utopia. I then noted that one concern with eliminating our ability to feel bored is that by making ourselves unborable we'd remove an important protection against the future becoming *objectively boring*. In response to this concern, I noted that (a) we have motives other than avoiding feelings of boredom that could drive us to create an interesting future, and (b) we could construct substitute mechanisms—boredom prostheses—to help steer us toward interesting options and experiences. This strategy—outsourcing our negative emotions to some external process or device—could also eliminate the instrumental need for pain and for many other unwanted feelings that we currently rely on as behavioral guides and guardrails.

I now want to bring up another concern that one might have about eliminating boredom. This concern is normative rather than instrumental. (Although I'll speak about it here with reference to boredom, it is worth bearing in mind that parallel concerns may arise if we set out to abolish other negative emotions; so many of the points that follow have more general applicability.)

Right, so let's analyze the concern that even if all the instrumental functions of subjective boredom could be subsumed by some clever technological replacement, this would not yet allow us to harmlessly remove our boredom propensity, because of the putative existence of a normative relationship between boredom and boringness. Feeling bored, somebody might say, is an *appropriate* and *normatively fitting* response to being in a situation that has the boringness property. If this is so, it may be intrinsically bad not to react in this way when one is in such a situation. (Additionally, the objectively boring conditions may also themselves be deemed intrinsically bad, independently of how we respond to them—we will come back to *that* issue in just a second.)

Everyone with me so far?

Student: Can you give an example?

Bostrom: Sure. So, on the view we are considering, if you are at a boring dinner party, then it is appropriate for you to feel bored. It would be normatively *inappropriate* to experience the event as engrossingly fun and stimulating. Only a dork would feel that way—and you don't want to be a dork! Thus, it is deemed bad for there to be a mismatch between one's objective circumstances and one's subjective attitude or response to those circumstances.

Maybe it's easier to grasp the intuition if you look at other possible cases of mismatches between our attitudes and our circumstances. For example, you might think it is bad to be jolly at a funeral, or to delight in other people's misery, or to take great pride in extracting a booger from your nostril. Some—but far from all—philosophers maintain that these attitudes are bad and not just because they might causally produce hurt or embarrassment but inherently bad.

If you've read Nozick's reflections on his experience machine, you may recall that he wrote: "we want our emotions, or certain important ones, to be based upon facts that hold and to be fitting. . . . What we want and value is an actual connection with reality.".[137] So Nozick endorsed a "fitting response" view as part of his attempt to explain why a life inside the experience machine would be undesirable. Quite a few contemporary philosophers hold similar views.

We are now looking at the implications of this view with respect to the question of whether utopia would be boring. And you can see that if we accept that there is this kind of normative linkage between feelings and reality, then the problem of getting rid of boredom in utopia might not be such a trivial thing to do after all. While technically doable, eliminating boredom feelings would incur an ethical cost by distancing us from the normative ideal that our attitudes should match reality…

Unless, that is, we can make utopia a place that is free from *objective boringness.*

*

Let's call the opposite of boringness "interestingness". (I know, it's not the prettiest word, but alas I haven't come up with anything better.)

What we must explore, therefore, is how much scope there is for (objective) interestingness in utopia.

Suppose that there is *unlimited* scope for interestingness. Then two nice things would follow. First, utopia need never "run out" of interestingness. It may be good in itself that utopia is interesting through and through or that it contains a potentially infinite amount of interestingness. Second, if we recalibrate our boredom proneness to such an extent that we never feel bored, this would then *not* need entail any mismatch between our subjective attitude (of ceaseless interest and fascination) and our objective circumstances (with their inexhaustible interestingness and aspects worthy of fascinated engagement). There could instead be a perfect match, which would be very convenient.

If, on the contrary, the potential for objective interestingness is limited, then not only could utopia only contain so much of the value of objective interestingness; but also (it would seem) we would eventually either have to experience feelings of boredom (undesirable) or else abandon the hope of satisfying the "fittingness" desideratum that our attitudes should match our circumstances (which, according to some moral theories, would also be undesirable).

(What about a third alternative: to recycle interestingness to make a finite supply last forever? The utopians could have their interesting experiences repeat themselves—but that may not be very objectively interesting; or they could die and let a new person take their place—but that is another kind of repetition, which may also not ultimately be very objectively interesting. More on this later.)

*

My inclination is to believe that we should *not* put much weight on the fittingness desideratum in the case before us, even if we assume that the "fitting response" views of some moral philosophers are correct.

This is not only because, in general, the fittingness of our emotional responses and attitudes to objective circumstances is only one type of value among many others; and I would say it's not among the very most important ones. It is also because I am particularly skeptical in the present case, on grounds that with respect to boringness/interestingness it is not clear that there *are* any sufficiently robust standards to apply as to what is a fitting response and what is not.

Criteria for attributions of objective boringness—and *especially* claims about *absolute* levels of objective boringness, as opposed to comparative claims about which of two situations is more boring than the other—seem to me extremely indeterminate. Yet without such an absolute calibration standard, there would be no fact of the matter as to what degree of subjective boredom-feeling constitutes an "appropriate" response to a given objective situation.

How interesting is Shakespeare?

Let me illustrate this with an example. Take somebody who has the education, intelligence, temperament, and life experience required for a deep appreciation of Shakespeare. Now ask: What level of subjective boredom would be a "fitting response" for this reader as she spends time immersing herself in Shakespeare's work? Is it fitting that she feels bored after she's read all his plays once? Or is it fitting instead that she begins to feel bored only upon completing a third reading?

Even if her readings and re-readings reach points of diminishing returns—provided that the texts continue to engage her, and that she finds the occupation to be thoroughly enjoyable, and that she occasionally finds some new nugget or perspective that at least slightly deepens her appreciation of the plays; and assuming also that she has nothing better to do: then would there really be anything "ill-fitting" about her failing to get bored with Shakespeare even after a lifetime of study? Or if that lifetime is *a thousand years* long?

I am not persuaded that there would be anything ill-fitting about that, that it'd constitute any kind of "failure to get bored in a situation where the appropriate response would be to have gotten bored".

If we do have an intimation that there would be something misfortunate about somebody spending her entire life reading and appreciating Shakespeare, I think it may reflect a sense not that Shakespeare's work is in itself unworthy of such dedication but rather that the reader is missing out on other things that life has to offer. But *that*, of course, is not a problem from our utopian perspective. If the *only* ground for it being objectively fitting to get bored with activity *X* at some point is that there comes to be available some on the margin even *more* objectively interesting activity *Y*, then there would be no cause for experiencing boredom. As soon as *X* starts to get boring, we simply switch to *Y*! Even momentary boredom could be avoided if a clever prompt helped a person switch her focus at the right time, just before *X* would have gotten boring.

And what if we start running out of new *Y*s to switch to? Well, then we recalibrate our boredom proneness so that the old *X* keeps holding our interest for longer.

Therefore, insofar as objective boringness simply tracks the opportunity cost of foregoing alternative more objectively interesting options, we need not worry: as far as *that* consideration goes, we could have a utopia free of boredom-feelings without having to become interested in things that are objectively boring. If we could only make comparative judgements, so that the "zero point" of interestingness is arbitrary, then whenever we began to run low on diversity, novelty, engagement, complexity (and whatever other elements contribute to keeping us interested), we could simply adjust our threshold for experiencing boredom. We could do so without engaging in any objectionable form of "wire-heading" that would make our feelings and attitudes discordant with our circumstances. Quite the contrary, in fact: the adjustments would be necessary to ensure that our feelings *remained* appropriate as the available reservoir of unconsumed objective interestingness was gradually being drained.

*

That said, I will admit there is still a concern that, as we consider longer and longer intervals, there may come a time when *all* the activities available to somebody become objectively uninteresting, because all novelty and interestingness has been used up. There need be only *moderately determinate* standards of objective interestingness for this to be a possibility.

Perhaps there is enough objective interestingness in Shakespeare's work to fill an entire human life, or a few lifetimes. But maybe the material *would* become objectively stale to somebody who spent five hundred years studying it. Even if they had been modified so that they didn't experience boredom, we might judge their continued Shakespeare studies to be no longer valuable (or at least much less valuable in one significant respect) once they have "exhausted" the Bard's work, in the sense of having discovered, appreciated, learned, and fully absorbed and mastered all the insight, wit, and beauty therein contained. We would then have definitively run out of Shakespearian interestingness, although we would be able to choose how to feel about that fact.

The 162,329th table leg

In his novel *Permutation City*, science fiction writer Greg Egan has a character, Peer, who has achieved immortality in a virtual reality environment over which he has complete control. Peer has modified himself so that he does not feel bored. Presumably in order to minimize the objective boringness of his existence, he has programmed an "exoself" to automatically change his passions at random intervals, ensuring that his life will continue to have some diversity and variation:

> "The workshop abutted a warehouse full of table legs—one hundred and sixty-two thousand, three hundred and twenty-nine, so far. Peer could imagine nothing more satisfying than reaching the two hundred thousand mark—although he knew it was likely that he'd change his mind and abandon the workshop before that happened; new vocations were imposed by his exoself at random intervals, but statistically, the next one was overdue. Immediately before taking up woodwork, he'd passionately devoured all the higher mathematics texts in the central library, run all the tutorial software, and then personally contributed several important new results to group theory—untroubled by the fact that none of the Elysian mathematicians would ever be aware of his work. Before that, he'd written over three hundred comic operas, with librettos in Italian, French and English—and staged most of them, with puppet performers and audience. Before that, he'd patiently studied the structure and biochemistry of the human brain for sixty-seven years; towards the end he had fully grasped, to his own satisfaction, the nature of the process of consciousness. Every one of these pursuits had been utterly engrossing, and satisfying, at the time. He'd even been interested in the Elysians, once.
>
> No longer. He preferred to think about table legs."[138]

A colleague of mine, Eliezer Yudkowsky, has written about the risk of running out of interestingness (he uses the term "fun") and explored some similar issues to those of today's lecture.[139] Yudkowsky finds the just-quoted Egan passage particularly horrifying:

> "I could see myself carving one table leg, maybe, if there was something nonobvious to learn from the experience. But not carving the 162,329th. . . . at *that* point, you might as well modify yourself to get pleasure from playing Tic-Tac-Toe, or lie motionless on a pillow as a limbless eyeless blob having fantastic orgasms. . . .

> carving the 162,329th table leg doesn't teach you anything that you didn't already know from carving 162,328 previous table legs. A mind that changes so little in life's course is scarcely experiencing time."[140]

Now, Egan's scenario does seem far from optimal. But *why* exactly is it so? How much of the problem is really due to Peer's life having a deficiency of objective interestingness?

One thing at least seems clear. Even if there is some disvalue in the putative misfit between Peer's subjective interest and the objective boringness of his preoccupation, *that* is not the main problem with this scenario. It would surely be *far worse* if, on top of everything else, Peer was also terminally bored as he stood at his lathe. If he is going to be making all these table legs, he might as well find his craft engrossing and enjoy himself while he goes at it.

Aesthetic neutrinos?

But there could be many other things wrong with Egan's scenario besides the fact that Peer is spending a lot of time on the allegedly objectively uninteresting activity of carving table legs.

The solipsism of his existence is one striking feature. If we imagine a whole society of Peer-like characters, who interact normally with one another but are collectively gripped by one great shared enthusiasm after another—imposed, perhaps, by a joint exoself (an "exocommunity"? aka "culture")—and who find in these serial fascinations a tremendous source of pleasure, satisfaction, and purpose; then the prospect immediately takes on a significantly sunnier aspect; although, of course, it is still not nearly as good as the best possible future we can imagine.

*

There is a tendency, I think, especially for the intellectual type who is likely to be reading and writing about these matters, whose self-image emphasizes the intellectual virtues, and who derives much of his or her pleasure in life from learning and problem-solving—this nerdy kind of person is, I believe, at risk of overestimating the value of "interestingness", and of forgetting that there are many other plausible values besides solving complex novel important problems. And these other values, even if they are centered on human experiences and activities, may not relate to repetition and the passage of time in the same way as the value of interestingness.

For example, while one might hold that the interestingness value that an individual can derive from the works of Shakespeare would be permanently depleted after several decades of study, what about the enjoyment value of *a nice cup of tea*?[141]

Drinking tea (or coffee, if you prefer) may not be a source of an intense flash of value, the way that an epiphany into some deep truth about human nature may be if discovery of such truths has interestingness-value. But it is quite renewable. The 162,330th cup of tea, on your 200th birthday, may not be less valuable than the one you had a century earlier. And whereas the supply of human-accessible profound truths might be limited, you can always put another kettle on.

*

Tea-drinking is a small pleasure with a smidgen of gustatory aesthetics mixed in. Or to use our terminology: positive hedonic valence plus some pleasing experience texture. Is it possible to scale this up?

John Stuart Mill, the English utilitarian philosopher, shares in his autobiography how he suffered a mental crisis in his youth. The young Mill was troubled by the thought that humanity would eventually run out of problems to solve, and that we would, as a consequence, be reduced to boredom, listlessness, and despair.

> "[T]he pleasure of music . . . fades with familiarity, and requires either to be revived by intermittence, or fed by continual novelty. And it is very characteristic both of my then state, and of the general tone of my mind at this period of my life, that I was seriously tormented by the thought of the exhaustibility of musical combinations. The octave consists only of five tones and two semitones, which can be put together in only a limited number of ways, of which but a small proportion are beautiful: most of these, it seemed to me, must have been already discovered… I felt that the flaw in my life, must be a flaw in life itself; that the question was, whether, if the reformers of society and government could succeed in their objects, and every person in the community were free and in a

> state of physical comfort, the pleasures of life, being no longer kept up by struggle and privation, would cease to be pleasures."[142]

Mill found the resolution of this quandary (and a remedy for his own melancholy) in the romantic poetry of Coleridge and Wordsworth. The answer, he concluded, was "to take refuge in a capacity to be moved by beauty—a capacity to take joy in the quiet contemplation of delicate thoughts, sights, sounds, and feelings, not just in titanic struggles".[143] He writes:

> "What made Wordsworth's poems a medicine for my state of mind, was that they expressed, not mere outward beauty, but states of feeling, and of thought coloured by feeling, under the excitement of beauty. They seemed to be the very culture of the feelings, which I was in quest of. In them I seemed to draw from a source of inward joy, of sympathetic and imaginative pleasure, which could be shared in by all human beings; which had no connexion with struggle or imperfection, but would be made richer by every improvement in the physical or social condition of mankind. From them I seemed to learn what would be the perennial sources of happiness, when all the greater evils of life shall have been removed. . . . I needed to be made to feel that there was real, permanent happiness in tranquil contemplation. Wordsworth taught me this..."[144]

Thus the contemplation of beauty, whether refracted through art and poetry or focused directly on our surroundings, is one proposal for how utopians might fill the hours of an endless summer.

By the way, Schopenhauer, too, saw disinterested aesthetic contemplation as one of only two possible ways to escape the reciprocating motion of want and boredom that powers our suffering (the other way being the path of the saint, which he conceived of as involving the self-abnegation of the will). Importantly, the objects that Mill and Schopenhauer had in mind as the focus of aesthetic contemplation were not necessarily or even preferably those of world-changing transformations, dramatic turnarounds, high stakes, or groundbreaking discoveries—not "interestingness" in a sense that depends on learning profound truths or solving pressing problems. Rather, they had in mind a beauty that is available equally in the small and mundane, in how the water flows in a brook, or the particular way that an oak tree spreads its branches, or that a workman's shoulders move when he is digging a ditch.

The true idea or aspect of any of these things could serve just as well as the most splendid cathedral. This is because the quality of an aesthetic experience is principally a function of how an observer looks at things rather than of the things being looked at. For a mind that has emancipated itself from slavishly processing all its experiences in terms of their instrumental significance and utility for the ego, and has freed itself instead to behold things as they are in themselves, objects for aesthetic contemplation are ubiquitous.

It is as if we had first experienced the world as cramped and tenebrous, because we lived in the dark and were scarcely able to see as far as our own hands; and then—the rising morning sun revealed to us that we've been cowering in an Edenic landscape whose dew-shimmering vistas extend beyond the horizon.

*

I think what these reflections point to is this. We might worry that the utopians would fairly quickly run out of interestingness in their lives, once they had read the great masterpieces and visited the Taj Mahal, the Notre Dame, the Grand Canyon, and some other top tourist destinations. Of course, superintelligent AIs could write us many more masterpieces and construct marvelous new buildings and landmarks; but one might still fear that the range of fundamentally different possibilities is quite limited—the way that the young Mill worried that composers would run out of beautiful melodies.

The severity of this threat depends on just how demanding our notion of objective interestingness is. If interestingness required *fundamental* novelty or *world-transforming* significance, then the utopians would run out of it quickly. However, the observations we've made about aesthetic experience suggest there is a form of "interestingness" that is far less demanding yet still eminently worthwhile. It is worthwhile in the sense of being able to provide the fabric for valuable forms of experience texture (by which I mean an additional factor of putative axiological relevance that can characterize one's experience stream, besides its hedonic quality).

I think, therefore, that we need to expand our conception of objective interestingness to encompass objects and ideas worthy of aesthetic contemplation. And these can include *little* ideas and *ordinary* objects, and small variations of what one has already experienced—in addition to the more eureka-inducing learning-opportunities and intellectual problem-solving that e.g. Yudkowsky appears to have had in mind.

*

Just as gazillions of neutrinos pass through our bodies every second without our noticing, so too might the world present us with countless beautiful things at every moment—which our minds are too coarse and insensitive to appreciate.

If, however, we tuned up our aesthetic sensibilities to a sufficient degree, we could register more of this pervasive beauty. We could capture enough to fill our conscious minds with a never-ending rapture of contemplated aesthetic significance.

*

I suggested earlier—and that was before we considered this expansion to the aesthetic—that the concept of interestingness is very elastic, lacking clear, hard, precise criteria for what is and isn't objectively interesting.

In our present expanded conception, this is even more the case. Aesthetic interestingness is *highly* elastic, if not altogether limitless.

Who really is to say that the interestingness of a lizard's rustling in the leaves, or of the ripples that a carp makes when it breaks the surface of a pond, is sufficient to fill ten minutes of contemplation with valuable experiential content but no more? Or sufficient only for one summer afternoon? Such that any modification of our boredom-proneness that would result in us finding lizard-rustlings and carp-ripples interesting for a longer period than this ought to be regarded as a discreditable form of wireheading, one that would produce an objectively disvaluable misfit between our attitudes and the objects to which they attend?

A hundred years of yellow

Admittedly, things get difficult if we consider more extreme cases of insignificance.

One *year* of contemplating the yellow of a duck's beak?

One *century* of contemplating the yellow of a duck's beak?

How long is too long?

Complexity in the observer

In a case like this, the key thing is not the external object but the internal response. If a given object is contemplated through a series of shifting mental perspectives and subjective modulations, which hold "the object" in place at the center of attention but enfold it in changing complexes of mental resonances, associations, and feelings, then even a very plain object—a patch of yellow, or even no real object at all, as an imaginary something would do just as well—could sustain a prolonged episode of objectively interesting contemplation. The question in such cases is *how much objectively interesting variation in our subjective experience* are we capable of generating? And how does the cumulative value of this experience scale with the amount of time we spend having it?

I see many of your faces registering confusion. Let me try again!

By *X* being *subjectively interesting* to Bob, I mean simply that Bob finds *X* interesting—he is fascinated by *X*, feels motivated to continue to do *X* or to pay attention to *X*, and so on. We can readily conceive of a person who finds counting blades of grass subjectively interesting and who would continue to find it interesting until he had counted all the grass on the college lawn. So that's what subjectively interesting means.

What *X* being *objectively interesting* means is less clear, but I'm trying to be accommodating to those who think that a life spent counting blades of grass would be seriously defective on grounds that this activity lacks the requisite complexity, challenge, novelty, diversity, and significance that would make *X* *worthy* of Bob (or anyone else) being so subjectively interested in it.[145] This may give you at least some rough idea of what this notion of objective interestingness is supposed to mean.

Ok? And now the point I want to make is simply that we can ask this question—whether *X* is objectively interesting or not—not only about *X*s that are external physical objects (such as paintings, books, and duck beaks) and facts (such as concerning the cardinality of the set of blades of grass on a given lawn) but also about *X*s that have mental components.

Let's say Bob is looking at a duck's beak. We can assume that the beak itself is not particularly objectively interesting. So if the beak is our object *X*, it would have a low interestingness value and it would be inappropriate for Bob to be highly interested in it. But suppose Bob happens to be a very sensitive and imaginative fellow. While he is staring at the beak, his intellect is more akin to a kinetoscope conjuring up a series of inner phenomena. Let us take a peek inside Bob's mind as this is happening:

> Initial visual impressions flood into awareness. The yellow of the beak contrasts with the orange of the animal's legs. The firmness of the beak is compared with the fluffier duck parts. A memory of some other water-related object casts a faint reflection into consciousness: a life vest, a buoy. A relationship again becomes salient: the beak (incorrectly assumed to be insensate) versus the fleshy living torso. A thought arises: how we incorporate into our notion of self not only those parts of our flesh in which we have sensation but also our nails, the enamel of our teeth, our hair, sometimes even external objects like a wedding ring or a phone. A further thought: how this applies not only to our bodies but also our minds—so much of our brain activity is inaccessible to consciousness yet is an intimate and inextricable substratum of our mind. Can information processes flowing outside our skulls similarly be part of our mind? A small jolt as we recall we were supposed to be focusing on the beak not ruminating on the extended-mind thesis. How could we render the beak in a painting if we were only allowed to use three brush strokes? And then this and then that.

In such a manner, Bob is having quite a party just by looking at this yellow structure. We may call this the *extensive* approach to beak-contemplation.

Alternatively we can consider an *intensive* approach, which would consist of trying to hold the mind still (something akin to Trāṭaka, or "yogic gazing"):

> No sideways movement but a deeper focus and immersion in just this one thing, this one very particular hue of yellow. Different lessons about consciousness being gleaned *en passant*—how our perception, closely examined, flickers and wobbles in and out of consciousness; how there is a universality in the particular when it is taken for what it is in itself rather than in its relation to the ego or some external concern; and then, beyond these travel-tales, eventually some state of awareness that is still and

> sufficient, that feels incomparably "more real" than the turgid confusions that entomb the everyday struggling mind. A place of clarity. A place of peace…

The point here is that, in either of these approaches, the extensive or the intensive, the relevant part of what's going on has very little to do with the duck's beak. The "action" is almost all in the mind of the observer.

Is that any clearer? Good.

*

What is most needful, therefore, if we wish to create a world of beauty, is not additional artworks, sculptures, poems, and musical compositions; but an increased capacity for aesthetic appreciation.

With enough such capacity, the rustling of a lizard or the yellow of a duck's beak can go a long way.

Without such capacity, it doesn't matter how large a collection of beautiful objects we accumulate. We would be like the guard dogs patrolling the Louvre.

*

But now, of course, we can ask whether this "action", which is in the mind of the observer, is objectively interesting or not. Does it constitute an inherently *worthy* occupation of our time insofar as we are concerned with filling it with interesting content? And *how long* could this activity go on before it gets objectively boring? (Which, to repeat, is a separate question from whether the person engaged in it would ever *feel* bored.)

If contemplating little things is interesting only because we (metaphorically speaking) can see reflected in them big things, then it doesn't really help the utopians that there are many little things. Once they've seen all the big things reflected, then they would be reduced to either just seeing smaller and smaller things reflected, or to keep seeing the reflections of the same big things over and over. In either case, they again have a challenge to sustaining the objective interestingness of their experience streams.

The roots of our desire for interestingness

Let me shift gears for a moment. Why do you think people are interested in interestingness?

I mean, it is easy enough to explain why humans tend to be interested in things such as food, sex, status, and health, but why interestingness? In particular, for those who hold that a boring future would be undesirable, or at least ceteris paribus less desirable than an interesting future—and who want an interesting future even if we postulate that they wouldn't *feel* at all bored in an uninteresting future: how might we explain this preference or value judgment?

I have a reason for posing this question, aside from generic curiosity. I think a better grasp of the etiology of the interestingness value might give us additional clues as to how it might fare in utopia. (Also, I want to put forward an idea about value formation generally, and this case will serve as a useful illustration.)

I will propose four hypotheses, not mutually exclusive.

*

> *The learning & exploration hypothesis*
> The value we place on interestingness derives from a kind of learning instinct, and/or an "exploration bias". We seek out situations that present us with significant new information and novel varied challenges, because doing so led our ancestors to acquire more knowledge and skills, which was adaptive in our evolutionary environment.

This hypothesis weakly predicts that placing great weight on the interestingness value should correlate with being young. Learning and exploration is a higher priority early in life, when we have not yet plucked as many of the low-hanging fruits of exploration and when we have more remaining lifetime to benefit from any knowledge and skills we gain. Older individuals should be relatively less keen on novelty and change—they may favor stability instead, which allows their existing skills and social assets to remain relevant.

*

The signaling hypothesis
The value we place on interestingness derives from a social signaling motive. We desire to engage in activities and to be in situations that will enable us to tell a good story about what we've been up to, because this increases our social status.

Some people might choose a strenuous adventure holiday or travel to some exotic location, which has a high level of "interestingness", because it allows them to post impressive updates to their social media feed, even though otherwise they might more enjoy staying at home and resting. The idea then could be that we (subconsciously) generalize the observation that certain types of experiences—"interesting" ones—tend to garner more social approval: as a result, we gradually come to pursue interestingness for its own sake, independently of whether we can see how it would earn us social approval on any specific occasion.

The signaling hypothesis predicts that activities will tend to be rated higher in "interestingness" if they have socially valued prerequisites, for example if they require special skills, virtues, or social or economic capital. This might help account for why spending the summer playing golf at an exclusive resort may intuitively seem to have higher interestingness value than spending the summer throwing tennis balls into a bucket in your backyard—even though, in terms of the basic mechanics of the activity, the two options seem roughly comparable. The fact that a future of golfing may to many seem fairly utopian, while a future of throwing-tennis-balls-into-a-bucket would not, could be then explained by many people having internalized the fact that the former activity has high social status and the latter does not.

*

The spandrel hypothesis
The value we place on interestingness is a side-effect or implication of the value we place on other things. If we value *X*, *Y*, and *Z*, and if a future that is too homogeneous or high in boringness would not have room for the kinds of complex and variable structures needed to instantiate *X*, *Y*, and *Z*, then future scenarios that are low in interestingness would appear undesirable.

For example, if we value the existence of a large number of humanlike creatures that interact with each other in open-ended ways and that gradually discover novel techniques for transforming their environment and for building ingenious and beautiful structures, then it could be the case that the only possible futures that would allow this value to be realized are pretty interesting.

Some axiologies seem more likely than others to imply interesting realizations. Hedonism, the view that pleasure is the only good, could achieve a high realization in an uninteresting future (and indeed may require such a future for its optimal realization) whereas, say, a balanced pluralistic axiology that also prizes individual autonomy and that respects certain kinds of historical path-dependency would more plausibly require interesting structures for its full realization.

The spandrel hypothesis might suggest that we should be especially likely to find scenarios "interesting" if they involve the expression or realization of other values—in contrast to, for example, a scenario in which some sort of giant loom weaves out some immensely complex and indefinitely varying and growing mathematical pattern.

*

Finally, we may seek the origin of our valuing of interestingness in the evolutionary usefulness of the aversive feeling of boredom as a regulator and balancer activity:

> *The rut-avoidance hypothesis*
> We tend to get bored if we keep doing the same thing for too long, especially if we don't see any positive results. This emotional disposition could be evolutionarily useful, not only as a mechanism for encouraging active learning (as per the first hypothesis), but also more specifically to prevent us from persisting in fruitless endeavors or getting stuck in situations that we've mistakenly estimated as more propitious than they actually are. A boredom propensity could also help us allocate time and effort more appropriately to the full range of our needs, by promoting task-switching after a suitable interval of engagement. This would be analogous to how you might get tired of eating the same food at every meal, a disposition which leads you to consume a more diverse diet that is more likely to cover your nutritional requirements.[146]

The rut-avoidance hypothesis might suggest that our judgments of interestingness would track such features as whether activities have progressive payoffs, and that they should prefer scenarios in which there is a relatively frequent cycling between activities. For example, if we compare the original version of Egan's scenario, where Peer spends a few hundred years exclusively focused on one occupation before moving on to the next, to a variation where Peer's occupations are interleaved, so that each week he spends some time in the woodshop, some time playing sport, some time watching films, and so forth—the rut-avoidance hypothesis would predict that we should find the latter scenario to be higher in interestingness (even if, at the end of Peer's life, he will have tallied up the same total number of hours on the same activities in both cases).

Intrinsification

It is easy to see how the spandrel hypothesis implies that an uninteresting utopia would be defective. Since a utopia that contained all other values would, according to this hypothesis, be interesting, an uninteresting future would necessarily be lacking in some values.

The other three hypotheses, however, may require an additional step before they can account for our intuition that interestingness is intrinsically valuable. We could interpret these hypotheses as, most immediately, furnishing explanations for why we should tend to be subjectively bored in certain situations and subjectively interested in other situations. But once we postulate that boredom can be completely dispelled—people in utopia could be made to be highly subjectively interested in *any* situation—the question comes to the fore why in that case the more objectively interesting is still to be preferred, even though objective interestingness is no longer necessary as a means of forestalling subjective boredom.

To explain this, I will propose a mechanism of value formation which, I think, has more general applicability and relevance. It deserves a name—I'll call it "intrinsification".

> *Intrinsification*: The process whereby something initially desired as a means to some end eventually comes to be desired for its own sake as an end in itself.

This kind of process may play a role, for example, in the development of moral motivations. Initially, as young children, we discover that treating those around us well—at least refraining from biting, scratching, or kicking them, or stealing or destroying their belongings, and so forth—tends to lead to social rewards: our parents and friends are nicer to us when we are nice to them. This gives us instrumental reason for respecting the interests of others. Later, this motivation migrates upstream: it becomes "intrinsified", and we begin to place final value on giving other people their due, and more generally on adhering to a moral code. That is to say, we become motivated to act righteously even in cases where we know that the original connection to social reward is severed; indeed even (if we become persons of moral integrity) in cases where doing what is morally right predictably leads us to suffer opprobrium and other disadvantages.

The actual psychology of moral development is undoubtedly more complicated than this simple sketch. I imagine, for example, that in humans the process of intrinsification is assisted by various specific emotions, inductive biases, attention-related phenomena, and other psychological and physiological tendencies, which serve as a scaffold within which the values develop that we hold to (and that hold us) as adults. And all of this, of course, takes place in a tight interaction loop with sociocultural factors, which themselves change over time due to both individual interventions and systemic developments in demography, economy, technology, and so on. But although I will not here try to pencil in all of this complexity, I still think we can get some use out of the basic idea of intrinsification.

*

We can observe the same phenomenon outside of the human individual, such as in the realm of institutional economics. For example, a state might initially establish a military and an arms industry as means to the end of protecting the country against foreign invasion. This simple instrumental rationale, however, may eventually intrinsify and become an institutional end in itself. The physical implementation of this may take the form of a military-industrial complex that drives policy and budget to itself, somewhat independently of the degree to which those measures are really needed to stave off foreign foes. This dynamic, in which some agent-like part is created in order to help a greater whole achieve some objective, and where the subagent then gradually develops its own agenda and begins to pursue its own goals even when they no longer serve the original larger objective, is very common. Not only in state bureaucracies but in many other organizations too.

*

Returning to the individual case, we may wonder *why* intrinsification should occur? Whereas in the organizational case we might explain the phenomenon by appealing to principal-agent problems, in the individual case it would seem that some different explanation is needed for why we would exhibit a psychological tendency that on its face may appear quite irrational. I mean, if something is desirable as a means to an end, why not keep valuing it for purely instrumental reasons? Why attach an additional value to it that persists even when the instrumental reasons no longer obtain?

One explanation for this is that human minds and characters are *translucent*. Other people can, to some degree, sense what our true motives and commitments are. For this reason, being able to intrinsify our adherence to a moral code, or our commitment to a cause or loyalty to a person, community, or system of norms, can be instrumentally useful to us. In some cases, only if the other is convinced that you truly love them will your advances be accepted; and the most effective way to convince somebody that you truly love them may be to truly love them. Similarly, people may be more likely to welcome you into their coalition or community if they trust that you will act honorably even in situations where you stand to personally gain from treachery and deceit.

In addition to this signaling explanation for intrinsification, there might be more "implementation-level" causes for the phenomenon. For example, it could be computationally expedient to represent some evaluative consideration or perspective by implementing it neurocognitively as a process that has some of the characteristics of a goal-seeking process with independent agency. In other words, basically the same reasons why we create government agencies and corporate departments to manage certain functions or to ensure that certain considerations and interests are duly represented—an environmental protection agency, a compliance department, a chief risk officer, a separate prosecutor and a defense lawyer in criminal trials, and so on. When things work well, this might be a more practical and efficient arrangement than, for example, having the entire process and decision-making be managed by a fully unitary actor, such as a dictator or a committee in which every member is individually equally responsible for representing all the relevant constituencies and considerations.

*

One notable feature of intrinsification is that the status of the elevated objective as an intrinsic value is not necessarily permanent. We can think of intrinsified values as *motivational flywheels*, which can get their initial impetus from positive reinforcements or by having momentum transmitted to them from other commitments and goals to which they are logically or statistically connected. Once charged up, the flywheels can keep spinning, and keep driving behavior, even if they are no longer receiving energy from their original instigator; and they can subsequently convey their accumulated momentum to new plans and subgoals.

It is possible, however, for intrinsified valuing to run out of joules. The young idealist, once burning for a cause, eventually exhausts his ardor, "burns out", especially if he receives no encouragement. The dieter, who starts off the new year with a strong commitment to slimness, even perhaps to the point of valuing this attribute for its own sake, may change their evaluation in February as a result of the dragging negative reinforcement from their calorie-craving palette and gut.[147]

Philosophers sometimes refer to those things that are (or, on more objectivist metaethical accounts, ought to be) valued for their own sakes as "final values". I don't particularly have an issue with this terminology, but we should bear in mind that such "final values"—although they may be ultimate within a scheme of axiological grounding, the last link in a chain of normative justification—need not be ultimate in any temporal sense. Psychological and cultural facts about what we value in this way—and, on some metaethical views, also facts about what is valuable in this way—may change over time. In this sense, final values come and go.

*

This is important when we are thinking about how to achieve utopia. For example, a natural idea is that we should defer the design of utopia to some lengthy process of deep collective deliberation: "the Long Reflection". It's been proposed that this deliberation should continue for hundreds or for millions of years.[148]

There is clearly *something* right about this. If we had an astronomically consequential choice to make, it would behoove us to think carefully about the options before coming to a decision. Perhaps we ought to dither for quite a while, especially if we can do so in existential security.

The predicament (or one of them—there are other difficulties with this idea) is that our values would be prone to change over the course of such an extended delay. This is not only because of the regular drift in our values that occurs with the passage of time as we go about our lives individually and collectively. An imposed condition of sociotechnical stasis, in which no potentially existentially risky or irreversible changes are permitted, would itself also plausibly alter our society and culture—and our values—in quite fundamental ways.[149] Furthermore, the proper kind of reflection might require big upgrades of our intellectual faculties and call for our becoming directly acquainted with a wide range of possible experiences, and so on.[150] Perhaps the ideal reflection would imply a radical enlightenment. That would surely involve a quite profound transformation of our psyches—and who knows what would happen to our final values along the way. I'm hoping to return to this topic later.

Critical playful spirit

Any questions up to this point?

Student: I'm a little confused. You first had all these hypotheses that debunked the idea that interestingness is really important—we just think so for various psychological or evolutionary reasons. And you said that actually interestingness is not necessary, since in utopia we could just engineer away boredom with drugs or whatever. But at other times you seem to be saying that interestingness is intrinsically valuable after all?

Bostrom: First let me say that it's good that you notice your confusion! Learning to notice one's confusion is a core skill for philosophizing—and a good well from which to draw original insights more generally.[151]

Now to answer your question: the four hypotheses were not meant to debunk anything. They were simply meant to explain why it is that we experience boredom in certain situations and why we might have come to form the intuition that a future lacking in interestingness would be undesirable. In and of itself, this kind of explanation says nothing about whether interestingness is valuable. Presumably, for *any* belief, emotion, preference, or evaluative intuition that we have, there is some causal story to be told for how we came to have it. But only in special cases—e.g. if the mechanism that caused us to hold a belief has no statistical connection to whether the belief is true—would the causal explanation tend to "debunk" or invalidate the explanandum.

The proposed mechanism of "intrinsification"—you could view that as an extension of the other explanatory hypotheses. It could help explain why we may have come to desire or value interestingness for its own sake, and not merely as a means for staving off the aversive subjective experience of boredom or as a means for achieving other objectives.

Now, as for where I myself come down on interestingness and utopia: we're still exploring the subject! I'm not really driving toward a predetermined thesis here. What do I know! Let's just see where we end up.

At the risk of committing an act of didactic overreach, I'll add that open-minded explorative reflection is another metacognitive strategy that I'd like to see more widely indulged.

Though I should warn that it can create a problem of social illegibility. Many people these days operate on a short leash. They stay on message at all times, then gradually lose the capacity to *think* off-message. In the terminal stages of this condition, they may even become incapable of recognizing that somebody else is not subject to the same limitation. If a freer spirit visits such a leashed person, this day approaching them from one direction, the next day from the opposite direction, the intellectual explorer may be received with confusion: "But what is your position? Where is your house? Do you belong on this side of the river or the other side?". And you say: "I am not talking about my dwelling (which is a rental in any case): I've been out exploring, and I just thought I'd share some of the cool things I've seen." And then they stare at you blankly. (And that's if you're lucky.)

I see there are more hands but unfortunately I think we'll have to move on, as there's a bunch more stuff we need to cover. You can try to pin me down after the class if you have further questions.

Scale exercises

Back to interestingness. We have some more analytic footwork to do.

Novelty and diversity seem important for interestingness. You may not think it is interesting to look at duck beaks. However, if you had never seen a beak, then seeing one for the first time might be somewhat interesting. Certainly if you had never in your entire life seen *yellow*, then seeing a yellow duck beak *would* be interesting. Presumably because it would be a very novel experience.

Even the most avid birder would get bored if she could only ever watch one particular bird doing one particular thing. But if she can watch various birds doing various things under various conditions, she has the makings of a lifelong hobby.

And Eliezer Yudkowsky could maybe see himself carving *one* table leg. Having done that, he could do something else, like making one drawing, cooking one meal, I don't know; but the idea would be that he could continue to pick new activities from some set, without his life becoming objectively boring, provided the set contain a sufficient diversity of activities that what he'd be doing at any given time would be adequately novel and non-repetitive that he'd learn something new from it.

*

From here, we can proceed to observe that the properties of novelty and diversity depend on the scale at which we're looking at things.

Suppose you bend down and look at a square inch of ground. Quite a diverse scene might meet your gaze at this scale. Here, a patch of dirt with an interesting shape. There, some yellow fibrous stuff twisting and bending expressively. Yonder, a green cylinder rising up to a great height. And there (Jiminy Cricket!) a huge six-legged monster with vast sweeping antennae, crashing in on our blithe pastoral.

You stand up and brush yourself off. Now you see a carefully groomed lawn. A monotonous sheet of green. Scarcely any diversity at all.

Next, suppose you climb into a hot air balloon. As you ascend, a stadium comes into view. And then—city blocks and streets and parks. Eventually suburbs, rivers, forests, fields, and a coastline, and ships dotted across the blue expanse. Quite a lot of novelty and diversity again at this scale.

Assuming that interestingness correlates with diversity, we thus find different amounts of interestingness at different scales.

*

But it gets yet more intricate: the *very same change* in the world can *increase* interestingness at one scale while *decreasing* it at another scale.

Imagine that on the opposite sides of a strait lie two cities, Solburg and Lunaburg. At the start, everybody in Solburg is a morning person, or lark. They get up early and go to bed early. In Lunaburg, it's the other way around: everybody is a night owl.

Then they build a bridge. People start commuting, mingling, migrating. After a few generations, the two cities have similar populations, each a mixture of 50% larks and 50% owls.

How did the bridge affect diversity?

At the scale of an individual city, diversity *increased.* Previously, the city had only one type of individual; now, it has two.

Yet at a larger scale—that of the region as a whole—diversity *decreased.* Previously the region had two types of cities, but now it has only one. A tourist visiting the area, who in earlier days may have crossed the strait in a ferry to catch a glimpse of two distinct populations and cultures, no longer has reason to do so, since both places now are the same.

Yet at an even larger scale, the bridge may have again caused diversity to *increase.* Suppose that, prior to the bridge, all cities in the entire country were either 100% larks or 100% owls. Post bridge, there is a new thing in the land: cities with a 50/50 mixture. So diversity at the scale of the country increased from having two types of cities to having three.

At a still larger scale, however, the bridge may have *decreased* diversity. Suppose that before the bridge, all countries had three types of cities: ones with 100% larks, ones with 100% owls, and ones with a 50/50 mixture—all countries except one, the one we are considering, which initially had only the first *two* types of city. When this previously unique country builds the bridge, it becomes just like all the other countries, and diversity at the scale of the world is reduced.

We thus see that the internally least diverse country could be contributing the most to world diversity, since it is such an outlier.[152]

Interestingness: contained versus contributed

Let's keep going.

Suppose you have an interesting book, and you buy a second copy of the same book. Do you now have "twice as much interestingness"?[153]

Or consider somebody who has an interesting life. Suppose you create a duplicate who lives an exactly identical life. (This would be tricky to do with a biological person who lives in an open changing world, but trivial with a digital person who lives in a self-contained virtual reality.) How much interestingness is there then? The same amount as if the life had been lived only once? Or double that? Or some other amount?

You might say that it would not be interesting to live a life that has already been lived. Of course, we're not talking about *subjective* interest—there could be as much of that as you please. In fact, there would be exactly as much subjectively felt interest the 500th time the life is repeated as the first time (since otherwise it wouldn't be a true replication). But *objective* interestingness: you might think your current life would lack objective interestingness if it were an exact repeat of a life already lived. Maybe the duplicate life would even be so deficient in interestingness as to drastically reduce its *overall* desirability—making it about as appetizing as a plate of pre-chewed food?

*

In the case of the book, I would say that each copy has the same amount of interestingness, but the second copy contributes nothing to the interestingness of your library given the presence of the first.[154]

We may distinguish between *contained* and *contributed* interestingness. How much *contained* interestingness is there in the second copy of the book? Same as in the first copy. How much interestingness does the second copy *contribute* to a library that already contains a first copy? Not nearly as much. You might say either that the second copy contributes zero interestingness, or that the added interestingness that the library has thanks to this title is offset by a loss of interestingness by the first copy due to its being duplicated by the second. But certainly one cannot in general increase total interestingness a thousandfold by taking an interesting object and making a thousand copies.[155]

We see, therefore, that in some cases a part contributes less interestingness to the whole than it contains in itself. It is also possible for a part to contribute more interestingness than it contains in itself. Suppose we discovered some old clay tablet with a strikingly expressive sketch of the person who invented the alphabet. This portrait would be quite interesting: more interesting than the sum of the interestingness of each individual stencil stroke of which it is composed.

*

What about the duplicated life? Consider two different questions we can ask:

A. How much interestingness does this life *contain*?
B. How much interestingness does this life *contribute* to the world?

The duplicate life contains as much interestingness as it would have contained if it were not a duplicate, but it may contribute less interestingness to the world. (It could also contribute *more* interestingness to the world—an otherwise ordinary life could be quite a lot more interesting if it were the *only* duplicated life in the entire universe. But we'll assume the case where duplicated lives are commonplace, so that there is no special interestingness arising from the very fact that it is duplicated.)

*

Suppose that we want to know how good a life is. Consider the following two different questions that one might ask regarding the goodness of a life:

1. How good is this life *for the person* whose life it is?
2. How much good does this life (directly, by its own existence, as opposed to via its wider causal effects) contribute to *the world*?[156]

The answers to these questions can come apart. For example, according to average utilitarianism, a life could be good for the person yet bad for the world. This would happen if the well-being of that life is high but not as high as the average level of well-being in the world.[157] More generally, unless the value of the world is a simple sum of the values of the individual lives it contains, we should not expect the answers to the two questions to coincide.

Suppose we decide that the question we want to ask is the first question, the prudential question: "How good is this life for the person whose life it is?". And suppose, further, that interestingness is one of the qualities that makes a life better for that person. How are we to think about this?

I think we need to be concerned with two factors in this case. First and most obviously, we have the interestingness that is contained in this life. This is independent of duplication. If having an interesting life is a prudential value, then the life that contains more variety, complexity, depth, coherent development etc. would be the more interesting life, and therefore the life that scores higher in this regard.

But in addition to this, there is a second factor that we may need to consider. One might hold the view that it is good for a person to be making a contribution to some larger whole—for one's life to achieve something, or amount to something, or participate constructively in some grander worthwhile enterprise. If one holds this view, then one may in particular also hold that *contributing interestingness* to some larger whole is one way to realize this prudential value and thereby to make one's life better for oneself. So in order to assess *this* component of how interestingness contributes to how prudentially good one's life is, we would have to look at how much interestingness the life "exports". Here the duplicated life would usually be at a disadvantage. And not just the exactly duplicated life, but any life that is too similar to a lot of other lives—they all contribute less interestingness to the world than a life that is more starkly outstanding as a one-of-a-kind.

*

Incidentally, the life that scores *the worst* on the contained interestingness criterion might score quite well on the contributed interestingness criterion. I mean, the very most boring life in the entire universe (if we assume that there is one life that is more boring than all others) is, ipso facto, distinctive and eminent in a particular dimension. It is therefore at least somewhat interesting from the external point of view, for reasons of the kind that we explored in the Solburg & Lunaburg thought experiment.

Let me illustrate this point with a personal anecdote.

I have forgotten almost all the lectures that I attended as a student, but one has stuck in my memory to this day—because it was so especially outstandingly boring. I remember trying to estimate the number of black spots in the acoustic ceiling panels, with increasing levels of precision, to keep myself distracted as the lecture dragged on and on. I feared I might have to outright count all the spots before the ordeal would be over—and there were tens of thousands. Memorability is correlated with interestingness, and I think we must say that this lecture made an above-average contribution to the interestingness of my student days. It was so boring that it was interesting!

There is a coda to this episode. You see, I don't have much natural talent for feigning interest. And the room was small, only a handful of people in the audience (which is why it would have been awkward to leave). The attitude of my soul was, alas, all-too-apparent to the lecturer—who was the departmental chair at the time. I later learned that he had blocked my admission into the department's PhD program.

Small people, Big World

Now let us suppose that you are Napoleon. It is then true that your life is quite distinctive among us Earthlings. But it is still not clear that your life is contributing any significant amount of interestingness to the world.

This is because chances are that you have been scooped—somebody else has already done all the things that you will do in your life. And I don't mean this merely in the loose sense that there have been generals and emperors before you. No, much more bitingly: it seems quite likely that somebody has done *exactly* the same things before, that there is not even the slightest scintilla of cosmic newness or uniqueness in anything you do or experience.

This, at least, is a consequence if we take the currently most favored cosmological models at face value. They suggest that we are living in what I've called a Big World: a world that is big enough and locally stochastic enough that it is statistically certain to contain all possible human experiences.[158]

For example, if there are infinitely many planets, and each has some independent small but lower-bounded probability of giving rise to any sequence of local state transitions (such as an exactly specified version of human history), then, for any such local transition sequence, with probability 1, it occurs multiple times throughout the universe. Infinitely many times, in fact.

The existence of infinitely many planets is likely if the universe is infinite in spatial extent, as it appears to be. Don't confuse *the universe* with *the observable universe.* When you hear a claim like "there are 10^{82} atoms in the universe", it is likely a mangled version of a true statement about how many atoms there are in the *observable* universe—but the part of the universe that is observable (from our vantage point) is an infinitesimal part of the whole thing, assuming we are in an open and singly-connected Big Bang universe, as the astronomical evidence indicates.

There is also some reason to think that there are other universes besides this one. The possibility of a multiverse, of course, further raises the likelihood that the Big World hypothesis is true.

In a Big World, snowflakes are not unique. Somewhere out there, far away, far far beyond even the remotest galaxies that we can see, there is an identical snowflake, right down to its precise atomic composition. Farther away, there is also an exactly identical "Napoleon". And farther away still there is an exactly identical "Napoleon" executing an exactly identical "Russian campaign" and ending up in an exactly identical villa on the exactly identical "island of Elba", and so on. (The campaign would not be in Russia, and the villa would not be on Elba; but they would be in places that are atomically identical to these sites on Earth.)

So we may say that there is nothing new under the sun, although it would be more precisely correct to say that there *are* things that are new under the sun, but they are all old hat in many other solar systems.

I'm harping on this, not because we know with certainty that we do live in a Big World, but because (a) it is quite likely that we do and (b) the implications are so striking. (But it is also possible that our basic way of conceptualizing possibilities apparently involving physical infinites is in some deep way flawed.)

*

When we consider things on the largest scale, therefore, we find that while the total amount of interestingness contained therein is great, our ability to contribute to it appears to be extremely small. And this would be true whether we are assessing our contribution in relative or absolute terms—that is to say, whether we are thinking about *what percentage* of the total interestingness we can be responsible for or *the magnitude* whereby the world is made more interesting owing to the fact of our existence. Either way, our role is negligible. And if the world is not only really big but canonically infinite in the way that Big World hypotheses imply, then we seem to be (if we construe ourselves as particular concrete individuals) responsible for a literally zero or infinitesimally small amount of total interestingness.[159]

In a way, you might say, this is reassuring. For if this is how things stand now, then at least we are not at risk of *losing* anything when it comes to our ability to contribute interestingness in utopia. That might otherwise have been a concern: that the gains in other dimensions of well-being that we can attain in a post-instrumental utopia would come at the cost of a reduction in our contributed interestingness. But if we are contributing nothing now, we won't contribute less later.

*

We should also consider the possibility that the Big World hypothesis is false, or that things are otherwise not as they seem. How much interestingness might we be contributing in that case?

If there is no multiverse, and if our universe is not too large, and if it is devoid of extraterrestrial intelligence: if, in other words, our planet is the only furnace in which the flame of consciousness has been lit—*then* the human phenomenon, flickering and faltering as it may be, does take on a kind of cosmic interestingness. On a dark enough night, even the faint glow of a firefly can stand out as a noteworthy sight.

And yet, even under these stipulations, there would remain the problem that while *humanity* may then be contributing a substantial amount of interestingness to the world, the case for us having an appreciable amount of *individually* contributed interestingness would still not have been made. The sober reality is that with over a hundred billion humans already having been born, and perhaps many more to come, it is hard for any of us not to disappear in the crowd—how interestingly different, really, can most of us claim to be from all the other burping lowlifes and sniffing highlifes, not to mention all the earnest midlifes out there?

An exception would be carved out only for the rarest and most extreme individuals. Maybe, if the universe were void of extraterrestrial life, Napoleon would make a (barely) perceptible difference to the total interestingness of the world. Maybe we could add a few other world-historical figures—some founders of religions, some great discoverers, a few peak cultural creatives. If our notion of interestingness places a premium on sheer eccentricity of belief and habit, we might add a little faculty of lunatics to round it off.

For the average king or prime minister, however, there would seem scant hope of contributing a significant amount of interestingness to the world, even if the Big World hypothesis were false.

*

Let us suppose—for the sake of the argument—that you are neither Napoleon nor any other world-historical figure, nor yet any (other?) sufficiently rare form of lunatic. Is there then *any* possibility that you could still somehow be contributing a humanly-noticeable fraction of the interestingness of the world?

I think there is such a possibility, though it requires a departure from the standard contemporary scientific worldview. We would have to consider more esoteric possibilities—ones that reduce the field of competition, from the infinite number of people that exist in a Big World, down past the billions of humans that we normally assume to have entered existence, down further to some still smaller and more manageable number.[160]

You might, for example, entertain a solipsistic speculation. If the external world is an illusion, and you are the only real person, then you would be responsible for a large share of total interestingness. You would be a subject of great originality—a most outstanding and remarkable figure!

(If you think this, per se, fact wouldn't really make your life that much better for you, well, then that's an upper bound on the amount of value you place on the interestingness contributed by your current life. Probably this means you actually attach little or no weight on contributed interestingness as a factor of well-being. Although it is theoretically possible that you might place great weight on contributed interestingness but just think that your current life is direly wanting in the relevant internal attributes; whereas if, for example, you became a planetary-sized superintelligence, you could gain enough complexity to make your interestingness-contribution to the world in the solipsistic scenario an important well-being factor.)

Variations of a solipsistic scenario could also work, where there are a few other real persons besides yourself. I'm not sure whether that view has been given a name. "Paucipsism"?

We can also consider simulation hypotheses, according to which the world you see is a computer simulation. It might then be possible that not all the apparent humans around you are simulated with a level of granularity that gives them conscious minds. You could be in a one-person or a few-person simulation, in which the majority of characters are NPCs.[161] But note that this would allow you to contribute a significant amount of cosmic interestingness *only if there are no other* simulations with a combined much larger number of "player characters" similar to you. For in *that* case, you would again be just another face lost in the cosmic crowd.[162]

Or perhaps I should say "just another microbe lost in the gut of some fabulous creature"—for if there be simulations, there be simulators; and it may be principally in *their* corridors and courtyards that the objectively interesting affairs of the realm are being transacted.

Parochialism

So far we've been considering whether you have *cosmic* significance, in the sense of your life making a significant contribution to the interestingness of the world as a whole. We found that you're probably too puny to make any perceptible difference at that level.

But what if we look at things on a more modest scale? At the scale of, for instance, *a community*, it is certainly much more feasible for an individual to have a fair degree of significance. If we go down in scale even further, say to that of an extended family, or a household, then having individual significance becomes the norm.

So if we sufficiently restrict the scope of the domain relative to which we evaluate our contributed interestingness, we shield our eyes from all those blazing suns that antedate us and that outshine us in all possible respects. And then we may appreciate the modest glow of our own candle. We may feel cozy in the understanding that at least *here*, within this local domain, there is nobody else like us, and also nobody else like the person we're sharing the table with in our little alcove.

We may well delight in such parochial uniqueness. Somebody who finds in their garden a bluebell: why should they not take joy in this pretty flower? the fact that it is growing right *here*—notwithstanding that there may also exist forests in which the ground is covered with carpets of bluebells?

*

If we go further, and restrict the scope of consideration all the way down to just our own individual life, we recover the concept of *contained* interestingness. We are then looking at how much interestingness there is within a life itself, without regard to how duplicative that life might be in relation to everything else that is out there.

*

A world-designer who took this constricted view of interestingness might fill the cosmos with identical copies of the most interesting life. (For the time being, we are considering interestingness only, not other values.) If one would *not* choose this option as a world-designer, then one is not accepting that the scope of the interestingness-consideration should be a single life. If one would also not fill the world with identical copies of the most interesting family, or identical copies of the most interesting community, then one is likewise rejecting the narrowing of scope to the family and the community levels. But then we get back up to the scale at which the problems we just discussed start to bite. At the scale of countries, civilizations, or the cosmos, it becomes very difficult or impossible for most individuals to stand out.

*

Suppose, as seems plausible, that if one wanted to maximize the interestingness of the world one would make a wide variety of different beings, rather than tile the universe with innumerable identical copies of a single (internally maximally interesting) being. Interestingness in this sense would be a global property of the world, and our ability to contribute to such global interestingness, now or in utopia, is likely very limited, for the reasons we have discussed. One could however maintain that our lives could still go better for us if we contribute interestingness on some intermediate scale. So it could be a consistent position to maintain that the world is better if it is interesting on a global scale, and that our lives are better for us if we contribute interestingness locally. On this view, the value of the world would depend on factors other than (and in addition to) how well the lives of its inhabitants are going for them.

In these lectures, however, we are mostly focusing on the question of personal well-being: how well could the lives of the post-instrumental utopians go for them? And especially: how well could our lives go for us if we become such utopians?

Time and becoming

Even if we restrict the scope of the interestingness value all the way down to that of an individual life—in other words, to what we have termed *contained interestingness*—we still are not in the clear. For we also face a problem in the temporal dimension. The need for diversity, novelty, variation—surely a central element in the value of interestingness—applies not only with regard to duplication in space but with regard to repetition in time.

This is the problem that Peer is grappling with—the reason he has had to resort to implanting a series of successive passions that controlled his desires and activities. Only thus could he avoid falling into a cycle of endless repetition, in which the value of diachronic interestingness would have been forsaken.

But while Peer's strategy can *delay* the onset of repetition and objective boringness, it cannot hold it off indefinitely. If we keep going for long enough, we eventually come down to a passion for carving table legs.

And from there, whereunto? A passion for sticking little pads at the end of table legs so they don't scratch the floor? A passion for unscrewing the cap of the glue bottle?

*

Although a great deal can be done within the space of human modes of being, the possibilities contained therein are finite. I am sure we could keep going considerably longer than the customary threescore and ten without depleting our reservoir of potential intra-life contained interestingness; but eventually it would run dry.

Obviously, the number of playable variations is enormous *if we individuate possibilities finely enough*. For example, if you have 100,000 hairs on your head, consider all the possible different braids you could make: I mean if we count as different any two braids that are composed of non-identical sets of hairs, or that have some difference in the precise pattern of intertwinement. Thanks to the power of combinatorics, even somebody like me, who has a much smaller number of hairs left, could form enough different braids to last me far beyond the heat death of the universe even if I made a new one every minute. There are even quite a large number of different ways of tying a simple necktie. One of my colleagues—the one whom I've never seen bored—has co-authored an article in this field, calculating that there are 266,682 distinct tie knots.[163] If we assume that you get a new tie each Father's Day, you could do up each one a different way for a quarter million years.

But of course, what is relevant for our investigation is the number of *interestingly different* permutations. In general, this number is very much smaller than the total number of different permutations. Even more restrictedly, we're concerned with the number of interestingly different permutations that could be combined into a single coherent human, which is presumably quite a lot smaller still. It seems not implausible that if we extended a human lifespan to, say, a million years, we would really begin to dredge the bottom of the interestingness reservoir.

*

Even if we live only to seventy, we seem doomed to experience a diminution of interestingness as we progress through life, at least if we measure things according to developmental milestones or major steps forward. Consider some of the things that happen in just the first year or two of life:

1. You enter into existence!
2. You realize that you have a body!

3. You learn that there is an *external world*! with *objects*! which *persist even when we are not looking at them*!
4. You discover that there are other people!
5. You begin to learn how to make sounds and how to move your limbs, allowing you to accomplish goals!

Cognitive upheaval upon cognitive upheaval. Epistemic *earthquakes* that upend the very foundations of our understanding of self and reality—magnitude 10+ events on the Richter scale of contained interestingness.

Compare this to an interval of the same duration later in life. A middle-aged person might reckon it an eventful year if they remodeled their kitchen or their dog had puppies.

Some of this slowdown could be avoided were we able to retain full vigor and zest as we aged, and if we enjoyed more conducive circumstances. But some of it is unavoidable, if the interestingness metric we use is something like "rate of personal development, of cumulative constructive life change, of attainment of qualitatively new levels of achievement, understanding, growth, and experience". Even with an optimal curriculum, perhaps along the lines of Peer's passion rota, we seem bound to encounter diminishing returns quite quickly, after which successive life years bring less and less interestingness.

The space of posthumanity

This stagnation could be postponed, though I think ultimately not averted, if we expand and enhance our faculties so that we can explore the *posthuman space* of possible modes of being. I am referring to that much larger set of possible ways of conducting and experiencing life—thoughts, perceptions, feelings, understandings, ways of relating, acting, appreciating, achieving, and aspiring—that is inaccessible to creatures with our present sorts of minds and bodies, yet which could be unlocked by progress in human enhancement technology.

I have argued elsewhere (and this view still seems to me plausible) that the posthuman space of possible modes of being contains riches that are beyond our wildest dreams and imaginings.[164] We may perhaps conceptually grasp them, but only in the most limp and abstract manner.

This should be unsurprising. There is no obvious reason to suppose that it should be possible to intuitively understand and vividly appreciate all these modes of being, when one is confined (as we currently are) to doing all of one's understanding and appreciating by means of a mind that is made out of about three pounds of meat.

Imagine a troupe of great apes sitting in a clearing and debating the pros and cons of evolving into Homo sapiens. The wisest of them articulates the case in favor: "If we become humans, we can have lots of bananas!".

Well yes, we can have unlimited bananas now. But there's more to the human condition than that.

*

So, on a timeline, we can expect to see an upwelling of interestingness in our lives when the posthuman realm opens up, especially if this technological transition unfolds reasonably rapidly.

Then, for a while, we are new again. Like infants opening their eyes afresh to the wonder of reality and beginning stumblingly to explore its affordances.

Each increase in capacity unlocks new worlds. Somebody might have lacked the gears to appreciate theoretical physics and high literature. They are given some cognitive enhancements. There is a click and a whirr, and the gates swing open.

If we keep upgrading our mental faculties, we eventually leave the human ambit and ascend into the transhuman stratosphere and thence into posthuman space.

I want to stress that it is not only intelligence that is involved here. All kinds of human limitations could be pushed back and expanded: lifespan, energy, emotional sensitivity and range, sensory modalities, creativity, our ability to love, our readiness for calm contemplation or playful sociality, special aptitudes such as for music, humor, sensuality, and so on, along with entirely new receptivities and generativities that we currently lack altogether. Just as was the case when some apes evolved into humans, many new sources of interestingness will likely come into view during this ascent transition into posthumanity.

*

How long could this epoch of increased interestingness last? The answer depends very much on how exactly we conceive of interestingness. Is the value rooted in *fundamental* novelty? Does it demand some minimum rate of gain in *general* capability? If so, we will run out of interestingness relatively quickly, as maintaining a constant level of interestingness would require a fast, perhaps exponential, burn rate of our growth potential. On the other hand, if *superficial* novelty and increasingly *narrow* capability gains suffice for interestingness, then we could keep going for much longer, perhaps for astronomical durations, without suffering any diminution in how interesting each successive day of life is.

Implications of three of the etiological hypotheses

To make further progress on this question, we can now revisit our earlier speculations about the origin of the value of interestingness. Recall that we considered four (not mutually exclusive) possibilities: the learning & exploration hypothesis, the signaling hypothesis, the spandrel hypothesis, and the rut-avoidance hypothesis. Let's explore what each of these would lead us to expect regarding the requirements for continuing a life at a sustained high level of contained interestingness.

*

The spandrel hypothesis asserts that it is only because and insofar as the realization of various *other* more fundamental values would entail an interesting future that we have the intuition that a future that is uninteresting would be deficient. The value of interestingness, on this hypothesis, is purely derivative. To understand the long-term prospects of these other values, we might do better to study them directly rather than merely look at the shadow they cast. That will be a topic for tomorrow.

*

The signaling hypothesis says that activities have interestingness value insofar as we have intrinsified their status-boosting propensities.

So we could list different such activities—yachting, directing movies, breeding racehorses, climbing Mount Everest, attending parties with beautiful and famous people, competing in the Olympics, performing rock concerts, governing a country or a large organization, and so forth. Then we can consider how these activities would fare in a plastic utopia. It seems all these activities, or close analogues, will remain possible and could be repeated indefinitely. Some would remain scarce (such as running countries or mixing with the famous) but others could become widely practiced (such as yachting or climbing tall mountains).

Of course, if everybody were climbing Mount Everest, it would cease to be a status-boosting activity. But the idea is that we might have intrinsified the allure of this activity so that it would retain (some of) its value even if it ceased to have the effect of making others impressed. Under ordinary circumstances, people might gradually lose their appreciation of activities that were once prestigious but have ceased to be so. However, using neurotechnology, the utopians could fix their psychology so that this would not happen. Notice that to someone who holds that climbing Mount Everest is intrinsically valuable (perhaps because they have intrinsified the instrumental value that this activity currently has as a status-booster), the prospect of becoming a person who no longer holds that this activity is intrinsically valuable may appear as a kind of *corruption*—a transformation which they would plausibly regard themselves as having reason to prevent from taking place.

Another potential problem is that some of these activities may somehow "lose their point" in a plastic utopia (as per yesterday's discussion of the four case studies). However, this seems less of an issue for interestingness per se, and is rather something we will need to consider when we look at how other values will fare in utopia—notably, purpose and meaning. We'll come back to that tomorrow.

The upshot is that the signaling hypothesis paints a relatively rosy picture of our prospects for being able to sustain high levels of interestingness even if we live for a very long time. One could in principle keep yachting, or listening to chamber music performances, or taking gallops around the grounds, or attending galas with finely dressed people—in perpetuity, if such were one's fancy.

*

Next, the rut-avoidance hypothesis. If the value of interestingness originates as an intrinsification of a desire to avoid the aversive feeling of boredom, and if our boredom-propensity is an adaptation to prevent us from persisting in fruitless endeavors and from getting stuck in over-rewarded situations, and to help us allocate time and effort to the full range of our needs by promoting appropriate task switching, then we should expect that what is required in order for our lives to remain interesting in utopia is that they offer a suitable variety of activities and settings.

This explains why the life of Peer immediately started to look more attractive when we modified Greg Egan's description to give Peer a portfolio of *concurrent* passions. Spending 90 years doing woodworking in the mornings, going for nature walks with one's spouse in the afternoons, and watching sports games with one's buddies in the evenings sounds a lot more appealing than spending the same 90-year period by first doing 30 years of nothing but woodworking, then a solid 30 years of nothing but spousal nature walks, followed by 30 years of sports watching. And, significantly, this intuition that the more chopped and mixed life is preferable may retain some of its grip on us even if we stipulate—as of course we should—that in neither scenario would any subjective boredom be experienced.

Admittedly, it is very hard to eliminate all confounding factors when we try to evaluate alternatives like these. For example, in the chopped and mixed variation, we might be imagining ourselves enjoying the interactions between our distinct spheres of interest. Perhaps we could be talking with our spouse about our struggles and achievements in the woodworking shop during our afternoon walks, whereas it might be a little bit of a struggle to imagine how we would keep a conversation uninterrupted for 30 years. Generally speaking, a lot of the objective interestingness that we might enjoy in our lives could come from the *interactions* between different pursuits and from the way they are refracted across our friends and associates. This consideration is another reason for mixing things up, separate from the fact that we might have intrinsified the value of mixing as a means of rut-avoidance.

In addition to introducing variation within the course of a day, utopians may also jazz things up by giving their lives texture and structure over larger scales. They could have phases, analogous to the distinctions we make between weekdays and weekends; but also seasons, holidays, projects of shorter or longer duration, career chapters, nested sets of aspirations, life phases, overlapping and interlocking, with different periodicities; "a time to cast away stones, and a time to gather stones together".[165]

If our longevity were greatly increased, we could use life-structures that extend over larger timescales than is currently possible. Maybe we could have periodic "rebirths": reformative transitions in which, while we keep an essential core, we "trade in" much of the incidental stuff and specific knowledge that we have accumulated over a century or two, in exchange for some profound new thing—such as a deep insight or an upgrade in some basic capacity, which then enables us to rebuild what we gave up, but better, and to go beyond it.

I don't know about you, but I sometimes fantasize about being able to go back to my childhood or youth and "redo" my life from that point, but with the benefit of hindsight and some of the things I've learned. I don't know exactly how that would work, but just having a single run-through of life seems so… I mean, one is missing so very much of what this place has to offer. And although the years may bring some modicum of understanding of the workings of the world, it tends to come late, often too late—making it seem like the only fruit that grows on the tree of experience is resignation. Wisdom withered on the bough.

(If some genetic engineer can fix this problem, I think it would be a good deed indeed.)

Implications of the learning & exploration hypothesis

Let's get to the last of our etiological hypotheses. We said that it traces the roots of the interestingness value to an instinct for learning and exploration, which has been intrinsified. We need to dig into this a little more deeply.

The core idea is that we are endowed with capacities for thinking, learning, and developing; that we have disposition and/or desires to use and realize these capacities; and that this leads us to value situations and life paths that involve the use and realization of these capacities.

How this plays out in utopia depends on which of these capacities we focus on. We can think in terms of a spectrum. At one end, we have capacities such as *visual perception*, which could easily find continuous fulfillment lasting for eons and eons. At the other end, we have capacities or potentials such as those for *neurocognitive development and maturation*, which are likely to culminate and then stagnate after a much briefer period of utopian burgeoning.

Perception is short-cycle. You see something, information dashes through your visual systems, a percept flickers in your consciousness, you've recognized an object: it's a bassoon. Then the neural activations that were involved in accomplishing this task of visual processing are reset. The job is over within the span of a second.

Memory operates on a range of time scales. *Working memory* is only slightly longer-cycle than sensory memory. Let's say you are performing mental arithmetic, and you are storing the intermediate results of the calculation in working memory. After several seconds, you have no more need of that content, and it's flushed out. You are ready to receive some new set of representations and maintain them in a high-availability state for a short period of time. This is all part of normal functioning: we wouldn't say that anything is "going wrong" when the content is dropped after the task to which it was relevant is completed. Working memory, like sensory memory, is grieflessly resettable after a short period of time.

But we also have memory capacities whose aim, it seems, is to store information indefinitely. We are not "meant to" forget how to ride a bicycle after we have once acquired the skill. Some elements of declarative and episodic memory also seem, in the ideal case, to be preserved permanently; and there is something at least slightly misfortunate when we forget. This means that if things functioned perfectly, we would keep accumulating ever greater troves of procedural and episodic memories.

We may or may not begin to exhaust the human brain's storage capacity after a hundred years, but *eventually* we would. In order to maintain perfect function indefinitely, at some point—you're gonna need a bigger head.

I don't think this in itself is a major problem. I put some further remarks on this in a handout, if you're interested.

HANDOUT 14
MEMORY STORAGE FOR REALLY LONG-LIVED PERSONS

The maximum number of bits we can remember grows linearly with brain size, so if we keep accumulating skills and experiences at the same rate as we do at present, we would just have to grow our brains by 14 deciliters every century (in reality much less, since we'd presumably migrate to a more optimized medium: but subsequently there would still need to be a linear increase in volume if we want to continue to accumulate long-term memories, though probably at a rate closer to 1 cm^3/century) for a human-level mind.

At some point we would get so big that the signal conductance delays between different parts of our brains would force some aspects of our thinking to slow down (those aspects that require integrating information stored in widely separated brain regions). This may already be a constraining factor today—the axonal conductance velocity in myelinated fibers is about 100 m/s, meaning a signal can travel up to 10 cm in a millisecond, which is roughly the maximal temporal firing resolution of biological neurons. If we used optical fiber instead, signals could travel at the speed of light, which is 300,000 meters per millisecond—suggesting a brain size limit of 300 km in diameter, the size of a metropolis. If we store a century's worth of accumulated long-term memory in 1 cm^3, this would let us live for more than 10^{22} centuries without forgetting any long-term memories, which seems plenty. (We'd also need to make some other complementary adjustments, such as creating a retrieval system that would let us find and use relevant skills and memories; but overall it seems doable.)

We could further increase the maximum size of the memory bank if we run the system more slowly, since that would increase the radius within which signal delays would be acceptable. If we live in virtual reality, and we slow it down by the same factor as we slow down our minds, we would not notice any difference.

Conversely, of course, if we insist on running at a higher subjective speed than that of a biological brain, the maximum size of an integrated mind would shrink correspondingly. For example, if we speed ourselves up by a factor of three million or so, we would be back to a brain that could fit inside our current cranium, although it could still have orders of magnitude more long-term memory than present-day brains, due to the use of more optimal computing and memory substrates. (Such a high-speed mind would also need a powerful active cooling system, unless implemented almost entirely using reversible computation.)

Instead of optimizing for living as long as possible with our current minds, we may instead prefer to make the cross-sections of our mental life bigger and more complex. Each second of subjective life would then require a greater quantity of computation, and the memories of such expanded minds would also use up more storage capacity. So this objective trades off against longevity. We could go big and die young, or stay small and live long. Plausibly we might want to do some of both—live orders of magnitude longer *and* have minds that are orders of magnitude more capacious—and this would be feasible for people in a technologically mature civilization.

*

Things get more problematic to the extent that we have a need for development and growth, beyond a simple linear accumulation of constant-sized life memories and motor skills.

How many more ontological earthquakes of a magnitude equal to discovering that you have a body or that other people exist can there be? If we are humble, we may surmise that the set of such fundamental truths that we are ignorant of outnumbers the set of those we know. Still, would we really think that there are *a hundred* such enlightenments left?

If understanding is compression, there is an upper bound to how well a certain set of facts can be understood: a finite string of bits can only be compressed so far.[166] To get more opportunities for compressing—assuming that we do not want to erase insights we have already gleaned—we need to keep accumulating data points. Long term, the amount of data we can receive and store grows at best linearly with material resources. We could thus potentially keep compressing at a constant or even polynomially increasing rate, until our civilization ceases to expand.

But presumably it matters not just how much we compress but also what and how we compress. For example, suppose that our data feed consists of measurements of the position and molecular composition of the astronomical bodies we encounter as we venture ever farther out into the universe. No doubt, there is a lot of microscopic and mesoscopic structure in this data, and thus opportunities for finding local patterns that enable us to usefully represent the raw dataset more compactly, using a smaller number of bits. But somehow the cognitive work involved in doing this does not seem very interesting. The information and the patterns that we would keep discovering in this way would eventually have only local relevance: they would tell us less and less about what's true elsewhere that we didn't already know, and they would cease to reveal any novel general truths or deeper levels of explanation. At this point (to revert to our Shakespeare analogy), we would be long past studying the characters and plots, and we would have advanced to a point where we are spending millions of years cataloging with ever-increasing precision the exact way each fiber in each page of the folios intertwine; and then the precise molecular shapes of each such fiber—and the most exciting thing to happen in the research lab during a given ten-year period (spoiler alert) may be when one day a new dust speck lands on the manuscript.

I suspect something similar will eventually happen with our exploration of mathematical patterns, although it might take longer. Sure, there are infinitely many to be discovered, infinitely many truths to be established, requiring arbitrarily difficult proofs. But how many are there that are really deep and fundamental? How many results of a similar level of profoundness as, say, Cantor's or Gödel's theorems? I would guess a very finite number.

*

Perhaps the most promising object of study, if we are in search of an inexhaustible source of interestingness, is ourselves: intelligent beings and the culture that we continuously generate together. Since we are changed by our experiences and since a culture can build on its past achievements, this constitutes a moving target which our individual and collective understanding might never be able to fully comprehend. The more that we boost our capacities of understanding, the more we may simultaneously accelerate our cultural creativity and complexity, so that the former never catches up with the latter—just as a sprinter may never catch his shadow, no matter how fast and far he runs. Even a "Jupiter brain" (to use a term from the old transhumanist lexicon—a computational megastructure with a mass comparable to that of a gas giant like the planet Jupiter) might remain intellectually stimulated and challenged in a peer group that includes other Jupiter brains, which keep learning and improving and creating at pace with its own advances in skill and understanding.

It might be that the idea that interestingness could in this manner be long extended derives its plausibility from our concept of interestingness having a social component embedded within it. Something to the effect that information that gives one an edge in understanding one's social interaction partners, or that improves one's relative level of understanding of objects that are of shared concern to others, is, thereby, ceteris paribus, rated as being more interesting. (Such a social component in our criteria for interestingness could have been deposited through a process similar to the one postulated in the signaling hypothesis, except the origination would not need be confined to a social signaling value but could derive instead from various forms of instrumental advantages tied to certain types of socially relevant information or learning.)

If our value of interestingness does contain such an intrinsified desire for socially relevant information, the consequences of this might spill over to make the minutiae of any other arbitrary topic count as "objectively" interesting, so long as they remain the focus of significant cultural activity. For example, progressively recherché patterns in prime number sequences might count as interesting if there are many mathematicians who are actively searching for them. The "objectiveness" of this kind of interestingness, however, may be of a somewhat weakened or qualified form. Although it would go beyond mere individual subjective interest—it consists in something more than somebody just happening to feel interested in something—it would nevertheless be an interestingness grounded in what seems like rather arbitrary cultural choices and dynamics. It should perhaps be termed "intersubjective" rather than "objective". But to the extent that there are other normative constraints on which directions a culture takes, or on the dynamics which ought to determine such choices, it might be that some intersubjective constraints are indirectly buttressed by more objective constraints.

*

Maybe. I'm unsure how this shakes out. If I had to guess, I'd say *if* we measure the objective interestingness of an era in terms of how many epistemic earthquakes of a high magnitude take place per unit of time, then objective interestingness will probably peak around the development of machine superintelligence. Depending on how steep the takeoff is, interestingness might then remain at unprecedentedly high levels for a decade or so, before gradually trailing off to levels far lower than we have seen even in relatively stagnant periods of human history. The most important things that can be discovered may by then already have been discovered.

However, this prognosis depends very sensitively on how interestingness is quantified. Different ways of measuring could allow for interestingness to plateau at an extremely high level, and to remain there indefinitely—at a level far above what we are currently experiencing in our individual lives or as a civilization. It is possible that our concept of objective interestingness is too confused and poorly specified for there to be any determinate fact of the matter as to which metric is correct.

Spirited kaleidoscopes

How much of a bummer would it be if we "run out of interestingness" in the epistemic earthquake sense, or if our consumption of such interestingness must asymptote to some level that is very low by contemporary standards?

Maybe we can get some grip on this if we compare such a hypothetical transition-to-an-era-of-perpetual-much-lower-interestingness to the transition we have already undergone individually, from our own infancy and childhood to our adulthood.

As we observed, for most of us, this transition involved a drastic decline in a certain kind of interestingness. We are no longer regularly making discoveries of the profoundness-level of object permanence or the existence of other minds. Even this very lecture series, despite my best efforts to put in place a justification for the exorbitant tuition fees that many of you are paying, or your parents are—and which we are compelled to charge in order to fund the construction of the university's new campus, which will be exclusively devoted to accommodating our exponentially growing administrative staff; not to mention the new ballpark with seating for twenty thousand, without which it would scarcely be possible to give a young person an adequate foundation—my efforts with this lecture series, I dolefully concede, will most likely fail to deliver the interestingness value once afforded to you by that colorful toy xylophone you received when you turned one.

Still, I'm hoping that your lives are currently going not too badly, and that even though in percentage terms you are not making as rapid cognitive strides as you did when you were even younger, you find compensation yet in the greater absolute level of sophistication at which you now operate.

And to the extent that our later experiences are tinged with nostalgia, it may be for altogether different reasons. We might have liked being taken care of and being able to spend our days playing. Being naive might have been good for us, not because it gave us more opportunity to learn and progress, but because it shielded us from harsh realities and allowed us to fit into a smaller, cozier, more human-sized lifeworld. Now we are stressed out, responsible, jaded, damaged, less vital; there is less fun and wonder.

The future used to dangle before us and above us like a magic veil, draped alluringly over creation in motley colors and translucent shadings. Now we instead see a corridor, lit by fluorescent lights, with numbered rooms, bills to pay, obligations to discharge—and we know that at the end lies the hospital, the hospice, and the morgue. Our parents, who used to be our loving protectors and maintainers of our little world, are helplessly withering away before our eyes, or already in the ground.

So if some part of us pines after the bygone days of childhood or the lost innocence of youth, there are plenty of sad reasons that could account for that.

*

What might it look like if we did attain a state of stagnation, from which further growth and development was not possible? How objectively boring would such a condition be?

I think we should not imagine this as tedious monotony, as getting stuck in an everlasting rut of triviality and grind.

Instead, such life could be like a living kaleidoscope, conjuring an ever-changing series of patterns that transform and modulate one another according to a fixed set of rules within bounded parameters. There is a sameness at a certain level, but also inexhaustible richness and novelty at other levels.

The posthuman kaleidoscope would be more intricate and complex than contemporary human life, perhaps by a margin similar to that by which the richness of our lives exceeds that of a literal plastic tube with a few glass beads and mirrors.

We could gape in fascination at the beauty of being, revealing itself in its endless forms and variations; we could step in and actively participate, like dancers with indefatigable legs, moved and moving according to a logic of wonder and appreciation—entrained by the harmonies and pulsations of full existential realization.

The scenic route?

However alluring such a destination, we may have reason to proceed toward it with less than maximal speed. I'm not here referring to the practical considerations that might arise, some of which we have alluded to already, such as potential tradeoffs between speed and safety.[167] What I have in mind, rather, is the idea that "the journey is the destination"—or, at least, that the journey can be an important and value-adding part of the total package.

Insofar as we are seeking to maximize the total interestingness value in our lives (whether contributed or contained), we should ask not "What is the optimal *state*?" but "What is the optimal *trajectory*?".

*

The best "state" is anyway not completely static. For example, mental activity and phenomenal experiences require brain *processes* to occur. A frozen brain state, or a mere snapshot of a computational state stored in memory, would not be conscious. So, strictly speaking, it has to be trajectories; and the only question is whether the relevant temporal scale upon which our values supervene is short (e.g. on the scale of a second or so, as may be the case for a moment of conscious experience) or long (e.g. on the scale of a potentially very long individual or civilizational lifespan).

*

If and when we manage to crawl into a condition of *existential security*, where we are protected from exogenous competitive pressures as well as from nature's slings and arrows, we might be well advised, from that point on, to take it slow. With all the time in the world, but only a limited number of significant things that we can do for the first time, we may want to carefully husband our supply of novelty.

This is especially true for *developmental* novelty, i.e. new levels of general capability that we could unlock through enhancement technologies. There are two reasons why these may need to be especially economized. First, because the total number of possible upgrades of our general capability levels (of some given percentage magnitude) is likely to be fairly small. (If you grew your brain by 10% every day, you would suffer a gravitational collapse within just a few years.) Second, because increases in general capacity enable us to more quickly deplete some domains of environmental challenge. For example, many games in which you or I may find a source of stimulating challenge, which could keep us delightfully entertained through many a winter afternoon, would be uninterestingly trivial to a superintelligence who, at a glance, perceives the optimal strategy. Perhaps, therefore, we should take our time to extract the fun and interestingness from these games first, before we enhance ourselves to the next level of cognitive capacity. In utopia, it might be a mistake to rush to grow up.[168]

*

I'm saying *perhaps*, because there are also non-developmental forms of interestingness—interestingness of the kaleidoscopic kind—which we could experience at greatly elevated (yet indefinitely sustainable) levels as posthuman superbeings. There are other values besides interestingness which we could more fully instantiate in our lives once we have greater capacities and are further developed; pleasure being but the most obvious one. It is therefore not at all obvious that delayed ascension is, all things considered, to be preferred. Posthuman life might just be *so* good that even if the path to get there should be quite scenic, winding through a charming countryside of meadows and orchards, it would be doltish to slacken our pace even a little or to stop to smell the flowers: perhaps we ought to just try to get there as quickly as possible.

Tessius [*whispers*]: Truckers call it "home speed".

Firafix [*whispers*]: Poetic.

Tessius [*whispers*]: But decorum goes out the window in juice bottles.

Bostrom: Still, the longer our lives will be, the more plausible it is that the life course that would optimally realize the panoply of our values would be one that sets aside some initial segment of our time for enjoying existence as less than maximally developed beings. This initial segment may be very short compared to the total span of a posthuman life, but it might be long compared to the duration of a typical human life.

Identity, survival, transformation, discounting

There is another reason why we might want to pace ourselves in our journey toward utopia, once we have attained existential security individually as well as civilizationally. This reason is more general: it pertains not only to the interestingness value but also to other values we may wish to realize in our lives.

Suppose that you suddenly became a Jupiter brain. If you were suddenly transformed into such a being, it is hard to see how you could avoid severely rupturing your personal identity in the process. It is challenging enough to see how such a transformation could be personal-identity preserving if it is done very slowly and incrementally; but compressing it into a subjectively brief interval may unavoidably tear the filaments of prudential concern that connect our present selves to the future subjects that we might hope to become.

Consider the following thought experiment:

> *Abrupt maturation*
> A four-year-old child goes to bed in the evening and wakes up the next morning a fully-formed adult. During the night, his body grew, his brain matured, and dreamless sleep instilled in him the skills and knowledge that normally would have taken twenty years to acquire. Neural patterns,

analogous to remembered experiences, were imprinted in his cortex, so that the person who wakes up is not entirely lacking in autobiographical or first-person-perspectival information.

By the way, note that this procedure would not necessarily imply that the person who wakes up in the morning is deluded about his past. We can suppose that the implanted memories come with flags identifying them as such. Although the person in the morning would thus not be remembering his own actual life (beyond the four years that he lived before this strange operation took place), he would have a database of quasi-memories sampled from a counterfactual life roughly similar to the life that he would have lived if he had continued to grow up normally, and which he could use for at least some of the inductive purposes that real autobiographical memories serve in people who have ordinary pasts.

Now, the man sitting on the side of his bed in the morning, stretching out his long limbs and stroking his stubble: he is in some ways the personal continuation of the boy who was tucked into the bed the night before. He retains the boy's memories, shares some of the boy's personality traits, and his body is a metamorph and outgrowth of the boy's body. Yet in some ways, it might also be said, he is a different person.

So we might ask: Did the boy simply experience an unusually accelerated growth spurt? *Or* did the boy die and get replaced with a vaguely similar young man?

*

Rather than posing the dilemma so starkly, as a choice between two radically separated possibilities, I think we do better if we take a more quantitative approach, one that focuses on "survival" and recognizes this to be a matter of degree.[169] Although I might sometimes use the term "personal identity", what I mean by this is not a relation that necessarily satisfies the mathematical criteria for identity (such as transitivity). Instead, I will use the term to mean a relation that holds when and to the extent that there are "grounds for prudential concern"—referring to the self-interested reasons we may have for being particularly concerned with what happens to certain future subjects in virtue of them standing in certain relations to our present selves (relations which probably involve forms of causal continuity as well as psychological similarity, and perhaps various other more circumstantial linkages as well). Such grounds for prudential concern can come in various degrees of strength. In principle we could have them simultaneously toward more than one future successor entity.

So let us reformulate the dichotomous question, and ask instead: In the case of the boy turning overnight into a man, is the degree of personal-identity-preservation much less than it is in the normal case where the boy grows into a man over the span of one or two decades?

*

I think probably it is. I'm not sure about it being "*much*" less; but it seems plausible that the degree of identity preservation would be significantly reduced when the metamorphosis is sudden.

And why might this be so? Several possible explanations spring to mind. (i) In a normal childhood one might have various projects that one has an opportunity to complete; not so in the abrupt maturation case. (ii) One also has relationships (e.g. to one's parents and friends) and social roles that are allowed to more smoothly develop and evolve; in sudden maturation these are all radically disrupted. (iii) In a normal childhood, one's later development is partly affected by the efforts and choices one makes along the way; in sudden maturation one lacks such agentive participation in the transformation that takes place. (iv) Normal childhood also allows one to *experience* the process of growth and development; in sudden maturation, this does not happen. (v) In normal childhood, many of the changes one undergoes may be viewed as themselves constituting the exercise of a kind of capacity for development or as being the realization and fulfillment of an inherent biological potential to grow and mature; but that would be a less natural way of viewing things in the case of abrupt maturation.

If we maintain that the preservation of personal identity is impaired in the case of abrupt maturation, then we should also suspect that scenarios in which normal human adults are abruptly transformed into Jupiter brains are ones in which personal identity is undermined to a greater degree than if a similarly profound transformation unfolded more... *adiabatically*—more gradually and smoothly, over a longer period of time.

*

So what do we have here so far? We have the idea that certain developmental or learning-related forms of interestingness could be maximized along a trajectory that is less than maximally fast: one where we spend some time exploiting the affordances available at a given level of cognitive capacity before upgrading to the next level. We have also the idea that if *we* want to be among the beneficiaries of utopia, we might again prefer trajectories that involve less than maximally precipitous upgrading of our capacities, because we may thereby preserve a stronger degree of personal identity between our current time slices and the time slices of (some of) the beings that inhabit the long-term future.

These ideas both favor a slower pace of ascension. Do we have anything on the other side—considerations that favor going faster?

Student: If we wait too long, we might go extinct. We could get hit by an asteroid.

Bostrom: Well, yes, and there are bigger risks than asteroids to be concerned about. But here I want to bracket those kinds of practical considerations. Let us suppose that we have attained a condition in which both our civilization and we ourselves as individuals have attained an adequate level of security—we've stopped dying left right and center, and we've attained a very decent quality of life. By "we" I mean everyone, including nonhuman animals and digital minds; and we've reached a post-instrumental condition in which our own efforts are anyway instrumentally irrelevant. I want to focus on the value question. Would the slower rate of ascension always be better then, or is there something on the other side of the balance?

Anybody? What if we *never* enhanced our capacities?

Another student: Then we'd never get to experience what life as a posthuman being would be like.

Bostrom: True. And even if we did enhance eventually, but we waited until the universe was just about to run out of steam, then we wouldn't get to experience it for very long, right? More generally, if we think that many values could be instantiated in our lives to a much higher degree once we become posthuman, then the longer we wait, the more of that higher level of well-being we would be missing out on.

Student: Yeah.

Bostrom: But the universe might have enough steam to keep chugging along for trillions of years, and much longer than that if we exercise a bit of responsible stewardship. So if we waited, say, a hundred million years: that would scarcely make a dent, in percentage terms, in how much time we would get to live in a posthuman condition.

Can you think of any reason that might push us toward going much faster than that?

Another student: What about time discounting?

Bostrom: What about it?

Student: In economics class, we are taught how to discount future benefits and costs to calculate their net present value. The rate is usually set to around 5% a year. Basically, what happens in millions of years doesn't matter. For the present value of a posthuman existence to be significant, it would need to happen much sooner than that.

Bostrom: With a discount rate of 5%, what happens even just a hundred years from now would matter less than 1% as much as what happens next year.

And what happens 10,000 years from now would matter less than 1% as much as what happens 9,900 years from now. Which seems a bit sus!

I think we need to be cautious in interpreting the temporal discount rate as an assertion about ultimate value. In its normal usage, the discount rate also serves as a rough proxy for a bunch of empirical considerations, such as opportunity costs given alternative investment opportunities (which might be tracked by the risk-free rate of return), expected inflation, consumption growth, or even the possibility of individual death or a complete collapse of the economic system. There are many practical reasons why we might prefer to receive a given sum of money today rather than being promised to receive the same amount at some future time.

But in addition to these practical factors, the discount rate is sometimes also supposed to incorporate a "pure time preference", reflecting a kind of irreducible human impatience—which is assumed to take an exponential functional form.

As a psychological description of human preferences, this seems like a dubious model. I would think, for example, that we are impatient in different ways over different timescales. Whilst we are in the grip of some immediate temptation, we might discount future enjoyments at a rate of several percentage points *per minute.* Somebody might prefer to eat one cookie now rather than two cookies in an hour. However, if the potential reward is pushed outside the span of immediate gratification, then perhaps we might discount it at a rate of a few percentage points *per year.* But even in that regime, the model's predictions are questionable. I certainly don't care a hundred times more about what happens in 9,900 years than I do about what happens in 10,000 years.

In fact, if I were forced to choose, I might prefer a life that starts out bad but steadily gets better rather than one that starts out good but steadily gets worse. Which would appear to imply a negative discount rate.

But it gets hard to know what to think of such cases. For example, we might intuit that we would experience greater total happiness in the upward-sloping life, since we would be able to look forward to things getting better. Yet we'd have to try to abstract away from such psychological effects if we want to determine whether we have a pure time preference over the distribution of intrinsic goods in our life. If we just ask ourselves this question directly, our intuitive judgment is likely to be confounded by implicitly assumed empirical correlations.

Let me put forward for your consideration the view that time is not fundamentally relevant in this context. Time is just a proxy for certain kinds of change which can separate distinct temporal parts of ourselves in such ways as to lead an earlier part to have a reduced level of prudential care for a later part.

Consider the following thought experiment:

> *Freeze*
> One day the spirit of winter, King Boreas, sweeps in and stops all change on Earth. Everything is magically frozen in place; all motion and all brain activity is arrested. The planet continues to circle the sun. After a thousand years, Boreas departs and everything is unfrozen and continues where it left off.

I think that if we learned that the scenario in Freeze were about to happen, it should have no impact on how we feel about what's to come. This shows that it is not the mere passage of time that modulates our prudential interest in the future.

A more plausible modulator of our prudential concern is the gradual attenuation of personal identity that tends, in ordinary circumstances, to take place over time. In truth, I suspect that this too fails to get to the heart of the matter; but it might get close enough to serve for the purposes of the present analysis. So let's consider what this identity-preservation consideration implies regarding the cases before us.

I suggested earlier that an extremely rapid metamorphosis (as in Abrupt maturation and in a scenario where we suddenly get catapulted into a posthuman utopia) may tend to rupture our personal identity. This would make the posthuman thrills less prudentially desirable to us now, as the posthumans who would be enjoying them would be us to a lesser degree. But I think the same consideration may also militate against an extremely *slow* metamorphosis. The problem with an extremely slow metamorphosis is that the normal background rate of personal-identity attenuation means that by the time the metamorphosis is complete our identity will have eroded too much, making subsequent well-being gains less prudentially relevant for us now.

Can you believe it, I've prepared a handout with a toy example to help explain the point.

HANDOUT 15

ON THE OPTIMAL TIME FOR TRANSCENDENCE

(In the following we will use some made-up numbers in order to illustrate a few considerations.)

Suppose that under normal conditions, our prudential connection with future stages of ourselves attenuates at a rate of 1% / year. And suppose that if we undergo "abrupt metamorphosis", there is an instantaneous attenuation of 90%. We could then have reason to slow down the metamorphosis if we can thereby make it less disruptive of personal identity.

Note that after about 230 years, normal erosion would have reduced our connection to less than 10% anyway. 230 years is therefore an upper bound on how much we would want to slow down the metamorphosis (in this simplistic model) if our only concern were to maximize the present discounted value of our posthuman phase.

However, we must also factor in that our human phase of existence has some value for us. Suppose, for example, that we thought that life as a posthuman would be at most twice as good as life as a human. Then—holding duration and other factors constant—we would never want to undergo a metamorphosis that imposed a greater than 50% additional erosion of our personal identity on top of the normal rate.

In reality, duration and other factors are not constant. If a human life is a hundred years long, whereas a posthuman life is a billion years long, then there would come a point at which we would want to take the plunge almost no matter how great the attenuation of prudential concern it would involve. When you're about to die anyway, you have little to lose.

Other significant considerations include the fact that the value of continued human existence is not constant. Even if we could fix the problem of deteriorating health, one might hold that we would eventually deplete some of the values of a human existence—for example, its interestingness value. This would make the leap into posthumanity progressively more attractive: the more we have exhausted the possibilities of a human existence, the less we have to lose by moving on to something new.

Another thing to consider is that the rate of erosion may be different for posthumans than for humans. Mature posthumans might be more tightly temporally integrated than humans, so that for posthuman phases of existence the rate of erosion of prudential connection is far less than 1% / year. This would also tend to make an earlier and faster transition desirable.

*

It is an interesting question whether it would be good for a child to never grow up, and to remain a child for an entire normal human lifespan. My guess is that the short answer is "no". But the question is not such a no-brainer as one might think. It is confounded in an intense tangle of empirical contingencies, which would need to be carefully unpicked; and the answer that would then emerge is probably quite complicated.

I will not attempt the unpicking. I will confine myself to making two observations.

First, the maturing of a human being tends to be a package deal: a whole slew of changes that go together, some for the better, others for the worse, still others for the simply different. We could become more eclectic. Maybe one can have the playfulness and the sense of wonder of a child, the passion and capacity for strong action of a young adult, and the tempered wisdom of an intact senior? Although there may be inescapable tradeoffs between some good attributes, yet a creatively integrated personality might have more room for apparently incompatible traits than you might think.

Second, one element that might make us think of the child who always remains a child as misfortunate is that a child has various capacities, dispositions, and potentialities, whose natural use and unfettered development leads over time to the child's transformation into an adult. It might—this is an idea we will touch on again tomorrow—be deemed bad when such things are thwarted. This judgment can be decoupled from any judgment about whether it is better to live as a child or as an adult: it is, rather, a judgment that certain things ought to be allowed to unfold and change according to their own propensities.

Now, in one sense, once we have reached maturity, at around the age of twenty, we don't have a propensity to continue to grow and develop physiologically. From that point onward, unfortunately, we commence a process of biological decline, slow at first but accelerating, eventually culminating in sickness and death. However, viewed from another (I think more justified) perspective, we could view senescence not as a fulfillment of our potential but as a factor that is external to our potential and that prevents it from being fully realized. The function of our eyes is to see; and the emergence of cataracts is an impediment not the attainment of a supreme excellence of this function. Similarly, the function of our brain is, among other things, to reason and to learn; and the onset of dementia is not the capstone to a lifetime of intellectual activity but rather the toppling over and ruination of everything that has been built. Similarly, stomachs should digest, hearts should pump blood, lungs should oxygenate blood, hips and knees should bend. If there is a natural telos in these things, decrepitude is not its proper fulfillment. And so aging, from this fulfillment perspective too, is a calamity, notwithstanding the heavy puffs of copium that cloud the topic.[170]

But I think we could make a case that the normative demands from such a fulfillment perspective require something more than the abolition of senescence. Even if we could remain in fine fettle and full fiddle indefinitely, without senility or loss of vigor, it is possible to view the ensuing condition as still in some sense stymied; as slightly akin to a child that fails to develop into a full-fledged adult—and that only a continued process of growth and development, enabled by the application of enhancement technology, all the way up to the fullest forms of posthuman realization, would constitute a complete attainment of our telos.

I don't wish to place too much weight on this particular argument, but it is there for your consideration.

*

We've gotten perhaps a little bit sidetracked, so let me summarize a few of the points we have covered regarding the question of timing.

The metamorphosis needed for us to access the space of posthuman modes of beings may inevitably entail some attenuation of personal identity. This would reduce—but, I believe, would far from void—the prudential desirability of attaining such access.

The attenuation of personal identity would be aggravated if the transition is abrupt. A very rapid transition would tend to involve several kinds of discontinuity, which would sever more filaments of prudential concern.

On the other hand, if the transition to full posthumanity is very slow, then the enjoyment of it would be pushed into the distant future, where its net present value may be heavily time-discounted. Even if we reject a pure time preference, the gradual erosion of personal identity that takes place under normal conditions would reduce the present prudential desirability of this prospect to us. (A very slow ascent would also reduce the time we get to spend at higher levels; but this is perhaps a lesser consideration, given how astronomically long the total available time appears to be in the standard model.)

Overlaid on these considerations are some further factors, which relate to the value of interestingness (and other values that share a similar structure): to wit, that in our current human existence some of the values it instantiates may eventually get satiated or the materials for their continuing instantiation may get depleted—for example, our lives might become uninteresting and repetitive after a thousand years or a million years of human existence. On the other hand, if we want to maximize the interestingness of our life trajectories, it could also be the case that ascending too fast would be suboptimal, since it would involve foregoing some of the opportunities for interesting activities and experiences that are suitable only for beings with human-level capacities.

Finally, we reflected that there is a "fulfillment perspective", which may recommend that some degree of continual forward or upward movement may be desirable, to allow for the unfettered realization of our capacities and potential for development and growth.

The upshot seems to be that an intermediate pace of transcendence is optimal. I hesitate to put a figure on it, but just to offer you something: Perhaps, given these considerations, the ideal trajectory for a typical human being might be something vaguely along the lines of what it would be like if, after we grew up, from infants to teenagers to adults, the development of our basic biological capacities didn't stop—let alone be thrown into reverse—but were instead de-arrested and allowed to continue toward ever greater heights, at a pace that is measured but still brisk enough for one year to make a noticeable positive difference.

Timesuits

Actually, I think we can do even better than this, by frontloading enhancements that protect against the identity-erosion that otherwise would result from the passage of time or the introduction of other desirable enhancements.

The starkest cause of identity-erosion is biological decay. Interventions that help stave this off would therefore be an obvious early priority.

Even if we abolish death and dementia, however, we remain exposed to several subtler kinds of erosion—mundane processes which, in our current condition, are continually separating us from ourselves, ripping us asunder temporally, bringing alienation between the time segments that make up our past, present, and future. Such prudential disintegration is occurring even during comparatively stable periods of life, such as middle adulthood, when there is little change in our basic capacities and our personalities and aspirations tend to remain somewhat stable.

*

One contributor to this is forgetfulness. Our bygone days, even quite recent ones, are now remembered only in the feeblest outline, if at all.

How much can *you* recall of what you did and thought and experienced on Tuesday three weeks ago?

Memory enhancement could stem the rate of such losses. And maybe that would help to slightly unify temporally displaced parts of ourselves.

*

Impatience is another cause of self-alienation: thus also something we might wish to attenuate at the outset of our metamorphosis. I'm not saying that we would necessarily want to eliminate temporal discounting altogether and in every respect—to immediately become beings who are indifferent to when things happen, who would prefer just as well to eat a cookie in a million years as right now. Such radical redesign of our volitional apparatus would itself constitute a disruptive and potentially identity-eroding alteration. But: a gentle tweak, a slight up-tuning of our tolerance for delayed gratification… that could be helpful in letting us develop a bit more slowly without bringing the present too far out of sympathy with the later.[171]

*

Another type of enhancement that we may have reason to frontload is the upgrading of our capacities for autonomous decision-making. Being able to make choices that are expressive of our true selves (insofar as such things exist), rather than being driven excessively by impulse, or controlled only by surface and happenstance, or by corporate interests and their advertising agents, may help us preserve personal identity better, by making the changes that we do undergo be to greater extent the products of our own values and volitions. This could bring the later stages of ourselves into a more vital communion with the earlier stages, from which they stem.

Parents often think of themselves as, in part and in a manner, surviving in their children. I expect this sense of vicarious survival is strengthened if the child shares or embodies some of the deepest values of the parent, and if the child's character has been partially shaped by its interactions with the parent and by the expression of the parent's character in those interactions. When our influence on an outcome filters through the sediment of our moral character, and then seeps through the layers of our active efforts and conscious experiences, in a richly interactive and participatory way, it becomes easier for us to regard the result that it shapes as being imbued with our own essence, and as being, indeed, a kind of extension or spinoff of our soul—a fruit dangling on our own bough.

The author of a book may have a similar cognizance, of partially surviving in their work—which would not be felt to the same extent by a person who hires a ghostwriter and thereby causes an equivalent book to come into existence.

This suggests that, when considering which enhancements to frontload, we should think of "our capacities for autonomous decision-making" in a broad sense: as including not only, for instance, the ability to make prudent decisions about various options for medical or technological procedures, but also a more general set of intellectual and emotional abilities for "self-authoring"—for authentically shaping our lives and development trajectories in an actively participatory way, together with other people we care about and have deep relationships with, as opposed to merely going with the flow, or being manipulated by market forces, of being buffeted randomly by stimuli that we've had little role in curating.

*

Another factor that can promote a sense of survival through our children is that we may live on in their memory and in their hearts. This factor is also relevant in the case of our posthuman metamorphosis. We have already mentioned memory, how fortifying it early on in the transformation process could help buffer us against erosion of personal continuity. We have also made some reference to the affection that we may feel toward our future selves—how boosting our patience and decreasing the rate at which we discount future benefits would let us proceed at a more leisurely and identity-preserving pace (and to smell more of the flowers along the way). To this we now add the further desideratum, that of a reciprocal sympathy on the part of our future selves. This is the mirror image of a future-directed care: a care directed at our past. We could nudge our development in ways that make it more likely that our future selves will care about our present welfare similarly to how we now care about theirs: that they may honor us and remember us with the fondness and gratitude with which we might, for example, hope to be remembered by our children—perhaps even to the point of establishing equality between our prospective and our retrospective cares.

I wonder if such intertemporally more unified selves will glance back at beings like us, such as we currently are, and pity our riven nature. One moment defecting against the other. Today blowing tomorrow's inheritance in sloth or lust or gluttony… tomorrow, grieving amidst ruined prospects, shooting arrows of hate and loathing and regret back at its predecessor—while plotting the same trick against the time-part next in line. Are we not like a hydra at war with itself, each head-brother sniping and snatching at the other manifestations of its being?

*

Somebody has a question.

Student: Yes. I wonder, if we really became temporally neutral… It seems a bit kumbaya. I mean, wouldn't that in itself be a very radical change—which would defeat the purpose of making sure that our personal identity is preserved?

Bostrom: Well, maybe. I think you are really asking two questions. One question is whether being temporally neutral would be good—would it, for example, make one (sort of) float above one's life rather than being fully immersed in it? The other question is, even if being temporally neutral could be good, would a transformation from the way we currently are into beings that are temporally neutral involve such radical change that it would severely rupture our personal identity?

I don't think we have the answers to those questions. However, we don't need to go all the way to full time-neutrality, at least not right away. We could just move a little closer. I would guess that if we were to look more closely at this, we would find that time-neutrality is not one thing but a cluster of distinct traits. For example, we might have one attitude toward our future selves when we step back and coolly reflect on our major goals in life, and some quite different attitude when we are in the midst of some activity and temporarily blocking out the future in order to be fully present. This suggests there might be opportunities to target any adjustments we make so as to circumvent some of the tradeoffs you point to. In some ways we may want to remain schismatic, but ideally without foregoing the benefits of intertemporal cooperation and communion that could be gained by becoming less impatient.

*

I'll mention one more consideration that is worth bearing in mind in this context, when we ponder the desirability of embarking on a process of transformation that could eventually lead us to become quite different from the beings we currently are: Love is not necessarily proportional to similarity. This applies also to self-love. We might stand in a deeper connection of sympathy with, and identify more strongly with, a future *better* version of ourselves—one that realizes many of our hopes and aspirations—than we do with a future version of ourselves that is more similar to the way we are now. Rather like it is possible for parents to take an even greater prudential interest in their children than they take in their own person.

Outriders

As with the individual, so perhaps too with the civilization: its most desirable trajectory need not be one that leads most swiftly and surely to the best state. The ideal trajectory might instead follow a more scenic route, one that is slower and more meandering. It would avoid excessive accelerations that risk injuring the occupants or smashing the traditions that connect us to our shared past. Plausibly, the best trajectory for us—individually and collectively—is one where we are not mere passengers, but in which we take turns behind the wheel, or in reading the map, or, at the very least, in discussing and selecting which places to go—even if this results in us having to back up from a few blind alleys; indeed, maybe even at the cost of us eventually ending up at a slightly worse destination than if we had let autopilot manage the entire expedition.

Slightly worse, maybe. *Not* horrendously worse. Humanity is obviously in need of adult supervision! I'm suggesting that we should perhaps be left to our own devices in cases where the downside is capped to the analog of a scraped knee or a wasted twenty-dollar bill. In cases however where a stumble means death or ruination, we should be grateful to anybody or anything that steps in to catch our fall.

(For the avoidance of misunderstanding: when I compare humanity to a group of children, I mean *all* of us. If anything, the relatively more mature and capable among us are the ones most likely to get us all into serious trouble. The same is true, to a still greater degree, of our formal and informal institutions.)

*

I have little to say here regarding the how-tos, as this lecture series is not focused on practicalities.

Perhaps one way to proceed would be to send out an advance guard of AI outriders that (only) secure the path for our own subsequent slower advance?

To carry out this mission, the outriders would need to gather some information about the possibilities ahead, to spot potential dangers. But they could confine their reconnaissance to what is necessary to accomplish the task. And what they do discover, they may use to secure a path, but they could refrain from telling us. That way, they would avoid spoiling the fun we could have in finding things out for ourselves later at our own less hurried pace.

*

The avoidance of catastrophes and the abolition of miseries. Life extension, healing, and the dispensation of second chances. Protective coatings, or "timesuits", to reduce the identity-eroding effects of the passage of time—we've discussed memory enhancements, strengthening of diachronic solidarity (prospective and retrospective), and improvement of our capacity for autonomous decision-making. Add some preliminary hedonic brightening and some deepening and intensification of our emotional responsivity, to help us appreciate the early steps of our journey.

These would be among the interventions prioritized in the beginning.

Later come the more profound transformations, which would involve the jettisoning of much more of our mortal ballast, and which would let us soar into the thoroughly posthuman realms.

A time to keep, and a time to cast away.[172]

Professor interruptus

Before I end today, I want to return to the stance of parochialism. We've already touched on this issue, but we didn't quite take things to their ultimate limits—the possibility of narrowing the scope of the interestingness value not only to an individual life but also to a single moment in that life. From this perspective, the optimal configuration for interestingness, contained and contributed, would be to instantiate a maximally interesting moment, as widely and frequently as possible throughout the cosmos.

What's that banging on the door? It appears there are people outside waiting to get in. Can someone check on that?

I thought Gastropods of Dagestan were going to be away on a field trip.

Student: Professor, it's a Political Science class. They say they've booked the lecture theater from half past.

Bostrom: Oj. If it's not one kind of slime trail it's another. Alright, we'll have to end it here. You have a homework question on the handout. See you tomorrow!

Assignments and assignations

Kelvin: I need to get ready for the funeral. See you later.

Firafix: See you.

Tessius: What is the assignment?

Firafix [*reads*]:

HANDOUT 16
HOMEWORK QUESTION

Consider the following three items:

1. A rattle
2. A rubik's cube
3. A radio telescope

Which is most interesting? Discuss.

Firafix: I can think about that while I'm getting my nails done.

Tessius: I can think about that while I'm on my date.

Firafix: Oh, you're going on a date?

Tessius: Yes, but as pertains to the cerebral forms of titillation, my expectations are of a humble order.

Firafix: Well, however that may be, good luck!

Tessius: You are not insinuating that I'm in need of *luck*, are you?

Firafix: Oh, no, well, no, I mean, good luck with the homework question!

Tessius: Nice save. See you tomorrow!

Firafix: See you!

Feodor the Fox

Epistle XXV

Dear Uncle Pasternaught,

I'll take this opportunity, just before tucking in, to write a quick thankyou for your hospitality and to say how great it was to see you!

I've made good progress and should be back at Pignolius's place within two days.

Although it was not the season for strawberries, it was nice to be back home and to spend some time with you. Hopefully next time we can try that hill!

Your indebted nephew,
Feodor

Epistle XXVI

Dear Uncle Pasternaught—

A terrible tragedy has struck. Oh Pasternaught—I will write down what happened. Maybe forcing myself to do this will help me collect my thoughts.

I was hurrying my steps, eager to get back to work on our great project.

But as I approach the residence, I get a feeling that something isn't right. An eerie absence of some of the normal background sounds? A smell in the air that isn't supposed to be there?

I come to our mushroom beds—and they are trampled up and wrecked! Now my heart is pumping and I feel a growing sense of alarm.

That smell in their air—it is pungent. I follow it upwind, past the clearing, around a thicket of bushes. And there he is, or pieces of his face and body, Pignolius! The ground all around is sticky; flies are buzzing.

I stare at the carcass. It is about 85% gone. Then a wave of panic hits me and I dash back to the den.

Inside I find Rey—oh what a beam of light in this darkness—and he tells me what has happened. A pack of wolves attacked the previous night. They attacked Pignolius. He tried to escape, but they caught up with him and killed him. Rey managed to get away and has been hiding here in the innermost cabin of the burrow.

We discuss the situation. The pack will probably return to finish off what remains of Pignolius. We decide that the site must be abandoned. It is no longer safe, now that the wolves know its location. We will depart at daybreak.

I still see the ghastly scene in my mind. I tell myself now is a time to be practical, only practical. We must wait here until tomorrow morning and then leave. That's what we must do, and I'll give this note to the pigeon when he stops by.

Feodor, your indebted nephew

Epistle XXVII

Dear Uncle Pasternaught,

We made it out, to a temporary encampment about a league away. Rey has been scouting the surroundings and has found a location for a new burrow. It is a place that is hard to find, barricaded on all sides by rocks or thorny bushes, so it should provide a measure of security, and there are good resources nearby. We will move there tomorrow.

My heart is heavy, like I'm carrying around a boulder inside.

Rey seems to be holding up quite well. I can't even think about what would have happened if he hadn't managed to escape.

The new location is far enough away that it should be relatively safe, yet close enough that we could easily make visits to the old place to retrieve the mushroom samples and other supplies.

Yes, this means that we are still intending to carry on the project. I was not sure. I feel responsible for having gotten Pignolius involved in this mad undertaking, and also for embroiling Rey and exposing him to risk. I have talked this through with Rey. He says that Pignolius would have wanted us to carry on; also that the cost has already been paid and that quitting now would not bring Pignolius back. It is logical.

But what about Rey himself? He says he can make up his own mind, and that he definitely wants to continue.

Oh Pasternaught, should I send him home? I believe I should. I'm just so weak at the moment, I can't bring myself to do it! I think we should be relatively safe here for the time being. Pasternaught, tell me what you think—can it be acceptable for me to allow him to stay for some limited period of time, and then maybe I can send him home as soon as things have settled down? I know what I'm hoping you will say, but you should be honest and tell me your real opinion. I will obey.

Your indebted nephew,
Feodor

Epistle XXVIII

Dear Uncle Pasternaught,

I feel I need to make a confession to get something off my chest, so I'm up writing this in the middle of the night.

We've moved into the new place. On our way there, just down by the creek which one must ford to get into the ravine where the burrow is located, we came across a doe. She wasn't moving. We passed her by and I didn't give it much thought.

But when we returned later in the afternoon, to get some supplies from the old site, she was still standing there, and I saw that she seemed to have an injured leg.

We continued on, and when we were returning home, with our jaws loaded with foodstuff and other items, she was still there. I suggested to Rey that we share some of the food with the doe, but he shook his head and maintained a brisk pace.

When we got home I brought the topic up again. Rey was adamant that we should not share. We have limited supplies and we will need all our surplus to maximize our chances of success.

We went to bed, but I was not able to fall asleep. I kept thinking of the doe. She was probably still standing there, hungry, and it probably hurt if she tried to walk on her leg. After maybe two hours of this, I got up, grabbed an apple from our storage room, and snuck out of the burrow. The doe was still in the same place. She seemed shy when I approached. I put the apple in front of her, and as soon as I turned around, I heard the crunchy sound of eating. A joyful sound! A flash of pleasure ran along my spine. I hurried home.

Now having written this and made my confession, I feel a little easier at heart, and I think I shall be able to get some sleep. Tomorrow I will have to tell Rey.

I know it will not be good; but each day has enough of its own sorrows, so I will not worry about that now.

Goodnight. Your indebted nephew,
F

Epistle XXIX

Dear Uncle Pasternaught,

I told Rey, and he was rather indignant. He explained to me—clearly it was trying his patience—that random acts of charity would solve nothing and that our project, as unlikely as it was, was the one singular hope that existed for this entire forest. The responsibility had fallen onto us. It was therefore not virtue but weakness to dissipate our advantage on unstrategic generosity.

I said that I would fast for the day to make up for it. He said don't compound one stupidity with another: that I was already skinny, and that I needed calories in order to contribute efficiently.

Well yes, of course I understood all this already. But now my pride was worked up. I argued that in addition to laws of nature there are laws of morality, and they must both be obeyed; they are equally necessary. I hadn't really thought about this before, but words now came out of my mouth to the effect that I will keep feeding the doe until it gets well. We have a fierce verbal altercation. We each go off in a huff.

I'm feeling righteous, but also worried about what this might mean for the project. I hadn't planned to double down on saving the doe in this manner.

After some ten minutes, he comes up to me. I try to read his expression but can't, though I get the sense that he's determined. He states that he "will sort this out" and turns around and leaves in the direction of the creek.

What is he going to do? Is he going to do anything to the doe? I decide not to follow.

A while later he returns. At first he doesn't say anything, and I don't ask, but my face must have been a question mark. Finally he says, "Okay, we will feed the doe. She has pledged that after she gets well she will spend the rest of her life helping us with our project."

Later I learn that he had given her a stark choice: starve to death by the creek or agree to his terms.

Technically he has told her a lie, or at least a falsehood; for in the heat of our argument, I had made it quite clear to him that I would keep feeding the doe no matter what.

But I must say that I'm relieved that we are over the impasse. How close we came to an utterly ignominious end! And it would have been all the fault of my stupidity and stubbornness. I shudder at the thought. I am grateful for having escaped being the cause of our downfall and thank my lucky star.

I have a renewed admiration for Rey's skills. I quietly resolve to let him take the leadership role in this project. I have precious little practical sense but I can just about mobilize enough awareness to realize that, at least when it comes to affairs of the world, he is the superior talent.

Your indebted nephew,
Feodor

Epistle XXX

Dear Uncle Pasternaught,

No sooner have I resolved to become a follower of Rey's lead than I must question the wisdom of my decision—

The mating season coming up, Rey informed me of his determination to "hook me up with a nice vixen" this year. I cannot see how the timing could be worse, the sorrow of Pignolius's killing still hanging over us and the tasks of our great enterprise demanding our time and attention.

He has also made it clear that he himself intends to go full monty and avail himself of the full extent of his charms to do "everything which is necessary, appropriate, or possible" with regard to the female sex this season.

Ought we not to at least postpone any thought of procreation until next year, when perhaps our project will be further along and less in need of continual pushing? We will have to talk about this further.

Clara (the deer) is able to take some steps on her leg and I'm hopeful she will make a full recovery.

Your indebted nephew,
Feodor

Epistle XXXI

Dear Uncle Pasternaught,

We've had a further conversation about the upcoming mating season, and I'm sad to tell you it has opened another rift between Rey and me.

I had thought that Rey's attitude was a kind of joke, or at worst a concession to base biological necessity; but no, it is allegedly founded on philosophy.

If he had simply told me that it is something he needs to do, for reasons he cannot control, I would not have had any problem with it. Indeed I would have been glad to offer him my help, had he needed any assistance in that department, which he obviously doesn't.

But instead of this simple and natural justification, he presents me with something quite different. He tells me that his planned actions are in service of the project! That since the goal is to overcome the evils of the world by breeding for cooperativeness, and since at present he and I are its main contributors, we should begin by breeding ourselves.

This doesn't sit right with me!

Another thing: Clara is now gathering food for the three of us. She is not permitted to go outside a certain area, which Rey has staked out, on pain of dire consequences.

I haven't heard her complain, and the area contains everything she needs. But she is a slave, there is no way around it.

I brought this up with Rey. He said she wouldn't be alive if he hadn't saved her. (*He*?!)

I have been trying hard not to fly off my handle.

I think the problem is that I've not been able to articulate clearly why I think what I think. If only Pignolius were still alive, maybe he would have been able to explain things in a way that makes sense. Now all I can say is just that I feel in my heart that the path we seem to be heading down is not the right one.

I'm ashamed that I have allowed a rift once more to open up between us, when *the one thing* that is obvious is that the two of us need to work together if we are to have any hope whatsoever.

Why oh why, Pasternaught, is it so hard for us to get along!

Your indebted nephew,
Feodor

Epistle XXXII

Dear Uncle Pasternaught,

You are full of wisdom and common sense. Thank you for your letter to both of us. We read it together, and we've agreed to the compromise you suggested. Rey will reproduce as he pleases and I will hold off until next year.

I had not reckoned on your desire to have grandkits from my side come and visit you in the future; but of course this is something to be done! I promise that if it is even remotely possible to accomplish this next year, I will not disappoint you. (This is something to look forward to for me, too.)

We told Clara that by end of next fall she will have worked off her indenture and will be free to leave.

Let us hope and pray that we can hang together long enough for the mushrooms to become less toxic. We will be the first to try them out, as soon as our stomachs are able to tolerate them. If our theory is vindicated, it should from that point onward become easier and easier to keep the peace.

The hope keeps me going; without it, I really don't understand how one could continue to bear it all. Evidently it is possible, since many people do, and without so much as a complaint. But if one looks? and sees? That might be the mistake I'm making—the mistake *I am*. But if I could be the instrument of something worthy somehow… If I could render myself into a vessel for light and love? I'm conscious that as I write these words I am making myself even more absurd, and I can only beg your indulgence. Unfathomable is the mercy we need.

Your indebted nephew,
Feodor

P.S. I'm reminded of something Pignolius used to say: "In the end it's always the same two options: be sad it isn't better, or happy it isn't worse."

Epistle XXXIII

Dear Uncle Pasternaught,

I woke early today and decided to take the morning off. I climbed up to the top of the cliff, and watched the sun rise.

I thought of the seasons changing; and the never-ending events, twists, pains, and complications that make up our existence, its alternating hopes and disappointments—its ever-renewing combinations of patterns: leaves budding, growing, falling off; rain falling and drying up; the winds blowing and blowing and blowing.

Insects hatching in countless numbers, and soon dying in exactly the same numbers. Not one escaping, not one missing.

Everyone has his fate.

This world, this life: it is strangely beautiful. If only it were much less terrible.

With love and prayers,
Feodor

FRIDAY

Postmortem

Kelvin: Hi Tessius.

Tessius: Hi guys.

Firafix: How was the date?

Tessius: It went according to expectations.

Firafix: So… you have a good answer to the homework question?

Tessius: It would blow your socks off.

Firafix: I don't usually wear socks.

Tessius: I'm pretty sure it would, though. Something at least was blown off when I came up with it.

Firafix: What do you mean—

Kelvin: I think the lecture is about to begin.

Pure pleasure

Welcome everybody. There are a couple of seats free over there, if some of you want to try to squeeze in.

I ran into an old friend the other day—he also sat in on yesterday's lecture—and he somewhat upbraided me afterwards for not placing sufficient emphasis on the hedonic dimension of value. I want to start today by remedying this.

First, a few words on terminology. The word "pleasure", unfortunately, carries certain connotations that can detract from the core idea. The word is sometimes used to refer quite narrowly to pleasant or voluptuous bodily sensations. Alternatively, it is sometimes used to refer to nerve-frazzling thrills or high-energy socializing, or even to suggest a lifestyle of indulgence and consumerism.

The problem is that such "pleasures" may be, but may also not be, genuinely enjoyable. There may be people who have at some point in their lives wholeheartedly devoted themselves to pleasure in one of these senses; and in many cases, they would have discovered that this in fact made them quite miserable. They have been peddled counterfeits—pseudo-pleasures with bitter tastes and hollow interiors. Having experienced disappointment, they may then very understandably come to the conclusion that pleasure is no good and that to propose it as a central component of utopian life would be wrongheaded. It might have a place, they will concede, but a limited one, like party balloons on a birthday or fireworks—not something one would wish to pervade the everyday as a general background condition.

But there is a different way of using the word "pleasure", which I take to be more fundamental. What I have in mind is the subjective quality of positive hedonic tone—roughly, the experience of an unmediated liking of the way things present in the moment. Pleasure, in this sense—the authentic article—genuinely feels good and fills our spirit with a warm affirming joy; and, whatever we may tell ourselves, some core part of us cannot help but *really* like it.

The circumstances most conducive to this experiential quality naturally vary considerably between individuals. In our contemporary lives, some might indeed find little crumbs of it in the party circuit; others, while pretzeling themselves on a yoga mat; others, while reading a good book with rain pattering gently against the window; others, upon discovering a kind note in their backpack from their crush. Some people experience pleasure under astringent conditions, such as when straining up a rock face or when enduring hardship for the sake of an ideal that they cherish. Some like sweet, some like sour. The notion to which I am referring is neutral about the cause and also as regards the accompanying thoughts and sensations.

Now, I assert that hedonics alone can go a *long way*. (And maybe it can go even further if we combine it with other valued subjective experiential qualities.) Pleasure is plausibly the most important thing about utopia. It would not be crazy to think it is the *only* thing that matters.

In these lectures, we spend an inordinate amount of time talking about other values; but this is only because they pose more complex theoretical challenges. In terms of their importance, they might perhaps best be seen as possible bonuses, as cherries and marzipan roses on the cake rather than its main substance.

When I speak of hedonics, I am also—but I trust this goes without saying—concerned about pleasure's opposite. I am saying little about suffering here, because the theme of these lectures is utopia not dystopia. But for the avoidance of any doubt or misinterpretation, let me again stress that when we make all-things-considered decisions about how to proceed into the future, the mitigation of suffering, especially extreme suffering, ought to be a criterion of the greatest and possibly paramount importance.[173]

On fools and paradises

People talk down the idea of living in a fool's paradise. But when one considers the nature of humanity, might it not seem that such a destination would be very suitable and desirable for us? I mean: If we are fools, then a fool's paradise would be exactly what we need.

Of course, being fools, we are not likely to recognize what is good for us. We might be more likely to try to clamber into an eagle's nest or onto some kind of icy pinnacle of truth and glory, because it is "higher". And then we would sit there freezing our butts off to the end of our days. That would be the sort of thing one would expect a fool to do when he could have been living happily in a fool's paradise.

Radically exotic beings

I am not actually asserting that hedonics is the only thing that matters; only that a future with great pleasure and nothing else going for it could already be very good. And we get an even stronger case that a future could be good if we add, to the positing of pleasure, the stipulation that it also has certain other properties that supervene on subjective experience. Yet still I'm not asserting that such an experientialist utopia would be the best one possible, though it could be very good indeed. If there are non-experiential properties that would add even more to the desirability of a future, we may as well seek to bring about those too. I think we should.

The relative importance of hedonics, non-hedonic experiential goods, and non-experiential goods may be different for different kinds of being. For example, I think that some nonconscious beings could have morally relevant interests—which might be based, for instance, on their preferences. Unless they particularly happen to have preferences that refer to experiential goods, then if *they* were discussing what utopia would be best for them, they might omit hedonic and experiential goods altogether from the specs. In so judging, it's not clear they would be making any mistake.

The question of what is good for somebody is not only often empirically difficult to ascertain: metaphysically, too, it can be unclear what the correct answer is. And for *some* types of being, it can also be unclear whether—or to what extent—the question of what is in their interest even makes sense. (This issue is explored in today's assigned reading.)

If there are such other beings, who have interests very different from ours—perhaps a fellow with a monomaniacal desire for paperclips—we ought, I believe, to try, as far as possible, to accommodate them (assuming we have a say in the matter). I have made some remarks elsewhere on how we might understand the normative questions involved in such contexts of radical difference.[174] Of course, moral and political questions of how to combine different wills and interests are not confined to far-flung speculative contexts; they arise all the time down here in our old terrestrial biosphere, too. Heck, they arise even within families—even within ourselves, sometimes.

However, we have enough ground to cover in this lecture series without venturing too far afield in these directions. We will mostly concentrate on the question of what the optimal continuations are, from our current starting point, for creatures like us.

Extreme parochialism

Yesterday we were discussing one putative value or value-ingredient: interestingness. We were investigating it from many angles, probing its nature and tracing out its entailments... until the intrusion of Politics, which, with a heavy clap on the door and an official claim to the venue, put a prompt end to our inquiry and forced our gathering to disperse.

Before we proceed to other values, I want to briefly pick up on a few points we didn't get a chance to cover yesterday.

*

You will recall that one of the issues we discussed was that if interestingness requires novelty, then we may be totally out of luck with regard to the value of interestingness, inasmuch as we are unable to achieve true novelty in our lives when considering things at a large-enough scale. This would be true even for figures such as Napoleon or Aristotle. If we pick a more average doofus, it would also be true on scales far smaller than that of the cosmos as a whole.

As a possible response, we considered parochialism: a stance of narrowing the scope of the domain within which interestingness is assessed down to that of a civilization, a community, a group, or even an individual, as required to inscribe our lives with sufficient local novelty to make them qualify as objectively interesting. The tales of our deeds and experiences at such modest scales may not reverberate down the halls of Valhalla, and they may scarcely even be talked about in our pubs and coffeehouses, but perhaps they get a mention around the kitchen table; or, failing that, at least that dutiful monologist inside our own heads can be counted on to cover the proceedings—with its never-waning pathos and headlining of every bagatelle (breaking news: that jerk just stole my parking space; breaking news: what is that click in my knee; breaking news: fly at large in the vestibule).

Yet even such parochialism does not reveal a clear path to high levels of interestingness in utopia, for in addition to the spatial scope of the domain within which novelty is assessed, we must also consider the temporal scope. And especially in light of the prospect of radical life extension in utopia—which of course is desirable for other reasons—it may prove challenging to sustain a high level of interestingness throughout the course of our individual lives. This at least appears to be the case if we measure interestingness in terms of something like "cognitive upheavals" (though it is *not* true—we should remind ourselves—if we measure interestingness in terms of "kaleidoscopically churning complexity").

Well, okay, but confining the scope to an individual life is not actually the most extreme version of parochialism that is possible. We *could* take an even more parochial view of things, and disregard not only the elsewhere and the others, but also the past and the future: thus confining the purview of interestingness to *an individual moment of an individual life*.

On this view, all that is needed to satisfy the desideratum of objective interestingness is that each moment of life contains within itself, considered in isolation, a superlatively interesting occurrence. We would judge each moment solely on the basis of its intrinsic qualities.

*

If we were to take this approach, then a natural next step would be to go looking for the most interesting moment we can think of.

For example, let us say that we could record the exact processes taking place in Einstein's brain at the very moment when he first begins to grok the contours of general relativity. This would be difficult to do in the case of a biological brain, but if Einstein had been a deterministic upload, or a simulation, it would be straightforward for anybody with root access to store a snapshot of his brain state just as this episode of momentous discovery begins. Having captured all the relevant computational parameters, we would then be able to repeat this episode—to replay it over and over again, the same Einsteinian brain processes unfolding exactly the same way in each repetition, and generating, we may assume, the same subjective experience, the same cognitive and phenomenal flash of insight, like a bright stroboscope.

We could put this recording on perpetual repeat. In the view of extreme parochialism, a life composed entirely of this aha moment, replayed over and over, would be outstandingly excellent, at least insofar as the value of interestingness goes.

For once we have attained perfection, why seek change? Any other state we could reach would be no better, and might be worse.

*

If we are seeking an interestingness-optimal state, can we find one that scores even higher than Einsteinian eureka?

Perhaps this question calls for research methods other than those that are sanctioned by academia. There is a team down by the river, they usually hang out under the bridge, that appears to be conducting relevant investigations, though I think they have yet to publish their findings.

It is not even clear what it was like for Einstein as he developed his insights. Perhaps the excitement of being on the path to an important discovery, and the pleasurable feeling of an intellect feeling itself strong and capable, and the sense of clarity, and the delight in curiosity-driven exploration: maybe co-mingled with these there was also a sense of strain, mental fatigue, and dissatisfaction with the remaining confusions? Who knows, even sensations of thirst and hunger, or some bodily ache, might have intruded on Einstein's contemplation while he was making his breakthrough. At a minimum, we would want to expurgate any such discomforts and incumbrances before making his experience the template for an indefinitely repeating pattern, or existential arabesque.

We must also suspect that our intuitions about the value of Einstein's experience are influenced by our appreciation of its external significance. We know that Einstein was the successful first discoverer of a deep truth about the physical universe. We also know that his theory went on to receive worldwide acclaim, and that many of the brightest minds in subsequent generations have put much effort into understanding his results. And we know that the episode of discovery formed part of a larger whole, the life of a man who had many admirable qualities. All these extraneous circumstances may conspire to cast an enchanting spell on the episode itself, making it appear higher in interestingness value than it really is when judged solely on its intrinsic qualities—which is how we must conceive of it if we take seriously the narrowly defined scope prescribed by extreme parochialism.

*

We may speculate that without strictures against repetition and in the absence of demands of external significance, then maybe the state of mind that would maximize interestingness—or, better, a combination of the value of interestingness and the even more important value of pleasure—would be some form of ecstasy.

What kind of ecstasy? If you could only have one experience forever, what would it be?

One candidate is the state described in Aldous Huxley's last novel, *Island.* In this novel, Huxley sought to present an alternative to the rather unattractive dilemma he had posed in his earlier book *Brave New World*—a third way that avoids, on one side, the raw, brutish, suffering-filled state of nature and, on the other side, the despiritualized society of Fordist consumerism with its shallow and soma-powered mass contentment.

The approach he takes in *Island* is to seek a fusion of the best of Western science and of Eastern Mahāyāna Buddhism. The inhabitants of his utopian community have opted for a selective form of modernization. They cultivate an enlightened, pacifistic, humanistic way of life that is aimed at facilitating the pursuit of humanity's final end, which Huxley (elsewhere) describes as "the unitive knowledge of immanent Tao or Logos, the transcendent Godhead or Brahman".[175]

The islanders' pursuit of spiritual awakening is strongly aided by the consumption of "moksha-medicine": a psychedelic entheogen prepared from yellow mushrooms. This substance has the capacity, when used under the right guidance, to elevate the user into a state of mind which seems at least momentarily to achieve the kind of enlightened awareness that the islanders view as the highest good:

> "'Luminous bliss.' From the shallows of his mind the words rose like bubbles, came to the surface, and vanished into the infinite spaces of living light that now pulsed and breathed behind his closed eyelids. 'Luminous bliss.' That was as near as one could come to it. But *it*—this timeless and yet ever-changing Event—was something that words could only caricature and diminish, never convey. *It* was not only bliss, *it* was also understanding. Understanding of everything, but without knowledge of anything. Knowledge involved a knower and all the infinite diversity of known and knowable things. But here, behind his closed lids, there was neither spectacle nor spectator. There was only this experienced fact of being blissfully one with Oneness."[176]

So, if each moment of experience had to be evaluated entirely based on its intrinsic qualities—or if we had to pick a single unchanging mental state to be in for our entire life—this kind of "luminous bliss" would be one candidate.

*

A technologically mature civilization would not be reliant upon harvests of mushrooms or toadstools, of course. Nor would it need to confine its search for enlightenment to rifling through the large but still quite limited set of configurations that some unreformed old encephalon is capable of being coaxed into. Even if we accept the additional constraint that personal identity must be preserved—though isn't it a little unclear why we would do that, given that many of the states of enlightenment which we are encouraged to seek, are said to involve, as a central characteristic, the understanding that the self is an illusion?—there would still be plenty of room to restructure our minds so as to optimize them more thoroughly for spiritual attainment, or for experiencing high and continuous levels of luminous bliss: rather than, as is currently still the case, for the task of gathering tubers and whatever else we needed to do in our environment of evolutionary adaptedness in order to survive and reproduce.

Such brain enhancements and brain modifications would let us access a large space of new experiences, and let us sustain selected experiences for longer without diminution of intensity or focus. We must imagine that, in this space, there are at least some experiences that would strike us as either far stronger and purer versions of the luminous bliss accessible to some people today, or, alternatively, as manifesting novel phenomenal qualities that are even more desirable than luminous bliss.

I say we "must imagine" this being the case—but that is not quite right. It is impossible for us to *imagine* many of these possible mental states, at least in any vivid, concrete, intuitive manner; since, by definition, we currently lack the neurological wherewithal for experiencing them. What I should say, rather, is that we should assign high credence to the hypothesis (which we can only abstractly comprehend) that the space of possible-for-us experiences extends far beyond those that are accessible to us with our present unoptimized brains. *Very* far beyond, I think. And out there, in that vast unexplored space of possible experiences, there may be some that are worthwhile to a degree that exceeds our wildest dreams and fancies. I think that's likely.

And if we are willing to venture further, into the space of the outright posthuman—though this might require sacrificing some measure of personal identity between the us-now and the successor versions of us-later—then the storms of ecstasy that become available may well be such that, if the phenomena can in any way be placed on a common scale of measurement, the luminous bliss that Huxley's islanders are occasionally fortunate enough to experience is like a mere weasel's fart by comparison.

*

I will mention one more candidate for the best state of mind to be in. Here I will be brief, because although this notion seems relevant, I'm wary of straying from my area of expertise and trespassing onto that of Professor Grossweiter. I refer you to his *Grand Theological Colloquium* for a deeper examination of this particular subject matter. In general, you may view the present lecture series as a humble footnote to some of the material he covers.

The notion to which I'm referring is that of the *beatific vision*. Thomas Aquinas says that this is a perfect happiness, the obtainment of which is the final end for a human being.[177]

In the beatific vision, one has direct knowledge of God. This is a kind of unmediated "seeing" of the divine essence. St. Paul says: "Now we see through a glass, darkly; but then face to face".[178]

When thus joined directly to God, the created intelligence finds the highest bliss. And since God's perfect goodness comes directly into view, the act of "seeing" is at the same time an act of loving.

A visit to the navigator's cabin

This may be a good place to pause for a moment and take our bearings.

Let us unfurl the charts…

On Monday, we set sail from the simplest conception of utopia ("walls of sausages"). We spent the day, and also all of Tuesday, zipping along near the coast, taking in a host of social, economic, and psychological sights. Finally, we surveyed the cape of ultimate limits to automation.

We then turned away from shore, heading out into deeper waters. By the end of Wednesday, we had arrived upon the sea of deep redundancy. Its placid expanse stretched around us, soothed and comforted under skies that seemed overcast in unchanging gray.

Here we confronted the redundancy concern, a reformulation and generalization of the purpose problem that had originally helped to instigate our expedition. The redundancy concern, remember, is that, at technological maturity (and thus in a highly plastic world), human effort becomes redundant—threatening to undermine the very foundations of many values, such as interestingness, fulfillment, richness, purpose, and meaning.

The redundancy concern carries a sense of foreboding: that we may find ourselves in the paradoxical situation of having reason to try to master the world while also having reason to hope that our attempts to do so will fail. We would be cast into the role of the seducer who desires only the chaste: who might find satisfaction in the pursuit, but never in the possession. Except that whereas the seducer can always proceed to his next target, we have only one world to conquer—after which we would be reduced, as it were, to a perpetuity of lawn-mowing and taking out the recycling.

Fortunately, the outcome did not seem so bad. Even if we got permanently stuck upon the sea of deep redundancy, its windless peace offering no propulsion—well, it's not as if there is anywhere else we needed to go. We could just drift along with the currents. We have supplies to last us a very long lifetime, and we would be free to engage in whatever make-do and merrymaking activities that we wish on the deck.

In fact, we'd have the makings to turn the vessel of struggle and suffering into the ultimate party boat. We could bring out from our cargo hold: hedonic valence, experience texture, autotelic activity, artificial purpose, and sociocultural entanglement.

Not only does our time onboard not need to be bad: it can be *unimaginably better than* the lives we bore back on shore.

Still, on Thursday, we set out to make a closer study of the redundancy concern and the values it places in jeopardy. Perhaps not all of these values need to be completely and irretrievably lost to us even in a perfectly plastic utopia?

We began with the notion of interestingness. While it would be trivial to banish subjective boredom from utopia, it was a far more subtle matter to determine the extent to which we could possibly realize objective interestingness. Basically, what we concluded was that (a) yes there are major limitations to what we could hope for in this regard, but (b) it's not as if our present lives are in every sense knocking it out of the ballpark in terms of objective interestingness, and (c) we could in utopia achieve a very high level of a form of kaleidoscopic interestingness. We reached many other conclusions too, which I will not recap here.

This brings us more or less up to the present. We are about to continue our investigation, now turning our focus to the other values that are imperiled in a plastic world. Our previous work on interestingness will stand us in good stead when we start on this, because all of these "higher" or more "ethereal" value concepts are to various degrees overlapping; and the clarifications and analytical apparatus which we developed in our study of interestingness will be helpful as we seek to understand these other notions.

If we want to keep up the nautical metaphor, we may express things as follows:

Out here on the open ocean of full redundancy, with no landmarks in sight—no practical necessities by which to steer—our only recourse, if we are to maintain our bearings over the longue durée, is celestial navigation. The higher or more ethereal values to which I have referred, such as interestingness, fulfillment, richness, purpose, and meaning: these are like guiding stars and constellations. The light we receive from them may be relatively faint (compared to that from the smiling sun of hedonic gratification), but it is important for maintaining our long-range normative course of travel.

The overcast sky makes it harder to see these astral objects. But if we pay close attention, we can sometimes peer through seams between the clouds at night; and then—if we temporarily dim the gaudy party lights and let our pupils dilate, we can still catch glimpses of the old empyreal canopy, and study how we might relate ourselves, and our direction of travel, to it.

Oh, and about the absence of wind: well, nothing hinders us from taking to the oars. Where there is a dearth of external forcing, our own volition is called upon to take a more active role. Generating some ergs on our own initiative wouldn't do us any harm!

Some remarks on metaphilosophy

So what potentially normatively relevant patterns can we see in the firmamental vaults? We looked at the interestingness value yesterday. What about other value constellations—such as meaning? People often ask about the meaning of life, or meaning in life. What would be the meaning of, or the meaning in, a utopian life?

And before we get to meaning, we have a few other canopy values to cover: fulfillment, richness, and purpose. (These, along with meaning and interestingness, help to mutually delineate each other.)

When we analyze notions like these, we must bear in mind that our ordinary concepts are not all that sharply defined. There is considerable arbitrariness in which dots we ascribe to which constellation—for instance, where we draw the boundary between fulfillment and purpose, or between purpose and meaning. And no doubt there are different words and ideas that we could use to conceptually organize the same firmament of value, which would group the stars into different motifs. While some such conceptual schemes would be more elegant and natural than others (or would better match ordinary linguistic usage), there could well exist multiple alternative systems that would each serve approximately equally well as navigational guides.

It might be that the more each alternative system is refined and perfected in its own terms, the more closely their practical upshots would converge. We may compare this to how different lossy compression algorithms produce different artifacts (patterns of distortion) at high compression rates, yet asymptote toward exactly reproducing the same original file as constraints on memory, bandwidth, and computational capacity are relaxed.

So I don't want to assert that the way I'm organizing things here is the only acceptable way of going about things. To the extent that a claim is being advanced in what follows, it is expressed principally in the sum rather than in each of its terms separately. I want to sketch a picture of what I see, and to do so I must draw some lines; but I'm not saying that each line individually corresponds to an objective truth, or is superior to all of the alternative lines that one might have drawn to create an equally good picture of the same scene. It is mostly the gestalt they convey when taken together that is significant.

To widely varying degrees, such holism may pertain in all areas of human inquiry. This was the position taken, for example, by W. V. O. Quine—though in my view he very greatly overstated the case (particularly with respect to our attributions of linguistic meaning). However, I think that the degree to which holism obtains is especially high in philosophy and ethics. This is one reason, I suspect, why it has proven difficult to achieve the kind of cumulative consensus progress, taking the form of a growing body of generally accepted truths, that we see in many other fields of inquiry where a divide-and-conquer strategy has been more fully applicable. Even within philosophy there is variability, with different subfields possessing lesser or greater degrees of holism—the topics we're about to discuss, concerning the meaning of life and such, being among the most holistic.

Fulfillment

These disclaimers out of the way, let us turn to fulfillment. This is a notion with a long history, stretching back at least to Aristotle. Many thinkers have sought to locate the good for a human being in fulfillment of one kind or another—fulfillment of our capacities, of the highest of our capacities, or of our aspirations, or of our true individuality.

*

Usually the conclusion has been that the best and most fulfilling way to live a human life, and the most praiseworthy one, is in fact to be a philosopher. This occupation involves the highest and fullest exercise of reason, which is itself, we are told, the loftiest and most distinctively human of all our capacities.

Far be it from me to dispute such profound wisdom! I would venture but to add a simple corollary, one which should be obvious and uncontroversial, yet which has received insufficient recognition: namely, the imperative of bestowing upon the practitioners of a profession of such unparalleled intrinsic worth all commensurate worldly compensations: high salaries, low teaching loads, long sabbaticals, short tenure tracks, and also awe-inspiring titles and honorifics, and free MacBooks. Not of course because philosophers *would care* about such trifles—a notion too preposterous to merit refutation—but because the enjoyment of these perquisites is (one must sadly condescend to acknowledge) necessary to make the surpassing value of our work legible to the wider community, and hence for enabling our university colleagues and society at large to partake slightly in the reflected glory of our exalted enterprise.

*

I will not attempt to survey the extensive literature on fulfillment or attempt to enumerate all the various roles that the concept has been made to play. The particular idea that I want to focus on here may be adequately evoked by quoting some passages from one comparatively recent contribution (by the political and legal philosopher Joel Feinberg):

> "[F]ulfillment so interpreted is often said to be a 'realizing of one's potential', where the word 'potential' refers not only to one's basic natural proclivities to engage in activities of certain kinds, but also to one's natural capacities to acquire skills and talents, to exercise those abilities effectively, and thus to produce achievements."[179]

> "[T]he most fulfilling [life paths] are those that best *fit* one's latent talents, interests, and initial bent and with one's evolving self-ideal (as opposed to one's conscious desires or formulated ambitions)."[180]

> "[Human lives] approach fulfillment insofar as they fill their natural allotment of years with vigorous activity. They need not be 'successful', or 'triumphant', or even contented on balance in order to be fulfilled, provided they are long lifetimes full of struggles and strivings, achievements and noble failures, contentments and frustrations, friendships and enmities, exertions and relaxations, seriousness and playfulness through all the programmed stages of growth and decay. Most important of all, a fulfilled human life will be a life of planning, designing, making order out of confusion and system out of randomness, a life of building, repairing, rebuilding, creating, pursuing goals, and solving problems."[181]

Feinberg contrasts this with the unfulfilled life of somebody who fails to realize their potential, whether because of poor health or lack of opportunity, or because of having squandered their time and talents. He says that even if such a life leaves the person with "the pleasure of his diminished consciousness, his soma pills and television programs, his comic books and crossword puzzles", there is nevertheless something unfortunate about his deepest nature remaining unfulfilled:

> "Now we think of that nature, with all its elaborate neurochemical equipment underlying its distinctive drives and talents and forming its uniquely complex character, as largely *unused, wasted, all for naught.* All wound up, it can never discharge or wind down again. In contrast, the life of fulfillment strikes us as one that comes into being prone and equipped to do its thing and then uses it up doing that thing, without waste, blockage, or friction."[182]

If we accept this intuition, that the fulfilled life is preferable to the unfulfilled life, then we have another desirable property to add to utopia: the lives of the inhabitants should, ideally, be *fulfilled.*

*

One might object that it is possible for beings to have natures whose fulfillment would be *bad.* For example, we have pests and plights and parasites, not to mention unreformed predators, whose fulfillment might seem to require injury and destruction of others. But we could simply say that in some of those cases it would be bad all-things-considered for those natures to be fulfilled, because of the pernicious consequences for others, even though it would be good for the beings themselves.

We could imagine a being who would only find fulfillment in self-harm. But if we have a pluralistic account of well-being, we could handle this type of example by saying that it could be bad on balance for such a being to be fulfilled, because even though it would gain some value from fulfillment it might lose more from the detriment to other components or contributors to its well-being (such as to its hedonic state or its health).

In practice, I hope that we could deal with many of these cases primarily by creativity rather than compromise; that is to say, that we could find alternative ways of achieving fulfillment of most natures which would not require the harming of self or others. The cat can play with a yarn ball rather than a mouse.

*

Compared to interestingness, fulfillment seems to be more closely tied to the specific capabilities and predilections of an individual—or, on some accounts, of a species.

If some creature, such as a jellyfish, has only a modest amount of "elaborate neurochemical equipment", then even a simple life might offer it complete fulfillment: allowing it to "use itself up fully", drawing on all its capacities, modest as those might be. The jellyfish's life may not contain much objective interestingness, but it could easily be maximally fulfilling (at least if we hold its nature fixed).

Conversely, a life stuffed full of objectively interesting activity could still be sorely wanting in fulfillment if there were some central capacity that went unused or some important disposition that were never given an opportunity to discharge itself. We can imagine some person who is put through some unique, complex, and consequential fate, who finds themself a central actor at the nexi of a series of world-historical events, and who leads, in such a manner, a life that is abundantly endowed with interestingness; and yet who is at the same time deeply unfulfilled—maybe because their nature had disposed them toward mathematical or monkish pursuits, which their tumultuous life never gave them an opportunity to engage.

So the two values, interestingness and fulfillment, are not coextensive. They do, however, overlap considerably; and many of the points we made about the former carry over, with suitable adjustment, to the latter. For example, much of what we said regarding the interplay between objective and subjective interestingness can be applied equally to the putative value of fulfillment.

*

One new issue that arises when we think about fulfillment is that, depending on how we construe our evaluation function, the value of fulfillment may be *satiable* in a way that the value of interestingness is not.

We can see this issue more easily if we lop off some of the complexities of the concept of fulfillment, and focus on a simpler concept: that of *fill*.

Let's say you have a bucket. You can increasingly fill it. But eventually the bucket is full. At that point, we might say the bucket has been maxed out with regard to fill.

From an axiological perspective, if we pretend that "fill" is a value, we could then ask several questions:

1. Which things, exactly, are such that it is good that they be filled? Only buckets? Or also cups, galoshes, etc.?
2. Is it good only to fill already existing buckets, or do we also have reason to make more buckets and fill them?
3. Is lack of fill bad or just less good than fill? For example, is it better for there to be just one bucket that is full than for there to be two buckets one of which is full and one of which is empty?
4. Is it more valuable for a big bucket to be full than for a small bucket to be full?
5. If we start with a bucket that is full, do we then have reason to make that bucket larger and fill it some more?

Analogous questions can be asked with regard to fulfillment, which you may ponder at your leisure.

*

Regarding the scope for fulfillment in an advanced technological condition, I want, however, to say this: it does not appear that fulfillment requires instrumental necessity. If somebody has a capacity for running, or for playing the mandolin, it seems perfectly possible for them to fulfill those capacities even in a world where there are cars and record players. This is good news, as it suggests that the value of fulfillment should be broadly realizable in a plastic utopia. (Feinberg does at one point make a passing mention of "achievement" in his characterization of fulfillment. We might wonder whether plasticity would undermine the possibility of achievement. But we'll hew off achievement from the notion of fulfillment, and discuss that instead under the rubric of purpose, which we'll get to shortly.)

*

As a lower bound, therefore, we should in utopia be able to accomplish a more complete and thorough fulfillment of the capacities we currently possess. There is a further question of whether it is possible to go far beyond this, and attain what we might term "super-fulfillment". Could we bring into our lives forms of fulfillment that are vastly more prudentially desirable to us now than the ordinary forms of fulfillment that we can hope to achieve (if we are lucky) in our present historical context?

I think that for such super-fulfillment to be possible for us would require at least one of the following two things to be true:

Either it would have to be the case that we already now possess some supremely important capacity or aspiration that cannot be fulfilled within our lives as they are currently lived but which could possibly be fulfilled in some radically different context. The example one might have in mind here is if we have some latent capacity to commune with God in a particularly intimate manner, and yet the fulfillment of which, while profoundly more desirable than other kinds of fulfillment, is unattainable in our current incarnation. Then a situation that permitted this paramount capacity to be finally used could offer the hope of super-fulfillment.

Or else—this would be the other alternative way in which super-fulfillment might theoretically be possible—it would have to be the case that (a) we could get major new capacities and fulfill them, and (b) it would be highly desirable for us now—in terms of fulfillment value—for this to happen, i.e., for us to acquire such new capacities and to have them fulfilled. While I'm confident that (a) is true, I am doubtful that (b) is. This is essentially question (5) that I wrote here on the blackboard: "If we start with a bucket that is full, do we then have reason to make that bucket larger and fill it some more?". To be clear, I *do* think it can be very desirable for us to obtain new capacities and to fulfill them; *but* it does not seem that the desirability of this prospect (for us now) derives from the value of fulfillment *per se*. Rather, there are *other* reasons why we should regard a future in which we attain and fulfill new capacities as prudentially desirable. For example, it could make our lives more interesting and enjoyable, and perhaps also more meaningful.

*

Feinberg seems to have had in mind humans and human lives as the objects of predication of fulfillment or unfulfillment. I incline to think we should also consider other types of being as capable of having fulfilled or unfulfilled lives, at least in an extended sense—including beings such as nonhuman animals, but also plants, organizations, traditions, and cultural movements.

On this view, we may well regard it as sad if a caged bird is never allowed to use its wings to fly, even if we suppose that the bird *feels* content in a small cage.

If we strain our sensibilities a little further, we can perhaps similarly detect something unfortunate about a promising new artistic school that never comes to full fruition because of the intervention of some external defeater—for instance a train crash in which all its leading lights are extinguished. For might we not attribute to an art movement many of the same properties by which Feinberg characterizes the fulfilled life of a person? A movement could be said to exhibit "inherent proclivities to engage in activities of certain kinds". It might be said to have a "life path" that either does or does not allow the full development of its "latent talents, interests, and initial bent" and that harmonizes with "its evolving self-ideal". A movement certainly can have a trajectory that involves "building, repairing, rebuilding, creating, pursuing goals, and solving problems".

Some might insist that these agentic attributes should be ascribed only to individual human practitioners, not to "the movement" itself. But why? An entity that comprises a group of individuals together with a shared culture and a set of interlocking interests and ideals, and that has a field of operation with an implied logic of achievement: such an entity can be more than the sum of its parts. It can manifest emergent forms of capacity, intentionality, and goal-directedness that are distinct from those of its individual constituents. To the extent that a collective can be usefully and illuminatingly described in such intentional terms, it would seem to be on a par with human individuals with respect to the question of whether fulfillment, and wherever value resides therein, can be instantiated in the course of its sojourn through the world.

Admittedly, the notion of fulfillment is vague and indeterminate in its application to entities such as artistic or cultural movements. But it is also so in its application to human individuals.

We must concede, however, that there might be important differences in degree in this regard—and more so as we consider cases at increasingly greater remove from the paradigm of individual human fulfillment.

For example, should we attach some slight disvalue, due to a thwarting of a potential for fulfillment, when an apple tree is cut down before it has had a chance to bear fruit?

What about a teddy bear that is never cuddled, or a violin that is never played? A firecracker that is never set off? A boulder on top of a cliff that never rolls down?[183]

At some point, as we move in this direction, we transition from an increasingly finely tuned ethical sensitivity into sentimentality, and from thence into fancifulness, and finally into sheer nonsense.

Suppose that, in the case of the boulder atop the cliff, we are willing to say that we are "helping the boulder reach a more fulfilled state" by giving it a push over the edge. But could we not, with equal justification, describe the same sequence of events as "harming the boulder by thwarting its soaring ambition and leveling its remarkable attainment of elevation"? Without any criterion to privilege one of these characterizations as more correct than the other, we must suspect that there is rather no underlying fact of the matter, and that the words that emanated from our lips have failed to express an intelligible proposition.

However, even though in the limit lies nonsense, I still think it would be appropriate to move some distance in this direction, once we enter a more utopically conducive era. Considerations that in today's world are rightly dismissed as frivolous may well, once more pressing problems have been resolved, emerge as increasingly important lodestars. This is part of what I alluded to earlier when I spoke of letting our evaluative pupils dilate in utopia. We could and should then allow ourselves to become sensitized to fainter, subtler, less tangible and less determinate moral and quasi-moral demands, aesthetic impingings, and meaning-related desirabilities. Such recalibration will, I believe, enable us to discern a lush normative structure in the new realm that we will find ourselves in—revealing a universe iridescent with values that are insensible to us in our current numb and stupefied condition.

Richness

Next up: richness. We might want a *rich* life, a life effervescent with living—

Student: Professor Bostrom! I think that's the fire alarm? Are we supposed to evacuate?

Bostrom: It's probably a false alarm. The notion of richness is quite closely related to fulfillment, and even more closely related to interestingness. There are some differences in emphasis, but our earlier discussion of those canopy values means we can be briefer here.

I'll introduce the concept of richness by quoting from a recent article in *Psychological Review* by Shigehiro Oishi and Erin Westgate. They write:

> "[W]e argue that a psychologically rich *life* consists of interesting *experiences* in which novelty and/or complexity are accompanied by profound changes in perspective. . . . Thus, on their deathbed, a person who has led a happy life might say, 'I had fun!' A person who has led a meaningful life might say, 'I made a difference!' And a person who has led a psychologically rich life might say, 'What a journey!'"[184]

The authors find that a significant minority of people across several countries report they would prefer a life that is psychologically rich to one that is happy or meaningful.

Student: Professor Bostrom, the sprinklers are on!

Bostrom: Yes, that is unfortunate. I think it's the new regulations from the University's occupational safety office: they must test the sprinkler system once a month. The opposite of a rich life would be one that is unengaging, monotonous, uneventful, one that does not involve much exposure to the variety of the human condition and does not produce growth in life wisdom and perspective.

How is this different from interestingness? As I said, there's a lot of overlap. But I think of richness as placing greater emphasis on active participation, on range and intensity of emotional and physical involvement rather than merely cold cognitive puzzling and perusal, and on the accumulation and integration over the course of life of a large and varied body of lived experience.

Richness is different from fulfillment too. The two are probably correlated, but they are not the same. You could imagine somebody who has a very rich life and yet remains unfulfilled. This would be the case in the example I gave earlier, of the person who is thrust into the midst of a series of historical events—especially if the person is not merely witness to these events but also a participant, and if, additionally, they have a dramatic and turbulent personal life, with many ups and downs, challenging relationships, tragedies and triumphs, different careers, and so on. Yet if they were really cut out to be a mathematician or a monk, and they never had an opportunity to pursue that strong calling—

Student: Professor Bostrom, the lights just went out!

Bostrom: The sprinklers must have caused a short circuit. I'm sure they are working on fixing it. Wait, it's Friday; they might have left for the weekend. But you can still hear me, right?

So, a rich life is one "in the arena". It's a whole-of-being wrestling match. It is loaded with impressions, expressions, and the feeling of being fully alive. Rainstorms, thunder, hail, and rainbows; sad and difficult times as well as joyous ones; strong loves and hates; a life of toughness and tenderness, trust and betrayal; a life of breaking waves, with foam and seaweed flushing over your face; life with a pulse, never dull and rarely easy. Life as something overwhelmingly real, heart-stoppingly scary and beautiful and bewitching—an adventure that leaves you drenched to your pip in lived experience. Life for better and worse.

Richness focuses not on avoiding imperfections or minimizing pain, but on creating positives—or, more accurately, on experiencing human life fully and participating in its creative process. It almost welcomes problems because they are a source of challenge and a way of intensifying and amplifying life. By the criterion of richness, it is certainly better to have loved and lost than never to have loved.

Somebody who has lived a rich life is, I think, unlikely to regret it. They may or may not be inclined to accept an offer to live their life over again; but I'm imagining the person who has lived a rich life as having a kind of satisfaction as they look back near their days' end: a sense that they wrung everything they could out of their allotment of years. That win or lose—and maybe in this context it is always lose in the end—they had their chance and their moment, and they left it all on the field.

Ah, the sprinklers came off. Things are looking up! There's a box of tissues here under the podium, still dry. Please help yourselves if anybody needs to wipe off your glasses or something. There's probably not enough for everybody but there are a few left.

Tessius: Shall I grab some for you?

Kelvin: Yes, please.

Bostrom: Now, I should state for the record that achieving a rich life in this manner can be highly *inconvenient*—potentially even hazardous to health.

For those of you who are not inclined to join the fray, I will point out an alternative path to psychological richness. This method involves getting yourself a nice recliner, a coffee brewer, and a stack of good books.

A sensitive reader (or watcher, or listener) can gain considerable experience by living vicariously: in some respects, *more* experience than one can get by venturing out there and garnering it firsthand—just like you can probably procure a richer and more varied diet by buying your food at the supermarket than by growing it all yourself.

Marcel Proust describes the summer days of his childhood, when he would sit in the garden of his family's country home and read:

> "[T]hese afternoons were crammed with more dramatic and sensational events than occur, often, in a whole lifetime. . . . [T]he actions, the feelings of this new order of creatures appear to us in the guise of truth, since we have made them our own, since it is in ourselves that they are happening, that they are holding in thrall, while we turn over, feverishly, the pages of the book, our quickened breath and staring eyes. And once the novelist has brought us to that state, in which, as in all purely mental states, every emotion is multiplied ten-fold, into which his book comes to disturb us as might a dream, but a dream more lucid, and of a more lasting impression than those which come to us in sleep; why, then, for the space of an hour he sets free within us all the joys and sorrows in the world, a few of which, only, we should have to spend years of our actual life in getting to know, and the keenest, the most intense of which would never have been revealed to us . . ."[185]

Such an ability to experience by proxy is one of the advantages bestowed on people of sensitivity and imagination. Others, however, may need to put their own hand into the flame in order to feel anything strongly and to persuade themselves of its vehement reality.

*

In a plastic utopia, the gap between experiences of fictional content and experiences of reality could be reduced.

One enabler of such a convergence is that reality could creep closer to fiction—what you dream about and wish for would be more likely to come true. If you want to know what it might be like to ride a unicorn, today you have to resort to fiction and imagination; in utopia, they might have bred unicorns, and you could experience riding one for real, assuming you could find one who was willing.

However, there are limits to the extent to which reality could be made to conform to fiction. For example, you could have a story about an emissary from an ancient alien civilization visiting Earth, gradually winning over people's hearts, and inspiring us to establish global peace and harmony. But if in reality there is no ancient alien civilization within travel distance, this story could not come true. Another blocker is that stories revolve around people, and there are many fictional scenarios that could not come true without rights violations that are unacceptable in utopia.

Most of the rapprochement between reality and fiction would instead come from the other direction: from fiction creeping closer to reality. Secondhand experience, and experience of completely artificial constructs, could become more similar to firsthand experience. It could become more immersive, more detailed, and more realistic in specific aspects in which realism is desired. Secondhand or synthetic experience could become more capable of inducing in the reader/watcher/participant the same psychological effects and developments as would have occurred if she had experienced the corresponding real situation. At technological maturity, books would be much better (because they would be produced by superintelligent authors); movies would be much higher quality (for the same reason, and because of better production tools); and virtual reality would be fully realistic and lifelike, or, when fantasy-like, could achieve an internal coherence and precision of rendering similar to that which characterize direct perception of physical reality. The main qualification here is the one we discussed on Tuesday—that it might be impossible to render fully realistic open-ended original experiences of deeply interacting with other minds without thereby effectively bringing those minds into existence.

Actually, I expect that virtual worlds will be experienced as decidedly *more real* than the physical world—more vivid, engaging, fruitful, relevant, and psychologically impactful. Many of us already spend more of our time and attention in the worlds of thought and imagination than we do focusing on the objects before us in our physical surroundings—and we find these mental conjurings quite sufficiently "real", much of the time, even within the limitations of our current primitive cognitive and technological methods of mise-en-scène. (And I'm not even referring to the simulation hypothesis here.)

*

The upshot is that prospects for richness are auspicious: the experiences of utopians could be very rich indeed.

This analysis assumes that richness is given a psychological construal—which seems like a fairly natural and convenient way to carve things up in conceptual space. Admittedly, one could instead choose to delineate the value of richness in such a way that it would require, for its instantiation, elements that go beyond the purely psychological. One could, for example, define richness to also involve aspects of *objective interestingness* or *fulfillment*; but since we have already discussed those values separately, it would be redundant to go over these elements again. Or to involve aspects of *purpose* or *meaning*; but those are values that deserve to be discussed in their own right—something we were just about to do anyway...

Purpose

In Wednesday's lecture we presented four case studies, of shopping, exercising, learning, and parenting. In each of these, we saw the potential for a great redundancy, and a foreboding of a generally "post-instrumental condition", which would commence upon the attainment of technological maturity, or shortly before, and in which human effort would be rendered obsolete and unnecessary.

A life without purpose might seem deficient. What is the point of all the hustle and bustle, we may wonder, if it is just so much busywork? Or, if the hustle and bustle were to cease, then what would prevent us from sinking into a deathlike passivity?

Exploring in this direction quickly leads to the question of the meaning of life. However, I want to postpone discussion of that for yet a while. The more we can hew off before we talk about meaning, the better our chances of being able to wrap our heads around whatever remains. I will therefore construe "purpose" here in a reasonably separated and unloaded sense. Purpose, we may say, is anything for the sake of which we make efforts and engage in activities.

*

We can then distinguish different "sizes" of purposes, or for-the-sake-of-which-thingies.

Small purposes we can call *aims*. For example, you go to the grocery store with the aim of buying the ingredients you need to make one of the healthy and delicious recipes in *Green Transcendence: The Complete Guide to Liquifying Your Salads* (which, you may be interested to know, is currently on sale).

Medium-sized purposes—which we can term *goals*—are ones that require undertakings of a more expansive scope. Graduating with a good degree from college would be an example of a goal.

Finally, we have large-sized purposes. These we can call *missions*. A mission is an objective so great and encompassing that it could motivate the devotion of an entire life, or at least a major portion of one. A mission could be something idealistic such as to make a significant contribution toward the eradication of some disease, or to lead a life that is pleasing to God; or it could reflect some less elevated aspiration, such as to accumulate a great fortune or attain a high political office.

Note that in order for something to count as a mission, it is not enough that its attainment is highly desired or would be very valuable. It must also be such that pursuing it properly would require deep lifelong dedication and striving. If somebody had the ability to cure cancer simply by pressing a button, she would have strong reason to press the button, and by doing so she would be accomplishing a great good; but she wouldn't really be in a position to make curing cancer her mission in life.

*

Now we can observe that all of us have aims, many of us have goals, but relatively few have missions. It follows that either relatively few have worthwhile lives or having a mission is not necessary for one's life to be worthwhile. I think the second alternative here is the plausible one.

What about not having *goals*? A person who lacks goals would drift through life, like a log floating down a river, without long-term plans and ambitions. We might say that such a person wouldn't be striving to accomplish anything, but this would be true only if we speak loosely; since they would presumably still have small purposes, or aims. Such a person would make their way into the kitchen *in order to* make an elixir. If they are playing a game, they may be *trying to* win. So there would be short-term objectives; and some efforts would be made, and some actions taken, for the sake of attaining those objectives.

Even if having a mission is not necessary for having a worthwhile life, many of us may be inclined to say that a life that is also without goals—the life of a drifter, with no pursuits beyond those prompted by a series of mere momentary aims—would be, in an important respect, impoverished. Perhaps we are also inclined to hold that a mission, although not *indispensable* to a life worth living, is nevertheless something that is capable of adding an extra good to a life—a good that is, in many cases, of sufficient value to compensate for the numerous sacrifices and inconveniences that usually are part of a mission-oriented existence.

*

But *why* exactly might one think it desirable for people to have purposes, and especially medium- or large-scale purposes?

We must separate out several possible grounds for such a judgment.

First, in the present world, there are still things that need doing. It is therefore desirable that there be folk who do those things. If nobody pursues a cure for cancer, we are less likely to get one. But this reason for celebrating purposes and missions does not apply in a post-instrumental condition.

Second, it appears that having a sense of purpose is beneficial for psychological functioning and well-being, by making one more resilient to hardship and less likely to become addicted, depressed, or bored. But this reason does not apply in a plastic condition, since with advanced-enough neurotechnology we could attain the same benefits, far more reliably and conveniently, by artificial means.

Third—and this is the relevant category for our investigation—there could be noninstrumental reasons for preferring a life of purpose to one without. I think we can distinguish two possible grounds for such a noninstrumental preference for purposeful life. We might prefer it (A) because the pursuit of a purpose *gives a certain content* to a life; or (B) because the pursuit of a purpose *gives a certain significance* to a life. Let us examine each of these more closely.

*

We'll start with (A). A life of purpose, especially one that has not only momentary aims but also goals or, better yet, missions, has a certain content. Specifically, it contains purposive activity. The execution of temporally extended projects requires the forming of an interlocking sequence of intentions, planning, internal resource mobilization, exertion, results-monitoring, course-correction, etc. One might judge that it is intrinsically good for a life to contain these elements—either the elements individually or when they are linked together into longer coherent chains of enacted intentionality.[186]

Here we can observe that, insofar as we are concerned simply with this value—i.e. that a life has these contents of purposeful activity—it does not seem important *why* the purpose whose pursuit generates these contents came to be pursued, nor *what* exactly the purpose is. What matters is only *that* some purpose is in fact pursued, and that the purpose be such that it organizes and motivates a suitable chain of goal-oriented activities extending over time, and perhaps that it exercises skills and calls forth conscious effort on the part of the subject.

Now, insofar as *this* is why a life of purpose is preferable, it seems the utopians are home and dry. One thing they could do is engage in autotelic activity. That is to say, they could undertake various purposive and effortful activities simply because it is intrinsically desirable to engage in those activities.

But what if the utopians (or some of them) happen to be psychologically constituted in a way that they just don't find such autotelic motives sufficiently compelling? Or what if it is judged—although upon what ground one might make such a judgment is quite obscure—that it is not as intrinsically desirable to be engaging in certain activities for autotelic reasons as it is to engage in them for instrumental reasons? Well, then there is another thing the utopians could do, which would take care of these additional requirements: they could create *artificial purpose.*

A utopian wishing to give herself artificial purpose could proceed as follows.

First, she would give herself some appropriately long-range goal. If she is psychologically capable of doing so, she could simply set herself a goal by an act of will. If she is incapable of doing that, she could use neurotechnology to spark in herself a strong desire to reach a suitable objective.

Second, our utopian would need to ensure that this goal is challenging for her to achieve and that pursuing it will require her to undertake an adequately prolonged, complex, and effortful series of actions. There are two ways for her to do this. One is to place herself in an environment in which there is no easy shortcut to the destination—metaphorically speaking, she would travel to some fenced-off area of utopia in which not all the affordances of technological maturity are readily available and within which, consequently, people are forced to rely substantially on their own capacities to make headway toward their goals. The other way, which achieves an equivalent result, is for her to select or modify her goal so that its achievement, by its very nature, precludes the use of (certain types of) shortcuts. Thus, instead of adopting the goal "Achieve outcome *G*", she could adopt the goal "Achieve outcome *G* without making use of methods *X*, *Y*, or *Z*".

Such strategies are not as bizarre as you might think. While contemporary neurotechnology does not yet give us the general ability to easily make ourselves strongly motivated to pursue any arbitrary goal, people do quite frequently resort to the other ways of generating artificial purpose. People challenge themselves by placing themselves in situations in which they are strongly motivated to push their bodies and minds to the limit, for example by going "off the grid" and arranging things in such a manner that they have no other way, other than via their own efforts, to achieve their goal of surviving and getting out intact. Same with the rock climber who is dangling halfway up a vertical wall: whatever motivational problems might have been plaguing them in their ordinary life have all vanished. Once one is in the exposed position, there is no room for second-guessing or for ruminating on whether the effort is really worth it: there is only the immediate perception of the overwhelming imperative of hanging on and not falling.

These are examples of the strategy of achieving purpose by placing oneself in a challenging situation. We are even more familiar with the strategy of achieving purpose by modifying a goal so as to make its achievement more difficult and thereby create the conditions for some interesting and desirable pattern of purposeful activity. For example, the recreational golf player does not set herself the goal of getting the ball to move sequentially between the eighteen holes—*that* could be too easily accomplished by picking up the ball and carrying it in one's hand. Instead, she adopts the goal of *making the ball traverse the course using only a particular narrowly prescribed set of means*—thereby imposing on herself what would be an utterly gratuitous inconvenience except for the fact that this is exactly what is necessary for her to realize the intrinsic value of golf-playing.[187]

Well, it is actually unclear why people play golf. The real reason might be to have fun—and if so, then, of course, utopia would offer the more efficient option of wireheading as a means of obtaining the pleasure. But if we assume that golfing has some value as a purposive activity, a value that goes beyond the pleasure it occasions, then we see how this additional value too could be obtained in utopia.

Climbing a rock or completing a round of golf might be objectives too limited to fully realize the value of purpose, especially if we think that the pursuit of bigger purposes (missions) has greater or additional value than the pursuit of small purposes (aims). But we can very easily imagine larger versions of the same kind of thing. If somebody sets their mind on climbing Mount Everest without the aid of oxygen bottles, this could give them a purpose that can be accomplished only by means of a long and difficult project that requires years of planning and practicing before the actual climb can even begin. For this strategy to work in utopia, one simply needs to embed a few extra constraints into the objective, to close off the multifarious shortcuts that become possible at technological maturity. For instance, we might have to define the objective to be to scale Mount Everest without supplemental oxygen *and* without upgrading our lungs or enhancing our red blood cells. Then we can proceed as before.

It therefore seems that nearly all purposes that are available to us now will still be available to us, with suitable amendments, in utopia. On top of this, we would have available to us many possible novel purposes which become accessible only with the development of new means and instruments. I mean: just as golf-related purposes were unavailable prior to the invention of golf clubs, and video game–related purposes were unavailable prior to the invention of computers, so too are there possible purposes that will become available only once technology further expands our equipment and/or our own capacities. We would consequently have no fewer—and would in fact have many more—possible artificial purposes available to us in utopia than we do at present.

*

Let us now look at reason (B), the second possible ground for why one might hold a life of purpose to be noninstrumentally preferable to one without. This was the idea that having a purpose endows a life with significance.

The thought here is that we may want our lives *to matter*. And the most obvious way for our lives to matter is if we accomplish some independently valuable result that would otherwise not have been accomplished.

The example of the person who pursues the mission of curing cancer is relevant. As a society, we might of course value the existence of this person instrumentally, for their useful contributions to oncology. But in addition to this, we might also judge that this person's life is made more desirable *for that person herself* if she succeeds in bestowing an important benefit onto the world. We might think that it is good *for a person* to have had a positive impact.

But note that if we do think that having a positive impact is in this way prudentially desirable, we face a potential predicament inasmuch as many things that need doing now will no longer need doing in a post-instrumental utopia. At any rate, they will no longer need doing *by us* (which is to say, by continuants of humanlike persons), since they would be better done by machine. It is true that we could, in the manner I just described, create artificial purposes for ourselves. However, we may wonder whether the achievement of artificial purposes endows our lives with the same kind of valuable significance as does the achievement of "natural purposes"—purposes that, as it were, "grow in the wild" from the soil of independently existing needs and problems, as opposed to being planted in little pots by ourselves merely in order to give ourselves something to do. This question warrants a closer examination.

*

At this point, I'm afraid, our subject matter is about to get a bit murky. I don't want to lose you in what will now become not merely literal but also a figurative darkness; so please do interrupt me if there is anything you don't follow. Let's think step by step.

1. Suppose that somebody complains of a lack of "real purpose" in utopia. They might say: "Sure, we could make up purpose, artificial purpose, but that's not the real thing. Artificial purpose would not give our lives true significance. In a post-instrumental condition, there would be no important function for us to perform—there would be nothing that we would *really* need to do. And so the seriousness of life would be all gone!".
2. To this, a number of responses can be made. We could deny that purpose, real or otherwise, has any noninstrumental value. Or we could assert that artificial purpose is a fully equivalent substitute for natural purpose. Alternatively, we could concede that real purpose can add some value to life that artificial purpose cannot, but that it is possible to much more than compensate for the loss of this value by the enormous gains in other values that can be realized in utopia.
3. Remember that we have already seen how artificial purpose can substitute for natural purpose in giving *content* to life. The remaining concern is whether the same can be said for giving *significance* to life. So, at worst, it is the significance-giving property of natural purpose, and whatever distinctive value may therein reside, that we are at risk of having to relinquish.
4. It is relevant to point out that our current human lives also appear quite limited in regard to significance. Many of the observations we made yesterday in our discussion of interestingness could be repeated here with respect to significance: our apparent triflingness from a cosmic perspective, and so on.
5. Even if we hold that the relevant scale on which to measure our significance is a more local one, it still does not follow that we are at risk on that account. Observe that we do not usually think there are vast

differences in the intrinsic value of a life to the person living it (to their "well-being", in philosophers' parlance) arising from the fact that some human beings contribute vastly more—millions of times more—objective good to the world than others do. In fact, many people contribute zero or net negatively (not necessarily through any moral fault). And some of these people, if they are happy and engaged in interesting hobbies and so on, nevertheless have lives that are good for them. In other words, the prudential (self-interested) desirability of their lives does not seem to be drastically impaired by the fact that they are making little or no net positive difference to the overall balance of goods and bads in the world.

6. But perhaps it could be objected that such "non-contributors" nevertheless currently do have a kind of local significance, in that their choices and efforts at least make a big positive difference *to their own lives*? Maybe if we were also to lose this local egocentric significance in a post-instrumental utopia, then our lives would become worthless?
7. Let us consider an extreme case, not so uncommon in the contemporary world, for instance in care homes, where a person has a very limited ability to have a positive impact even on their own life. Consider a person who cannot chew their own food or brush their own teeth, who requires expensive continual medical support, and who, let us suppose, is also unable to bring much joy to other people (perhaps because all the people they interact with are strangers who are mostly indifferent to them and are looking after them only because they are paid to do so). This hypothetical person, we are thus supposing, contributes net negative value to the rest of society. (Please let not the cold words I use to describe this case deceive. I think we should do more, not less, to support persons who are in such a condition. Our fundamental need and worthiness of receiving love, respect, and support is not preconditioned on our ability to make useful contributions to society. Which should go without saying. But there are, sadly, some people who are eager to misunderstand on purpose. Not that I hold much hope that anything I say can protect me from such folks and their aspersions, alas.)
8. Clearly, this individual faces a lot of challenges. But is it nevertheless *possible* that they could have a good life? I think yes. To see this,

however, we need to clear away some potential confounders. To make the case as unambiguous as possible, we could imagine the person in our example to be suffering from no pain or discomfort or anxiety from her medical condition, and no guilt from being a burden onto others. We should imagine, rather, that she experiences high levels of positive affect and effervescent joy, that she takes delight in observing the world and its wonders, that she is sensitive to beauty and humor, loves listening to music and does so with great understanding and appreciation, and so on and so forth—although we must also suppose that these positives are *not* the product of her own choices or internal efforts at regulating her attitudes, but rather are simply the effect of her spontaneous and effortless responses to experiences that her caretakers arrange for her. With these stipulations, I think the person in our example could be having a great life, despite her being utterly insignificant, or even negatively significant, insofar as we measure significance by the causal impacts she has on the world around her.

9. Now as for our utopians, not only could they have all the good things in this person's life—to unprecedented degrees, thanks to being able to benefit from physical, emotional, and cognitive enhancements far beyond the current human level—they could also benefit from having amazing active experiences via a pursuit of autotelic activities or artificial purposes. *So what* if this comes at the price of no longer having the kind of positive significance that most people have today? I mean, like what… the significance we gain by duly paying our taxes? stimulating the economy through our Amazon orders? Or, perhaps, the contribution we are making to the global conversation by reposting memes on social media?

*

You all with me so far?

Student: I have a question. I agree that my life probably doesn't gain much significance from my Amazon purchases. But I want to think that I'm significant on a more personal level. To my family, for example. And my fiancé. Don't you think it would be sad if there was literally nobody who cared, nobody to whom one was significant?

Bostrom: Good. Yes, that would seem sad. I was actually just going to get to that.

Let me first make a preliminary observation. While we want it to be the case that at least some other people care about us, and while we want to care about at least some other people, this does not seem to *require* that we have significance to each other in the sense of having the ability to take actions that affect each other's well-being. For example, you could care about somebody you know you will never see or interact with again. A relative might have departed to a new world, and even if you were sure that you would never be reunited, and that no mail service or other communication could ever be established, you might still be very much concerned with how things are going for them over there, thinking about them often, and hoping and wishing that they are doing well. So *that* kind of significance is not at issue here, right? No reason the utopians could not have that. It requires only that they care about each other, but it is not based on causal impact or the ability to be practically useful.

Now, lest you read different things into my words than they were intended to convey, let me hasten to clarify that with the example I just gave, and with some of the other examples that have come up earlier, I do not mean to suggest that utopian life would necessarily be a solitary affair. No! We are just trying to take things one step at a time. Obviously, in utopia, we would be able to continue to interact, communicate, do and experience things together, and generally enjoy each other's company.

In fact, we could, if we choose, establish much closer relationships with other people in utopia than is possible at present. We might, for instance, establish higher-bandwidth communication links between our minds, or use other psychotechnical means to facilitate openness, trust, and intimacy. We'd have means far more effective than alcohol or the primitive empathogens that exist today. I'm not making the claim that we ought to embrace all such technologies wholesale and indiscriminately, and merge ourselves into some sort of hivemind or borg; but the *possibility* would certainly be there. I am quite agnostic as to how much and what kinds of such enhanced social closeness we ought to seek in utopia. Perhaps this is something that would vary over time. It would seem like a tad improbable and suspiciously coincidental to suppose the amount and quality of connection that we currently have with each other to be exactly optimal—under either present or future conditions.[188]

One more preliminary remark. There is a theory that people feel more alienated from one another in modern societies because we are less forced to rely on friends and family for the essentials of our survival than was the case before contract law and police forces and welfare systems—and maybe also because we less frequently encounter life-or-death situations in which we discover who our true friends are. If this theory is correct, and if utopia carries the historical trend further, making us even less reliant on personalized sources of support and more able to depend on the state or advanced technology to keep us safe and to ensure that our needs are met, would this then mean that people would feel still more alienated in utopia? No. Because, if necessary, feelings of alienation could be easily banished with mature neurotechnology.

So, we could *be* closer to one another, *feel* closer to one another, *interact* more closely with one another, and remain significant to one another in the sense that we could *care* about each other's welfare. Let's deposit these valuables in the bank. And then we can venture out again to see if we can identify and secure still further values.

This brings us back to the main line of our inquiry, where, you will recall, we were asking specifically about impact-based significance at technological maturity. And your question about personal relationships is relevant here. It provides an important element in what I referred to earlier, back on Wednesday, as "sociocultural entanglement". This was the fifth and outermost perimeter in our multilayered defense. Cultural and interpersonal complications might provide us with purposes in utopia beyond those which we may create for ourselves individually by setting ourselves challenging goals.

*

I was going to approach this by first discussing a sort of engineering-fix, which I call "the gift of purpose", which I have here on a handout. I was going to read it, but that's a little difficult now… Maybe if I stand over there.

Yes. By the faint green light of the EXIT sign we shall proceed!

HANDOUT 17

THE GIFT OF PURPOSE

Let us suppose that it is good to have a purpose, and that you want to help out a poor friend of yours who is lacking one. We will assume that your friend cares about you to the extent that at least one of the following is true:

I. Your friend cares about your preferences.
II. Your friend cares about your well-being.
III. Your friend cares about your opinion of him.

To give him a gift of purpose, all you have to do is to establish a suitable linkage between your preferences, your well-being, or your opinion of him, on the one hand, and his actions, on the other.

For example, suppose that your friend cares about you in the sense that he desires that you get what you want. Then you would form a preference that he achieves *G*. If you find it hard to simply form a preference for something like that merely by willing it, you could use some psychotechnical means to create that preference in you.

Alternatively, suppose your friend cares about you in the sense that he wants your well-being to be high. You would then arrange things so that your well-being is higher if your friend does *G*. Exactly how to do this would depend on which account of well-being your friend endorses. If, for example, your friend believes that wealth is helpful to achieving well-being, then you could sign a contract with a third party that obligates you to pay a fine unless your friend does *G*.

Finally, if your friend cares about your opinion of him, you could simply commit to regard him more favorably if he does *G* (again, resorting to neurotechnology to aid you in this, if necessary).

You have now provided your friend (or maybe, rather, you are his mentor or coach, or parent) with a reason to do *G*. In order for this to amount to giving him a purpose, you need to choose *G* appropriately. If *G* is something he could too quickly and easily accomplish—such as patting himself on the head—then *G* would be for him a mere aim, which presumably does not bestow on him much purpose-value. So *G* needs to be something that necessitates a larger project—complex activities, efforts extended over a long span of time, bringing into play multiple skills and talents that your friend may have, and perhaps requiring significant emotional involvement and dedication. Having a sufficiently strong instrumental reason to pursue such a *G* would then give your friend a fine purpose—an ambitious goal, or even a mission.

Of course, it would be no good if the most efficient way for your friend to accomplish *G* were simply to press a button that causes a robot assistant to perform the actions necessary for the achievement of *G*. That would obviate the need for your friend to make any efforts himself, which would defeat the purpose. Nor would it be any good if the most efficient way for your friend to accomplish *G* were for him to swallow an enhancement pill that made it trivially easy for him to accomplish *G*. Therefore you must define *G* in a way that precludes the possibility of purpose-destroying shortcuts. The most straightforward way to do this is by baking into *G* itself the postulation that only certain kinds of means are permissible. So *G* might have the form:

> *G*: Achieve outcome *X* using only means from set *M*.

The admissible means would not include shortcuts such as ordering a robot assistant or popping an enhancement pill. The combination of *X* and *M* would be chosen so that achieving *G* provides a suitably engaging challenge for your friend. For example, if your friend might like to have a purpose related to the intellectual challenges of playing chess, *G* might be something like:

> *G*: Achieve a victory against the chess engine Stockfish at difficulty level 7, using no computer aids to assist you during training or during the match, and using no cognitive enhancers or other means that go against the spirit of this challenge.

Since your friend desires your preference to be satisfied (or your well-being to be increased, or your opinion of him to be high), and since you have arranged things so that the only way that he can satisfy this desire is by accomplishing G, he now has a purpose.

Wielded in such an explicit and reductive form, this method of generating purpose seems quite hokey. But if applied more subtly, with refinement and finesse, and inside the context of a suitable cultural embedding—then maybe not so much?

Schemas for esteem-allocation are a core feature of culture. Cultures leverage our esteem-craving to incentivize us to pursue a wide range of projects, including projects that would otherwise seem arbitrary and pointless. The pursuit of such culturally sponsored projects is a source of deep and fulfilling purpose in many people's lives.

If we take our golf example: think of how ridiculous it must appear—if we adopt a detached perspective and consider this activity outside the context of a culture that values people who develop skill in this particular endeavor—in fact, how disturbingly close to insanity it must appear, to devote the prime decades of one's life to the project of improving one's skill at hitting a small ball with a club into a series of narrow holes.

And think of all the person-years of learning, training, striving, sacrificing, and overcoming, all the toil and effort and ingenuity, the sleep foregone, the inconveniences endured, the troubles incurred, not just by athletes but likewise by artists, writers, actors, wealth-accumulators, soldiers, scholars, dieters, fashion-followers, and so many others—and how much of these exertions and sacrifices are fundamentally driven (though it is not always admitted or recognized) by a desire for social approval and esteem! This is often true even in the case of careers that have a solid claim to social utility. Like that of a healer—how many doctors entered their profession because they didn't want to disappoint their parents? All these people have been given a gift of purpose.

*

Blaise Pascal wrote: "All of man's woes come from his inability to sit still in a room."[189]

Well, much trouble would be avoided if we did sit still in our rooms—but also much life.

And even in solitude, our social aspirations scarcely leave us alone, as we continue to evaluate ourselves based on internalized representations of what others would think.

For better or worse, we are pretty socially entangled. It would be a highly invasive procedure (involving risk of substantial damage to our personal identity) to remove all of these volitional dependencies, even if we wanted to.

It is true that in a plastic utopia *some* of our current reasons for seeking esteem would no longer apply. For instance, we may no longer need esteem to make money or obtain other material advantages. Nor would we need esteem in order to feel good about ourselves—neurotechnology. However, we may continue to seek esteem for its own sake.

We may wonder whether perhaps our desire for esteem would be *weakened* if esteem did not confer these auxiliary benefits that it does today? However, we must also take into account that many of the other desires we currently have, which compete with our esteem-craving for the control over our minds, would also fall away as motive factors in a plastic utopia, as they would become trivially easy to satisfy.

It is perfectly conceivable, therefore, that the desire for esteem may come to make up a *larger* share of our remaining motive drive in utopia.

Might we then witness, shortly after the intelligence explosion, a vanity explosion? Or are we perhaps already so close to maximally vain that there is little room for a further increase?

*

Let's keep moving.

10. From the point of view of the person who receives a "gift of purpose", the given purpose could be regarded as real and genuine, since it exists independently of their own volition. It is an objective reality to which they must accommodate themselves, not a postulation that they can choose to make or not to make.

11. Might one nevertheless object that this kind of gifted purpose is of a less-than-sterling grade simply on account of it not arising completely independently of human agency? I think not. The man who runs away from a tiger and the man who runs away from an axe murderer seem on a par insofar as the value of purpose goes. Likewise if we imagine these two men embarked on longer-range projects, whether those be rooted in natural or in social realities: for example, one planning his escape from a desert island, the other from a prison colony.
12. There are, however, a couple of related concerns one might have about the value of gifted purposes, even if the mere fact that such purposes derive from the preferences and choices of other people is not in itself disqualifying. First, one might worry about significance claims that are based on success in zero-sum games. Second, one might worry about purposes whose roots lie specifically in somebody else's desire to help us achieve purpose. Let us consider these in turn.
13. First, zero-sumness: can a purpose qualify (as providing whatever value that having a purpose can provide) if it consists in trying to achieve success in a zero-sum game? The case against would be that efforts in zero-sum competitions have a kind of global futility that might seem inimical to real significance. On the other hand, we normally regard the purposeful strivings of, for example, competitive athletes as having significance-value. If several athletes compete in the Olympics, and one of them takes the gold, we might say that the winner achieved something of extra significance that the others did not (though they share in the lesser achievement of having qualified for the competition). This would seem to imply that the zero-sum nature of the activity cannot be a disqualifying characteristic. If this kind of zero-sumness is not incompatible with significance in sport, it may not be so in other contexts either.
14. Against this one might retort that an athletic competition as a whole is positive-sum. It produces net positive value, not because of which particular athlete is triumphant, but because of the competitive activity providing entertainment to its participants and its audience. Well, yes, but we could make a very similar claim for activity that flows from "gifted purposes" in utopia; such activity could also be positive-sum. Not, it is true, by virtue of it providing enjoyable entertainment—for *that* is something that could be more efficient-

ly provided by technological means—but by virtue of it bringing non-self-generated purpose into the life of the recipients. Under the assumption that having such purpose is good, it is hard to see why this contribution would not count toward making the arrangement positive-sum, just as the contribution of fun can make competitive sports positive-sum.

15. Second, there was the concern about purposes whose roots lie specifically in somebody else's desire to help us achieve purpose. Maybe one might think that those purposes are not as good as ones that arise otherwise? Pursuing purposes given to us merely so that we have a purpose—might that not seem just like so much busywork that lacks genuine significance? Once such a purpose has been created, we may have reason to try to achieve it; yet the whole arrangement might seem a bit like a sham—like digging a hole just in order to create a need to fill it. Which seems—absurd?
16. But we might just say: "That's life!". Once there is life, there are needs; once there are needs, they must be met. This might also seem a bit like a sham. Everything would be easier if there were no life: nobody digging holes, nobody needing to fill them. Yet we are where we are—and maybe we still have some significance, even if it be a merely local and perhaps even somewhat absurd kind of significance.
17. If our Creator made things the way they are partly in order to give us purpose, would this purpose then therefore be defective? Many people believe the opposite: that if there had been no Creator, or if our world and our lives bore no relation to the Creator's intentions, then our lives would be *less* purposeful and significant, not more. But that is an issue you would have to take up with Professor Grossweiter.

Let me check if you are following so far?

*

Ok, either you are still following, or I lost you so far back that you don't even have any questions. Maybe you aren't even there anymore? Well I don't let that stop me! Professor Grossweiter gets paid either way.

Student: Will this be on the exam?

Bostrom: Ah, the bottom line! No, I don't think it will.

Second student: But what about my purpose?

Bostrom: Oh, I see. Very good. Well, I suppose for the tuition fees you are paying, it seems only fair that you receive some purpose in return. Alright, it will be included in the exam.

Third student: What have you done?!

Second student: I've leveraged Professor Bostrom to give you the gift of purpose.

Third student: But why!

Second student: Why not?

Third student: Griefer!

Second student: Prig!

Bostrom: Order! Order! The anonymity of darkness is bringing out primitive tendencies. But in the name of Rudolf Clausius, let us try to keep the entropy at bay for a while longer.

Evidently, it is possible for gifts of purpose to be unwelcome. In this respect they are no different from other gifts, which can sometimes irk, for example by imposing obligations. Caveat dator—there is a bit of an art to it. In any case, purpose-giving is an option available to utopians.

*

For the sake of balance, I should perhaps also offer a few critical remarks about purpose.

I suspect that many of you have grown up in a culture that extols purpose and celebrates the mindset and lifestyle of the striver—somebody who works hard and seeks to become a success in life, or better yet pursues some vaulting ambition and gives it their all. In part, this value scheme may be an inheritance from the Protestant work ethic, although I think it can also derive support from other sources and traditions.

It is worth reminding ourselves of the existence of alternative perspectives on these matters. For example, there is an outlook found in ancient wisdom traditions and religious teachings, which either opposes or at least would tightly circumscribe such a valorization of human volition. For example, there are important strands in Eastern religions such as Buddhism, Jainism, and Taoism that emphasize the desirability of nonattachment or even the extinction of desire. Christian spirituality likewise has often recommended against attachment to worldly goals and aspirations. To the wise of these perspectives, it might seem quite a peak of perversity if, having somehow succeeded in attaining a condition in which we might have no or few remaining unfulfilled desires, we were then to proceed to deliberately engineer another set of new and more-or-less arbitrary desires just so that we would be compelled to continue to strive and toil in order to satisfy them: what, they might wonder, could possibly be a clearer demonstration of the madness of the modern Western mind than that it should endorse such a proposal as profound philosophy?

Hinduism, however, presents a more complex picture in this regard. On the one hand, it recommends a path of spiritual liberation through nonattachment similar to those espoused by the other Eastern traditions. Yet, on the other hand, it also presents us with the idea of Lila, or "divine play": the notion that the gods make sport by voluntarily imposing on themselves limitations and constraints in order to engage in playful activity within the mortal realm—an expression of their freedom and spontaneous creativity (the result of which is the reality that appears to our senses). Perhaps this suggests a model in which might deliberately adopt new purposes in order to keep the game going, but where these purposes are held and experienced in a more lighthearted and playful manner than glum and compulsive cravings that drive much of human existence in our present condition.

*

I will also offer another observation: that even if the appearance is correct that our human lives currently have only a modest and merely local significance, it might still be argued that we have too much; and that what we should be hoping for is to become *less* significant, not more.

Less significance means less responsibility, less opportunity to screw up.

I think we may already be beyond the optimum level of significance, relative to our current capabilities. I mean, look at it this way. You have been put in charge of *an entire life's worth of human conscious experience*: your own. This human life is at the mercy of your dictatorial powers during every waking hour of its existence. What an absolutely fearsome responsibility you have!

If I had to guess, I would say that the average adult maybe ought to be responsible for about one year of human life. After this period, things ought to be restored to some acceptable condition if they have messed up. Maybe the most mature and worldly-wise among us could be responsible for one decade of their own life. But to be responsible for *an entire human life*—and some would think without even the possibility of a do-over at the end—well, that is just too much.

*

It might be worth spending a moment to reflect on the etiology of this putative value of purpose, as we did for interestingness. How could we explain, in causal terms, why we have ended up valuing purpose (to the extent that we do)?

I'll put forward three hypotheses, not mutually exclusive:

> *The usefulness-of-effort hypothesis*
> We begin life with simple goals. After a while, we notice that if we apply ourselves, we are more likely to achieve these goals and to get the associated reward. We come to positively evaluate the striving itself because it is useful as a means to achieving a wide variety of goals. Eventually, the broad instrumental value of striving becomes intrinsified into a value pursued for its own sake.

This would be analogous to money. We start out not caring about it; then we notice that it is robustly useful as a means to many ends, and we come to desire it for instrumental reasons; finally, some people begin wanting it for its own sake and become misers.

Here's another hypothesis:

> *The innate drive hypothesis*
> We have innate psychological drives for activity and effort (alongside other drives, such as for rest and relaxation). We might also have some innate mechanism that wants our activity and effort to be directed toward endorsed objectives, and especially toward long-range internally ratified goals. In order to give this drive an outlet, we need purpose. Without

> purpose, the frustration of this drive is experienced as malaise—a kind of internal pressure or enervated restlessness, which might also manifest as an unpleasant weariness or inability to mobilize the organism's resources and vital energies. Purpose then comes to be valued as a means to forestall this unpleasant condition. Eventually this means becomes intrinsified into something that is valued as an end in itself.

We can also consider a cultural explanation:

> *The cultural hypothesis*
> To various extents in different cultures, purposeful exertion is lauded. Like other socially reinforced behaviors, purposeful exertion is then instrumentally useful—not only because of whatever specific objective the actions might accomplish but also simply because the effort is made and is seen to be made. This instrumental value of purposeful exertion then becomes intrinsified. (We may be especially likely to intrinsify values that are celebrated in our culture and among our peers, or are held in high esteem within groups whose acceptance we seek.)

This hypothesis raises the further question of why a culture should have come to extol purposeful exertion in the first place. A functional explanation might focus on how it helps societies thrive and prosper. A signaling explanation might focus on how it constitutes a hard-to-fake indicator of other positive traits, such as health, energy, and opportunity. But the signaling explanation could also involve reference to more historically contingent factors. In some social contexts, it might be that indolence and aimlessness sends a more positive signal, for example by indicating that an individual is so talented, wealthy, or otherwise privileged that he or she does not need to put forth much effort. Many other potential explanations could also be furnished. Probably the real story of why any society comes to hold the values it does is messy and complex.

The usefulness-of-effort and the innate drive hypotheses may predict less cultural variation in the degree to which people on average value purpose, though they could still allow for plenty of individual variation (for instance in the readiness with which different individuals intrinsify an initially instrumental value). Biological factors are no doubt also influential: something like dopamine or testosterone levels (and receptors that are sensitive thereto) probably have a big effect on how appealing a life of strenuous activity appears to an individual. However, to the extent that there are large differences between different societies in how much they valorize purpose, it is likely that the cultural hypothesis has an important explanatory role to play. Philosophers may need to take heed lest they overinterpret either an idiosyncratic personality trait or a local cultural fixation as a universal truth about human value.

*

To complete our inquiry into the value of purpose, let us now suppose the most challenging case: that while there is value in having purpose, this value is entirely voided if, as we may say, the purpose has been generated on purpose. In other words, let us assume (for the sake of the argument) that purposes that we either set ourselves or artificially induce in ourselves for the sake of realizing the value of having purpose or for the sake of enabling active experience do not contribute anything to the value of purpose; and let us also assume that purposes obtained via a "gift of purpose" are also to be regarded as worthless. Under these assumptions, is a plastic utopia then doomed insofar as the value of purpose is concerned?

That is not so clear. I suspect that even under this most stringent standard, in which only (what we'll term) "practical purposes" pass muster, there might still be possibilities for us to secure some purpose-value.

There are two directions we could scan for such practical purposes. We could look for practical purposes where what matters is only that something be done: we'll call these *agent-neutral purposes*. Or we could look for them where what matters is that something be done by some particular agent (or agent-type): we'll call those *agent-relative purposes*.

*

First we will look at the possibility of agent-neutral practical purposes.

When we discussed some related issues, back on Tuesday I think it was, we concluded, in essence, that there was no hope for us good ol' humans of keeping pace with the machines, in terms of ability to perform practically useful tasks. However, we might wonder whether it would perhaps be possible for us to avoid obsolescence if we are willing to enhance ourselves?

Clearly, for this to be an even remotely tenable proposition, something far beyond anabolic steroids or cognitive stimulants would be required. Even thoroughgoing genetic reengineering of the human organism would be completely inadequate here, since biological substrates are fundamentally limited in terms of power density, strength, computational speed, and many other basic parameters.[190]

Let us therefore consider some more radical potential upgrades. The most natural first step would be to upload your mind.[191] You could then increase your mental speed by moving to a faster computer. Having been digitized, it should also be easy to increase the number of neurons that you have, and to add new types of processing units as well as high-speed interconnects with external digital infrastructure. By means of such augmentations, you could become superintelligent.

I think we don't need to fuss too much about the non-mental parts of your being, since we seem mostly already reconciled to the reality of being surpassed by machines (and by many nonhuman animals) in terms of strength and speed and such. But if you insist on trying to maintain your competitive edge not only in mental but also physical capacities, then for a body we could give you, for example, a distributed network of advanced nanotechnological actuators, enabling you to grab and manipulate individual molecules—multiple Avogadro's numbers of them at a time; or, when needing to handle larger objects, your nanoscale actuators could morph into larger appendages strong enough to let you rip trees out of the ground.[192] You would also be able to leap tall buildings in a single bound, etc.

Then would you be able to keep up with the machines?

Color me skeptical. In fact, it is questionable both that we *could* keep pace with the machines and that, even if we were able to, we would *want* to. But there are several issues to disentangle here, so let's take them one by one.

The first thing to consider is what kinds of (agent-neutral) practical tasks there are that need to be performed in a mature civilization. Suppose, for example, that at technological maturity all infrastructure has reached such a high level of perfection that it doesn't need any maintenance or active operation—everything is self-healing and runs on its own accord like self-winding Swiss clockwork. There would presumably be some level of intelligent processing embedded throughout this infrastructure, but perhaps it could all be fairly low-level and routinized? In this scenario, most tasks are most efficiently performed using simple automated processes operating according to mostly precomputed plans whose execution requires little or no higher cognition or creativity.

We might imagine this as an extreme continuation of the deskilling trends we can observe historically in some areas of the human labor market. Once upon a time, a cobbler had need of some intelligence and resourcefulness to make a shoe. At some later stage of development, a shoe-producing worker might be employed in a factory and assigned the task of picking up soles from a box and placing them onto a conveyor belt: something which requires a lot less skill. Perhaps at technological maturity, all productive activities will have been optimized and decomposed into subtasks that do not require more than insect-level intelligence to carry out? Then the problem is not that humans are not capable enough, but that we are *too* capable, and therefore inefficient. It would (if nothing else) be energetically wasteful to employ a human mind to perform a task for which a simple processor, like the chip in an old-fashioned pocket calculator, would suffice.

Well, this is *conceivable*. However, I think it is likely that there will continue to be at least some practical need for higher levels of cognition for billions of years to come. I have put some examples of tasks in one of the handouts, which you can peruse later.

HANDOUT 18

SOME HIGH-LEVEL TASKS AT TECHNOLOGICAL MATURITY

The following examples of "high-level work" are task-domains which, even at technological maturity, plausibly cannot be routinized or automated by cognitive systems significantly less sophisticated, creative, or generally capable than human minds. (This list is not claimed to be exhaustive or definitive, merely suggestive of some possibilities.)

- *Physical expansion,* at least until the accessible parts of the universe have all been settled and optimized. Upon arriving at a new resource, cognitive work would be required to adapt existing plans to local circumstances—determining how to settle and bootstrap civilizational infrastructure in an optimal way, given the precise distribution of materials and factors such as temperature, pressure, radiation, etc.
- *Dealing with rare stochastic events.* Even optimal processes might have certain error rates, and occasionally those errors might compound in ways that cause unique problems. Sophisticated cognitive processes may be needed to detect, diagnose, and solve these problems.
- *Preparing for interactions with aliens.* Whether or not aliens exist, figuring out how best to interact with them might be regarded as important; and even though research into this topic would be subject to diminishing returns, it might retain a sufficiently high expected value to warrant ongoing investment; and this research may require high-level cognitive processing. (Understanding inaccessible superintelligences could also be very important.)
- *Policing and coordinating internal civilizational activities.* If civilization consists of a simple repeating pattern, such as a uniform grid of hedonium boxes, then there may be less need for this; but if we have a more multifarious and evolving civilization, with many independent loci of development, then there could remain a need for complex coordinating activity—for example, policing local developments so that they don't start some process that leads to a spreading corruption, or more general political work. This activity could require high-level cognitive processing.
- *Cultural products and attainments.* If it is viewed as important that certain cultural artifacts be produced by the civilization (whether for its own sake or so that members of the civilization have access to them) then a need for advanced cognitive processing may continue indefinitely, provided that either there

are an unlimited number of attainable steps on the quality ladder or some steps have an unlimited or growing number of products/attainments such that having or producing more of them is better. For example, one could imagine an indefinite sequence of ever harder-to-prove but still interesting math theorems; or an ever-increasing range of ever-more exquisite aesthetic experiences.

- *Ongoing cultural processing*. It is also possible to imagine an unending demand for creative cognition if we suppose that there are certain cultural artifacts or responses whose value depends on an ever-changing context. Suppose, for example, that there is value to an artwork or an experience that aptly sums up the prevailing zeitgeist, but that the creation of such artworks (or other independent developments) also perpetually change the zeitgeist, thus introducing the opportunity for new artworks and experiences that encapsulate the new zeitgeist; and so on.

So let us assume that there are at least some agent-neutral practical tasks which, even at technological maturity, call for high-level creativity, intuition, and advanced problem-solving. The question then is whether we, by undergoing uploading and radical enhancement, could remain competitive at those tasks. These could be tasks such as, for instance, analyzing the characteristics of possible alien civilizations, or dealing with rare cascading failures, or optimizing infrastructure plans according to the details of local circumstances. Could future versions of ourselves possibly be as good at such tasks as advanced AIs?

Student: I have a question! What if humans could perform those tasks by working together? Even if no individual can do them, that's no different from today, when no one person can build a smartphone or a jet plane. But we can still do those things by working in teams. Perhaps in the future, teams of enhanced humans, using much better collaboration tools, could solve even these super-hard problems that a technologically mature civilization has to wrestle with?

Bostrom: Well yes, very possibly. However, the question here is not about what we can *do*—individually or collectively—but about what we can *do efficiently.*

What if we could perform these tasks, but not efficiently enough to be economically competitive with machines designed for the purpose? More resources would then have to be expended if we do the tasks ourselves than if we outsource them. If we nevertheless do the tasks ourselves, we would be acting on some other motive than simply that the tasks be performed. We consider such motives elsewhere; but here we are focusing on whether we could remain practically useful at technological maturity in the same robust sense in which, for example, a car mechanic is useful at present. We would not be practically useful in that sense if our doing the tasks ourselves is more expensive and more wasteful of resources than having them done by machine.[193]

I believe that the most efficient method to accomplish these practical tasks will be by using AI systems designed for that purpose. If that is correct, then the only way that we could remain competitive on these tasks in the long run is by *becoming* such AI systems. Is *that* something we could possibly do?

*

A lot here depends on which criteria we adopt for personal identity. There's some set of tasks that need to be performed at technological maturity, and some set of possible minds and bodies that are optimal for performing them. Some of the tasks in question seem to require high levels and general forms of intelligence. Could we-as-we-are-now be transformed into some being that has a mind-body type that is optimal for some subset of these practical tasks, while preserving our personal identity?

In the present context, the relevant sense of remaining "the same person" is that some significant degree of prudential concern for the future entity would be appropriate from the perspective of present self-interest. Note that without this requirement that personal identity be preserved, the question becomes uninteresting. I mean, let's say that it would be possible to "transform" ourselves into a task-optimal system by arranging to have ourselves disassembled into nucleons and then for these to be reassembled into elements that are used to construct an optimal machine: in no way would that help show that *our* labor could remain practically useful at technological maturity.

*

I actually think it is not unlikely that for at least some of the functional tasks that would need to be performed at technological maturity, we could perhaps in principle be able to become the type of being who would be most efficient at accomplishing them.

In making this tentative suggestion, I am assuming a fairly expansive conception of personal identity: one that allows for a person's character to undergo rather fundamental change over time and for them still to count as "the same person" they were before—provided that the transformation is sufficiently gradual and continuous, and perhaps provided also that it satisfies some additional constraints, such as that the change to some degree involves the person's own autonomous choice.

One argument for such an expansive conception of personal identity—one that doesn't require much by way of qualitative similarity between two person-segments imputed to belong to one and the same person—is that we normally hold that a young child is the same person as the adult they grow into, even though both the physical substance and the morphological and psychological characteristics of the adult are quite different from those of the child. If such a radical transformation is consistent with the preservation of personal identity, then it might also be plausible that we could remain "the same person" even if, upon having reached adulthood, we continued to grow up, incrementally, in a long slow series of small self-chosen steps, over time transforming ourselves into some kind of AI-like program that is optimal for performing some functional tasks in an advanced civilization.

This might seem like good news, from the point of view of us being able to secure the kind of purpose that comes from having raw practical utility even in a plastic utopia. However, I think most of us would, upon closer inspection, find this path toward remaining employable to be unappealing.

One problem is that even if we *could* become a being who is practically useful in utopia, we might not actually *want* to be such a being. A useful being would be tightly optimized for some set of tasks. It would not expend memory or computational resources on matters unrelated to those tasks. Yet much of what we value in life may be irrelevant to the basic instrumental tasks of a technologically mature civilization (such as the ones I've listed in the handout). Be it childhood memories, personal friendships, a love of music, the enjoyment of food, and so on, or really most any kind of leisure activity or idleness: such ingredients, which we might regard as important for our eudaemonia, would quite probably be dispensable, and therefore wasteful, and therefore not part of a task-optimal system, and therefore would likely have to go by the wayside if we were to decide to try to stay in the race.[194]

Furthermore, to the extent that some time and resources *would* need to be spent by a task-optimal system on such "frills"—for instance, because the task to which it was devoted involved trying to model possible alien civilizations to whom some such extravagances might be important—it is not clear that *the way* the task-optimal system would spend that time and those resources would correspond at all closely to the way that we would choose to spend them if we were not subject to the requirement that our operation must be functionally optimal and we were instead directly pursuing the goal of having a good life. For example, maximally efficiently analyzing how an alien culture might relate to music could well involve cognitive activities quite different from those that we would engage in if we were actually ourselves appreciating and enjoying music.

Consequently, while it may be possible for us to become task-optimal systems, and while it is *conceivable* that some task-optimal systems would maintain some of the faculties and behaviors that we currently value in ourselves or that we would independently wish to develop for the sake of improving our well-being, it is far from certain that these conditions do in fact obtain—and it would be quite surprising if they did so to anything close to a eudaemonically ideal degree. If task-optimal systems preserved eudaemonic functioning only to a very limited degree, then it is doubtful that whatever extra value they might derive from possessing the particular kind of purpose-value that requires having raw practical utility would be sufficient to compensate for the sacrifices in other values that transforming ourselves into such systems would entail.

A second problem is that the *path* toward becoming a task-optimal system might be either unavailable or unappealing, even if the endpoint itself were not objectionable. For example, once there already exist task-optimal systems for performing the relevant tasks in a technologically mature civilization, then your opportunity to be practically useful has presumably passed. There would be no economic advantage in refashioning yourself to become efficient at performing some instrumental task when it is already being accomplished by systems optimized for performing it. Even if demand were to increase over time, creating a need for additional task-optimal systems, it would still not be efficient for you to transform yourself into such a system if, as seems highly likely, there were more expedient ways of producing task-optimal systems—such as by copying the existing systems, or in any case producing more of them via methods that are not constrained by the need to preserve your personal identity.

To have a chance of being task-optimal at technological maturity, therefore, you would probably have to start your transformation early and proceed at close to maximal speed. You would have to be among the first to upload, and then adopt further enhancements and augmentations almost as soon as they become available. You would have to ruthlessly eliminate any inefficiencies. As soon as technology permits, you would have to delete all parts of your mind that are not useful for the tasks to which you intend to dedicate yourself. Such a rush toward specialized perfection would involve additional sacrifices on top accepting that the being you ultimately become will have jettisoned much of what is valuable in life. We could compare the case to that of a human child who is hothoused from early infancy and trained with maximal rigor for the sole goal of being made into the best possible mathematician, pianist, or gymnast—at the cost of foregoing all the enjoyments of a normal childhood, and with the eventual result of becoming an adult prodigy who is stunted in all areas except their one designated field of excellence. The difference is that, in the case of a biological human being, there are limits to how far such an approach can be taken before it becomes counterproductive even on its own terms (and beyond which any further strictness or narrowing of focus produces burnout, dysfunctional rigidity, psychiatric problems, or rebellion, rather than additional gains in the targeted ability); whereas, with minds that are increasingly the product of engineering, it is plausible that a far more unbalanced and monomaniacal focus on a specific task would continue to be the most effective way of achieving the highest possible level of task performance.

Even if you took the most single-minded and uncompromising approach to reshaping yourself for optimal performance—one that pays no heed to quality of life during the optimization process or subsequently—it still may be impossible to remain competitive with machines built de novo for the purpose. In fact, I'd say it's likely that the optimal contribution one could make, at technological maturity, to the accomplishment of practical tasks, would not be by improving one's own capacities and then working on those tasks oneself, but instead by donating first all of one's financial resources and finally also the matter making up one's own body and brain to the construction and operation of systems optimized for performing the tasks in question. In other words, at technological maturity, your greatest practical usefulness would be as feedstock to the machines. Not quite "the dignity of labor" of yesteryear's proletariat—but you could perhaps aspire to the glory of having your atoms used to form a segment of cooling pipe in a datacenter running an AI that calculates deployment trajectories for mining equipment.

One of my friends—the never-bored one—confided to me many years ago that he wants to become an information transfer protocol. You know, like the TCP/IP standards upon which the internet runs, or the DNA code, or the Roman alphabet. Lock-in effects can make such protocols extremely durable. It might be tricky even for an advanced civilization to coordinate its way out of the local optimum of a globally suboptimal standard. We remain stuck with the (allegedly) slow QWERTY keyboard layout, for example, many decades after people stopped using mechanical typewriters; and a few countries even continue to cling to the imperial system of measurement units.

So my friend reckons that if he can become a new standard, he could then enjoy great longevity. Well, now we can see that this destiny might not only give him great longevity (which should hopefully be readily available to everyone in utopia) but also great practical usefulness, and hence purpose. I mean, how useful is the TCP/IP protocol? Very. So we might imagine some more advanced information transfer protocol of the future, perhaps a compression algorithm that requires sentient mental activity to compute, and that my friend could end up as something like that. You scoff—but compared to some other forms of "immortality" that people have coveted, such as having their likeness imprinted on a postage stamp, this fate could have a stronger claim to constituting an actual kind of survival and preservation of personal identity!

I'm speaking half in jest and half in earnest. There's more that could be said, but we better move on: we still have a lot more ground to cover.

Student: May I ask a question?

Bostrom: Fire away.

Student: I'm a little confused here. I thought the key benefit of getting to a post-scarcity civilization would be that we could do all kinds of things that aren't "task-optimal" but are still fun. Like, maybe I want to design and ride in my own spaceship, even if it isn't quite as fast or efficient as one made by superintelligent AIs. Why do we still have to optimize everything, once we have enough and more than enough to cover all our needs? Maybe that's a stupid question.

Bostrom: In my experience, when somebody asks what they think of as a "stupid question" it is usually a question that many people in the audience were secretly hoping somebody else would ask. Maybe all lectures should be held in the dark for this reason! That might also help with the opposite problem—the "clever questions" that get asked not because anybody wants to know the answer but because the questioner looks good for having posed them. Though on second thought… I'm not sure what would happen to my profession if that incentive were swept away. So perhaps it's best things are kept as they are.

Well, the utopians *don't* have to optimize everything! Certainly not in that sense. Oversimplifying a bit, our preset rhetorical situation is something closer to this: In a plastic utopia there are many things we want that we could have; many values that we could realize to very high degrees. That is great! Among the good things about the postulated condition is that we could have plenty of time to enjoy our hobbies. In fact, the condition would have so many truly good things about it that the interesting question becomes whether there is anything of value that we *can't* have there? And in particular, is there anything of significant value that we have now but that we would necessarily forsake in utopia?

It might seem like caviling to focus attention on any possible deficits of utopia when it would overflow with so much obvious good. But I'm not here to deliver a pep talk. Our goal in these lectures (or one of them, at any rate—in truth there is more than one goal) is to sharpen our analytical tools and to gain a deeper understanding of the topic we are studying; and to that end it is expedient to pay special attention to *the contours* of utopia—its possible limits, exceptions, and other complications.

This is where our discussion of purpose comes in: purpose being an example of something one might think has value yet might appear to be undermined in the postulated condition of radical plasticity. At this point in our discussion, we have already seen how certain kinds of purpose could in fact be secured in utopia; and we are now inquiring into a kind of purpose that seems especially elusive—the kind that requires our efforts to have "practical utility". In the course of investigating whether we could have this special kind of purpose in utopia, we have just observed that, among the practical tasks that fall under the rubric of "agent-neutral", while there are some such tasks that would still need to be performed in utopia, it would probably not be efficient for us to be the ones performing them. What this means is that so far we have not shown that purpose requiring practical utility is something we could have in utopia.

We did note that there might be some scenarios in which we possibly could, just barely, continue to be practically useful at technological maturity in the performance of functional tasks. Those scenarios, however, have other features that would likely make them undesirable—including that they might involve foregoing free time and hobbies. So the thrust here is not "let's give up on having free time in utopia" but rather "we should probably not rely on this method of securing practical purpose, even if, just conceivably, it might be possible to do so".

Is that clearer?

Student: Yes, I think so.

Bostrom: We certainly wouldn't want anybody to lack the time to build their own spaceship, if that's something they want to do.

I guess I should also explicitly state at some point—I might as well do it here—that even if some particular form of utopian life (purposeful or otherwise) could be shown to be more prudentially desirable than other forms, that would not imply that the more desirable form ought to be forced or foisted upon anybody! Questions of political philosophy, including issues such as state paternalism, or the boundary between individual versus collective decision-making, or distributive fairness, or the proper form of government—these fall outside the scope of these lectures.

(For what it's worth, I tend to imagine that there probably ought to be a big role for individual autonomy and self-determination, combined with respect and compassion for a wide range of different types of being and accommodation of many different types of interests. If you have a different or more definite conception of what the ideal political order would look like in a technologically mature society, feel free to substitute that in your own imagination! But one way or another, somebody or something would ultimately need to confront the value questions that we are exploring in these lectures.)

*

Okay, so much for the idea of remaining useful in the performance of agent-neutral practical tasks in a technologically mature civilization. Now let us consider whether we could remain useful in the performance of agent-relative tasks.

These would be tasks in which what needs to be done constitutively involves the active contribution of specified agents. It is *these agents doing it* that is needed, rather than simply that it gets done. The case that we are interested in is one where the agent-relative task needs to be done *by us*. (The "us" will typically refer to some specific human individual, or an enhanced continuer of a particular human individual; but in some instances it might instead refer to humanity in general or to some designated group—the basic reasoning is the same either way, and the meaning should be evident from the context.)

*

An example of agent-relative purpose will make it clearer what I have in mind.

Consider this value: honoring your ancestors. What is required to realize it? The prescriptions vary between cultures, but let's assume it requires that you sometimes remember your late father, that you think of him fondly and appreciatively, that you cherish the time you got to spend together, that you treat his remains respectfully, and that you continue to take his preference into account. For instance, if your late father always emphasized the value of honesty, then one way of honoring him would be to try to conduct yourself honestly, even after his death and in situations where doing so may be inconvenient.

These requirements, by their very nature, cannot be met by outsourcing the requisite sentiments and actions. Even if you could construct a machine that ruminated on the merits of your late father, and experienced fond feelings for him, and behaved with impeccable honesty in all its dealings, it would not fully satisfy. The value of honoring your ancestors demands that you do these things yourself.

The purpose that this value gives you is not artificial or arbitrary, and it is not gifted or intentionally generated for the sake of giving you a purpose. It is also distinct from the kind of purpose that comes from goals that one sets oneself in order to have certain active experiences, such as the strenuous sensations of scaling a mountain. In the case of the climber, his purpose lacks an external grounding, its justification being only that it enables the activity of pursuing it. Whereas for the person who is honoring his ancestors, his purpose does have an external grounding. You are not doing it (we can presume) in order to generate the pleasure of an uplifting honoring experience, or even because you think that your life goes better if it contains some quantity of honoring-activity. Rather, you are doing it because you think that your father deserves to be honored by you—owing to who he was, or what he did for you, or the special way he was related to you. This external grounding should make the purpose fully legitimate: which is to say, fully capable of giving you whatever kind of value that having a purpose can provide, even on views that are very restrictive in what kinds of purposes they regard as capable of undergirding this value. It's in every sense a "real" purpose—nothing fake or finagled about it.

*

Having found one instance of practical purpose that could remain relevant in a plastic utopia, we can look around for others. I think we will be able to find many more, although of course this depends quite sensitively on what theory of value or well-being we adopt. For example, if you think that only pleasure, or only mathematical insight, has value, then there would likely be no opportunity for you to have any practical purpose in utopia—though then again: if only pleasure or mathematical insight were valuable, purpose would not be valuable, so you wouldn't be missing out. But generally speaking, the more pluralistic your inventory of values, and the more that at least some of your values are tied up with complicated patterns of human behavior or preference or society or history or spirituality, the more plausible it is that you will find plenty of practical purposes in utopia.

Discovering these purposes will mostly have to be left for the utopians themselves. I will merely gesticulate indistinctly in some of the directions wherein such purposes may be sought.

First, we can generalize the example I just gave, that of honoring your ancestors. You may also honor fallen comrades, benefactors, and historical heroes. More broadly, you may find it valuable to honor or continue to follow and adhere to various traditions. This category of purposes, which can be quite wide, might serve to structure and constrain our otherwise potentially quite amorphous existence in a plastic utopia, giving us many things to do that we cannot outsource.

Second, we may have reason to follow through on earlier commitments and projects that we have (perhaps implicitly) undertaken. In some cases, this could require that we pursue them in the spirit that launched them, which might involve confining ourselves to only certain classes of means: again preventing wholesale outsourcing.

Third, we have the broad category of what we might term *the aesthetic*, where the expressive significance, and therefore the value, of what is done often depends on how it was done and by whom. This is most clearly seen in some works of modern art, though it applies in a diffuser form far more generally. I would say this: We should not think about this in terms of museums, art studios, concert halls, or public architectures, but rather in terms of how to live beautifully. Every moment of life offers rich opportunities in this regard; and a beautifully expressive gesture can as easily take the form of a refraining as one of a doing. We could develop a whole aesthetic of Noes, which precisely by refusing easy ways of achieving makes possible the glory of achieving in harder ways—makes possible a greater Yes.

Fourth, we have the domain of the spiritual or supernatural. In parts, this category might overlap with the preceding three; but we also have here the distinct possibility that there might be important connections between what happens inside the bubble of utopia (which might be internally organized so as to maximize our positive affordances) and what is the case outside of that bubble. A being in the divine sphere might have preferences and powers over what transpires inside our bubble, and in particular might be distinctively concerned about what we do ourselves as opposed to what we bring about vicariously via automation or other forms of technological indirection. If this is the case—or if at any rate we are not in a position to exclude the possibility of it being the case—then we may identify practical purposes in a plastic utopia that flow from such considerations. (Purposes deriving from transcendental sources could be either agent-neutral or agent-relative.)

*

It might appear—with the possible exception of the fourth category—that the values at stake would be relatively minor, and that the purposes they could underpin would therefore be correspondingly weak. But we should recall the remarks I made earlier about pupillary dilation. I suggested that it would be appropriate, once starker and more immediate needs and moral imperatives have been satisfied, to increase the apertures of our evaluative lenses so as to allow fainter normative considerations to come into sight and prominence. If this is right, then it could well behoove us, as we migrate onto utopian grounds, to begin to regard tasks like honoring, following through on implicit personal undertakings, and various forms of complex aesthetic expression as quite serious and important—as calling upon us to make considerable personal investments of time, effort, attention, and selective priority, and as thereby potentially giving us multitudinous purposes that are quite sufficiently real to give us much of whatever value there is to be had in having purpose.

*

Hmm, I'm not sure whether our discussion about purpose has been autotelic but it has definitely been bradytelic…

I should wrap this up. In conclusion, while we will likely become incapable of contributing to agent-neutral practical tasks at technological maturity, there are various kinds of purpose that will remain available to us (summarized in one of the handouts). It is true that in a plastic utopia, we would be able to get along very well without making even the slightest effort. We *could* have not only room service but also mouth service, esophagus service, mitochondria service—really, all-inclusive full-package "life service", with a five-star rating. However, if a condition of aimless indolence is not to our liking, we may either self-generate some suitable purpose or induce one by technological means—either for the sake of giving us active experience or simply in order to have purpose. Alternatively, we may be given purpose by somebody else (or by some cultural system). Or, if we prefer that our purposes not be created arbitrarily just in order so that we have purpose, then we could turn to the plausible sources of natural purpose that I described, particularly ones deriving from various agent-relative tasks that are inextricably entangled with our own agency and for that reason not outsourceable, and possibly also to purposes of a supernatural or religious derivation.

HANDOUT 19

SOURCES OF HUMAN PURPOSE IN UTOPIA

Artificial purpose

(deliberately created, either for its own sake or to enable active experience of autotelic activity)

Self-imposed

- Resolving to adopt a goal
- Neurotechnologically inducing a goal
- Placing oneself in a challenging situation

Given

- By other individuals
- By some collective or cultural process

Natural and supernatural purpose

(derived from some independent external motivational grounding)

Agent-neutral

High-level tasks that remain relevant at technological maturity, e.g.

- Physical expansion
- Dealing with rare stochastic events
- Preparing for interactions with aliens
- Policing and coordinating internal civilizational activities
- Cultural products and attainments
- Ongoing cultural processing

[Note: Humans may be uncompetitive on these tasks, even if highly enhanced in identity-preserving ways. Even if we could remain competitive, it may involve unacceptable costs. However, if additional constraints are imposed on use of machines—deriving for instance from agent-relative purposes—then it is possible that human purposes could be located in these task areas.]

Spiritually or supernaturally mandated attitudes and performances

Agent-relative

- Honoring people, traditions
- Following through on commitments and (possibly implicit) undertakings
- Aesthetic expression
- Spiritually or supernaturally mandated attitudes and performances

Alright, that's it. Time has run out, as it is wont to do in the end.

I have a bunch of copies of the assigned consternation—I mean the assigned reading here. I'll put them right outside the door, and you can pick it up as you leave.

Lastly, a reminder that the final installment in this lecture series will take place tomorrow. It will be open to the public, but everyone in this class will get in for free, and there should be reserved seats available. See you there!

A fair deal

Firafix: I've got copies. "The Exaltation of ThermoRex", hmm…

Tessius: Thanks.

Kelvin: Thanks. Now we have two options.

Firafix: What are they?

Kelvin: One is to go home and change into dry clothes. Then we could all meet afterwards.

Firafix: And the other?

Kelvin: —

Firafix: The hot springs again?! You are like a duck.

Kelvin: I am merely presenting options.

Firafix: What say you, Tessius?

Tessius: Fighting water with water—seems galaxy brained to me!

Firafix: Ok, you guys have it. But you'll have to tell me whether my analysis of yesterday's homework question is any good. And we pick up a bag of pears on the way.

Kelvin: Deal!

The Exaltation of ThermoRex

Part I

1.
Once the country's top law firms had taken their many pounds of flesh, there was still plenty left on the immense carcass that was the estate of Herr von Heißerhof.

Heißerhof, the country's leading industrialist, had bequeathed his vast fortune to a foundation established for the purpose of benefiting a particular portable electric room heater. We will refer to this room heater by its brand name, "ThermoRex". Heißerhof, who'd developed a reputation as being a bit of a misanthrope, had often been overheard saying that ThermoRex had done more for his welfare and comfort than any of his human companions ever had. The heater, he maintained, had always been faithful to him, keeping him warm through many a winter month when the northern winds howled around his castle; furthermore, ThermoRex had never plotted against him or sought to extract any personal advantage: conduct which, Heißerhof said, elevated it on the ladder of merit far above any man or woman he had known—including his own two children, one of whom was serving an eight-year sentence for a series of sex offenses, the other, plunged even further in von Heißerhof's estimation, having married a union organizer.

A suit was brought by the would-be heirs to have the will invalidated. So overpowering was their spite, however, that they insisted—against the recommendations of their legal team—on seeking a complete annulment on the basis of the testator having been *non sana mente.* Their case collapsed when the court discovered that Heißerhof had, in the period during which he composed his testament, actively and successfully managed an industrial consortium with operations across twenty-two countries; not only that, he had also remained a world-class bridge player up until the very end. In fact, his demise had occurred during the semifinals of the world bridge tournament, his cold stiff fingers, it was subsequently found, clutching a King-high Straight Flush (a hand which, in all likelihood, contributed to his death, as it was conjectured to have been the reason he had delayed seeking medical attention at the onset of his fatal heart attack).

Consequently, following the conclusion of the legal proceedings, ownership of all of von Heißerhof's assets was transferred to the legal entity "Foundation to Benefit the portable room heater ThermoRex serial number 126-89-23-79-81".

A complex institutional mechanism, many years in the making, now clicked into action. A lengthy "operating manual" was unsealed from Heißerhof's *Nachlass*, with detailed instructions to his executors. Several interlocking organizations were incorporated, in a variety of onshore and offshore jurisdictions, each with its own carefully defined objectives and bylaws. The key officers had been handpicked by Heißerhof—individuals he knew he could trust to carry out his wishes; but in any case the legal structures had been so cleverly contrived, with multiple overlapping and mutually reinforcing checks and balances, that it would have been difficult even for a cabal of conspiring insiders to subvert his intentions. Even from beyond the grave, Heißerhof's spirit remained firmly in control—and it was absolutely determined to ensure that the resources he had spent a lifetime amassing would be used, entirely and exclusively, for the benefit of ThermoRex.

2.
At the center of this construct was a twelve-member Board of Trustees, composed of Heißerhof's closest associates. In their twenty-three hands (one had a hook) rested the responsibility to dispose of a fortune, which, it was rumored, was of a size sufficient to bail out the insolvent pension system of a medium-sized country.

"But how do we do this?", asked the Chairman. "Let's give it another read through."

With furrowed brows the Trustees read. The text was quite emphatic about the Foundation's goal: it was to "work for the benefit of", "promote the interests of", and "facilitate the general flourishing, well-being, and ideal functioning of" ThermoRex. To the extent that these objectives were unclear, the instructions required the Trustees to "act as if they were motivated by a selfless love for this unique and wonderful being, ThermoRex".

—"Ideas? Who wants to go first? Günther, what do you think?"

Günther Altman, a silver-haired gentleman in a three-piece suit, was the oldest member of the Board and had known Heißerhof since their school days.

—"I move that we order in some cognac," Günther replied. This proposal was met with general approbation.

After the liquid had been served and imbibed, the conversation flowed more easily, and several decisions were reached. These focused on ThermoRex's physical safety. The device would be placed under around-the-clock armed guard. A report would be commissioned from a top engineering firm to evaluate risks from flooding, earthquake, fire, and power surges, and to recommend measures to reduce these hazards. This work was to be completed on an expedited timescale.

Satisfied with their progress, the Trustees retired for the day.

3.
Next morning, the twelve reconvened, and the Treasurer asked to have the floor. One of the younger members of the group, he had a focused look and wore round wireframe glasses. He explained that he had run some numbers, and based on the average rate of return on Heißerhof's investments in recent years, it was likely that the value of the Foundation's assets—after subtracting

the outlays agreed to the previous day—had *increased* since the start of their deliberations. In a sense, therefore, they had not made progress, but had instead fallen further behind on accomplishing their objective. If they were ever to discharge their fiduciary duties, they would need to think and act on a far more ambitious scale.

The Board quickly decided on some upgrades to the measures adopted the previous day, such as increasing the size of the guard unit from four to eight men. But it was gradually dawning on them that in order to make even a dent in the task with which they had been entrusted, it would be necessary to widen their purview beyond physical protection. They needed to find ways to positively benefit ThermoRex—to elevate its welfare above the baseline level.

—"What does a room heater *want*?", pleaded the Chairman. "What does it *need*? *Think! Think!*"

—"Maybe it wants the room to be warm?", someone suggested.

—"Or the whole building," another chimed in. "It wants its house to be warm."

—"Why not the whole Earth, then?"

—"Well, that would be too much. We can't afford to heat the entire planet. Besides, it would not be responsible to impose a higher temperature on everybody else."

Back and forth the conversation went, and glasses and cognac were brought in again.

—"Do you know how a space heater works?" said one of the Trustees, who had previously run the R&D division in one of the conglomerate's subsidiaries. "There is a thermostat inside. It measures what the temperature is, and if it is below the setpoint, it activates the heating element. Once the temperature reaches the desired level, or maybe one degree above, the heating element is switched off. But you see, all that matters is what the thermometer says—ThermoRex has no way of knowing what the *actual* temperature is anywhere else. So we could keep it in a small room, or in a closet even: and so long as the temperature always matches the setpoint, it will be perfectly happy!"

—"No, no, that argument proves too much!" another Trustee objected. "By that reasoning, we could equally say that you only care about what happens inside your brain. But I, at least, care about other things as well. Things outside myself. Including things I might never find out about."

—"What do you mean?"

—"Well, for example, I would not want my wife to have an affair with her tennis instructor. Even if I would never find out about it. I don't want my life to be based on a big illusion."

—"But if you'd never find out, and she behaved toward you exactly as she would if she were not having an affair, it wouldn't matter!"

—"That is a very French attitude, coming from you, Heinz!" somebody interjected.

—"But it *would* matter!" said the objecting Trustee. "It would matter to me."

—"I, for one, agree," said the Chairman. "I don't believe that Herr von Heißerhof would have been pleased if we made a cuckold out of his ThermoRex!"

The lively discussion continued throughout the day. By evening, although the Trustees had not attained unanimity on the criteria for thermostat flourishing, they had reached agreement on some points.

The first was that if there were two views, one according to which it did not make sense to speak about thermostat flourishing, the other according to which it *did* make sense, then they should condition their efforts on the latter, since if the former were true it wouldn't matter what they did or did not decide to do. More generally, they should try to benefit ThermoRex according to as many different theories as possible, at least for theories that had any appreciable support. Theories that were more plausible or that enjoyed wider support would be given greater weight.

The second point of agreement was that they should not rely solely on their own efforts, but should also seek the advice of external experts. They were actually quite eager to delegate the work of philosophical analysis, because although they had enjoyed the intellectual sojourn (which reminded them of their student days), they were on the whole more comfortable operating closer to their métier, which, for all of them, lay in the art of conducting business.

The Board spent the remaining days of their executive session mapping out a set of initiatives for soliciting ideas and opinions from a wide range of outside sources, including not only philosophers but also poets, engineers, scientists, and theologians, as well as "ordinary men and women". It would involve research grants, opinion surveys, focus groups, citizen juries, and essay contests. Funding would be lavish and implementation fast-paced and coordinated by an elite operations team.

4.

These investments had been made, but had not yet had time to fructify, when the Foundation was confronted with a crisis.

A weekly magazine (which, because its staff had degrees from the oldest universities, regarded itself as the fount of good sense on all issues around the globe) had published an editorial calling for the Foundation to be dissolved and its assets transferred to a public trust. It argued that the objective of promoting the welfare of a room heater was nonsensical and frivolous. This charge was soon taken up by an avalanche of other pundits, each one coming up with their own favored way of spending the assets that they proposed to confiscate.

The Trustees convened for an extraordinary session. The adverse publicity, they assessed, threatened the welfare of ThermoRex, and therefore the Foundation's mission, in several distinct ways.

First, there was the obvious: the polemics might gain enough traction to provoke a legislative response. This was regarded as the least serious of the concerns. Foundation officials had strong connections in government, and it was felt that senior people in relevant departments could be relied upon to squelch any untoward initiative. In any case, ThermoRex's assets were not actually owned by the Foundation itself, but by a group of offshore trust companies linked by complex contractual arrangements. Short of a coordinated international effort, which did not seem a realistic prospect, these safehavens were beyond the reach of national courts and parliaments.

Second, the animadversions could harm ThermoRex's interests in less formal ways, such as by impeding the ongoing program of external consultation, or by making it harder for the Foundation to recruit talented officers. There was also a concern that the negative sentiment could stir up hoodlums, who might attempt to commit acts of vandalism against ThermoRex.

And third—this was a more subtle consideration, but it gradually came to be recognized as the dominant concern—there was the possibility that it would be a bad thing *in itself* for ThermoRex to be publicly maligned. At least, this consideration did not seem that much of a stretch, *given* that ThermoRex had any welfare interests at all: and *that* was a premiss the Trustees felt they had to accept as the working assumption for their entire endeavor.

To assist them in dealing with the crisis, the Foundation retained the services of Abracadabra Communications, one of the country's top public relations firms. Their response, it was decided, should focus on two messages:

- Von Heißerhof's wealth had been acquired through honest toil. He had started at age six, selling apples he'd picked from his grandmother's backyard. From such humble beginnings, he had created a business that had provided tens of thousands of jobs and brought billions of tax revenues into the national treasury. Heißerhof had worked notoriously long hours, continuing his toil well into his eighties. Legally and morally, it was his money to spend, and nobody else's business how he chose to spend it.
- Across society vast amounts were spent each year on products that were harmful (such as cigarettes, alcohol, unhealthy food, coal mining equipment, cluster munitions, and envy-inducing status symbols of all kinds). It was unfair to single out for rebuke a private project which was, even if arguably wasteful, at any rate *harmless*, when these positively harmful practices were being widely condoned and indulged (including, they would not refrain from pointing out, by many of the Foundation's most vociferous critics).

Whether this counter-salvo would have been effective may never be known; because the next day, it came to light that a celebrated striker on the national soccer team had said something offensive and the incident had been caught on video. Several teammates came forward in support of the striker, which only further incensed the censoriat. Thanks to Soccergate opening a new outlet for the bilestream of public rancor, interest in the Heißerhof estate story quickly dried up.

5.

The Foundation acquired all the properties neighboring the building in which ThermoRex was housed. They hired an on-site fire marshal, an electrical engineer, and two dozen additional guards, including a canine patrol unit. They paid all personnel with site access at double the market rate to make them harder to bribe. They commissioned a private security firm to perform background checks and to continually monitor all staff. Then they hired a second security firm to keep an eye on the first.

And still the spend rate was too low. All the obvious ways of benefiting ThermoRex had turned out to be disappointingly affordable.

The Trustees pinned their hopes on the big consultation push they had initiated. And indeed, after wading through the results, they did identify some promising leads.

There was an interesting cluster of ideas that centered on building up ThermoRex's presence as a cultural phenomenon. For example, one proposal that caught the Board's attention was to develop educational modules on various ThermoRex-related topics—electrical engineering, thermodynamics, industrial design, supply chain management, business history. Award-winning film directors would work with leading scientific authorities to create teaching materials that were both engaging and informative, and famous actors would be hired to do the narration. Market research would be used to determine which versions worked best for different audience segments. Textbooks would be printed, lab kits manufactured; and everything would be given away for free. Schools and universities could then be lobbied to integrate these resources into their curricula.

—"And here is how this would benefit ThermoRex…"

The proposal had been put together by an interdisciplinary team of educators, and its lead author, a philosophy professor, had been invited to present the ideas to the Board. She now advanced to her final slide, which progressively disclosed a series of bullet points.

- *ThermoRex would come to have significant beneficial effects on the world.*

"Through its Foundation and with your assistance, ThermoRex would provide excellent educational services, which is a public good."

- *These beneficial effects would closely reflect and derive from the specifics of ThermoRex's nature.*

"You can argue, and some philosophers have indeed argued, that achievement is one of the things that makes somebody's life go better for them. We think that achievements that provide significant positive contributions to the world, and that are more closely linked to the achiever's personality and derive from their own distinctive skills, assets, or character traits, may be especially valuable in this regard, making the achiever's life more meaningful and overall more worthwhile."

"Now, somebody might object," she continued, "that ThermoRex is not consciously achieving these outcomes; and, well, that is true. But if, for example, a poet gets into some state of inspiration, and straight from the bourne of her unconscious mind upwells a great poem, maybe while the poet is asleep, we would still count that as a valuable achievement. So we don't think that the lack of conscious effort and awareness is a complete showstopper here."

"There is more. This initiative would also provide the following benefits."

- *ThermoRex would become more famous.*
- *ThermoRex would become better understood.*
- *ThermoRex would become widely and justly praised and appreciated.*

"These can plausibly also be regarded as prudential goods, contributing to somebody's objective flourishing. In summary, we think a strong case can be made that it would be well-being-increasing for ThermoRex to be the wellspring of all the learning that this program would enable!"

The Trustees found these arguments persuasive enough to approve the requested funding. They also greenlighted a number of similar initiatives, which aimed to enhance ThermoRex's stature as a public figure by means of prosocial contributions that reflected the benefactor's unique style and character.

As these seeds sprouted, they were carefully tended to by the Foundation's functionaries and amply watered by its affluence.

Over time, they grew into a nice grove, under whose boughs (which became laden with cultural reference points, symbols, memes, and stories) congregated an energetic flock of humming, buzzing, and chirping followers—colloquially known (tongue-in-cheek) as the Cult of ThermoRex. These fans organized activities throughout the year, culminating in an annual carnival, which was eventually made into a public holiday by an act of Parliament.

Part II

6.

Years had passed, and the Trustees were gathering for their annual retreat in Heißerhof's old castle to review progress and to plan future steps. Autumn colors dressed the alpine slopes which could be seen through the windows, and gentle rain drizzled over the sturdy stoneworks.

The Chairman opened the proceedings. Much had been achieved, he said, yet there was a sense that existing activities had reached a culmination point. The ThermoRex cult had established real roots in popular culture and no longer needed to rely on the largesse of the Foundation for its sustenance. While it might make sense to continue to provide some subsidies, it was not believed that increasing their level would be helpful—excessive liberality could risk undermining the spontaneous expressions of public affection and support that ThermoRex currently enjoyed (and which were hypothesized to contribute more to the heating element's well-being than "bought love" generated from paid advertising and the like).

The other main operating expense for the Foundation was in providing for the physical needs of ThermoRex, particularly security. In this category, too, the Foundation saw diminishing returns. Hiring still more security guards would not significantly increase ThermoRex's safety, when the current provisions were already more than adequate; and it would increase the number of people who could potentially be bribed or corrupted by a hostile actor. Expanding the security perimeter was an option, but it was not clear that it would provide any practical benefit; and it would risk antagonizing the residents who would be displaced in the process. Only some minor security improvements were recommended, and these were not costly.

—"Next I was thinking that we could hear from our Treasurer. Eduard, would you care to give us a picture of how we are doing on the economic front?"

Eduard, who still wore the same wireframe glasses but now had gray hair around his temples, proceeded to give an update on the Foundation's financial position: it was very solid. Owing to a strong performance by the asset management team, helped by a recent bull market in the industrial sector, the size of the endowment, adjusted for inflation, had about doubled since Heißerhof's passing.

—"Thank you, Eduard."

—"So," the Chairman concluded, "I think that if we reflect on these facts, it is possible that we might find that the time has to consider broadening our views on certain topics."

The meaning of this remark was not lost on anyone in the room. It meant that the Chairman was coming around to the view that the Foundation should begin funding proposals from "the third cluster". The question of the third cluster had been a contentious one from the beginning, reflecting a deep philosophical divide among the people chosen to represent ThermoRex's interest.

The "first cluster" comprised measures to satisfy ThermoRex's basic needs, such as for electricity and physical security. These were generally uncontroversial and had been prioritized in the early days. The "second cluster" was geared toward ThermoRex's emergence as a member of society, and included measures to help ThermoRex make contributions to the common good and to receive due recognition for its unique nature and its good works. Most of the Trustees supported investing in second-cluster projects; and even those who were not convinced that social accomplishments truly benefited ThermoRex saw little reason to oppose these outlays, since they weren't doing any harm and the Foundation could easily afford them.

The third cluster, on the other hand, had met with resistance from several of the Trustees, who adjudged it potentially injurious to ThermoRex's interests. This third cluster consisted of proposals to *enhance* ThermoRex in various ways, to improve its basic functionality or endow it with novel capacities. The Chairman had been among those who were skeptical of such interventions, and the Board had made a decision to postpone any forays into enhancement territory until some unspecified later date—which seemed now to have arrived.

Admittedly, it was not strictly true that the Foundation had hitherto done nothing to alter ThermoRex's physical makeup. Several such interventions had been undertaken. One had involved the replacement of a missing button; another, the removal of dust and dirt from ThermoRex using pressurized air; yet another, the reattachment of an electrical certification sticker that had come loose from the heater's power plug. Each of these operations had been conducted with the utmost care. In the case of the loose sticker, for example, the Foundation had hired a distinguished neurosurgeon to affix the sticker with the same adhesive as had been used in the original production run (as confirmed by spectrometric analysis by two independent labs).

It could plausibly be maintained, however, that these earlier interventions had been merely therapeutic. They had pursued the comparatively modest aim of restoring ThermoRex to an earlier undamaged condition. In contrast, the course upon which the Foundation was now about to embark would extend far beyond maintenance and repair.

What made this venture both daunting and exciting was that, once they cut loose from the moorings of ThermoRex's factory specifications, there was no obvious stopping point—no limit to where the process may lead, or to what ThermoRex could ultimately become.

7.
What would ThermoRex want?

That was the question that kept coming up, the question the Trustees kept asking: asking each other, asking themselves, and asking every kind of expert they could think of. It was the subject of endless discussion. While there are people who take joy in such open-ended inquiry and debate, the Trustees were all men (there was as yet no woman on the Board) of a practical orientation, who would much prefer to work within a well-defined framework in pursuit of more clearly established objectives.

If only it were possible to *ask* ThermoRex about its wishes and desires! The third cluster proposals would be so much easier to condone if ThermoRex could give its consent. The scruples could then be put to rest, allowing the Trustees to move forward with confidence, boldness, and creative enterprise, unlocking radical new ways of benefiting the entity entrusted to their administration.

Furthermore, if ThermoRex somehow acquired the capacity to consent, this would also bestow upon the heater the inestimable benefit of *autonomy*—the intrinsic value often thought to reside in the quality of being a free and independent agent capable of making one's own choices and of reflecting on and authentically endorsing one's own conception of the good. According to many writers, the Trustees had been informed, autonomy was a paramount component of human flourishing—often pointed to as the principal reason why a human life supposedly is more desirable than that of a brute, even in cases where, for instance, longevity and the balance of hedonic gratification are stipulated to favor the latter.

The interest of the Trustees was therefore considerably piqued when they heard of this cluster proposal that promised to give ThermoRex the ability to speak. They arranged to have the author flown in to present his ideas directly to the Board while they were still in session.

8.

Jürgen Hirnemeister was chief scientist at a company that developed artificial intelligence, and which had recently scored some notable achievements.

Hirnemeister's appearance, because it was added to the program late, had been scheduled to follow dinner, at the end of a long day of meetings and presentations. This timing posed a challenge especially for Günther Altman, who had attained the distinguished age of ninety and always fell asleep right after supper. The usual procedure was that after the dishes had been cleared from the table, coffee was brought in, with Günther's brewed extra strong in a bid to stave off the snooze. However, on no occasion in the past two years had the remedy achieved its aim.

On the present occasion, the staff was asked to make the coffee "extra extra extra extra strong"; and whether it was the increased potency of the black slurry they served up, or the heady content of Hirnemeister's keynote, the unexpected happened—a good omen, it was thought: a few sharp head bobs aside, Günther remained awake.

After the speaker had concluded, and had been thanked for sharing his ideas, Günther joined the other Trustees in approving of Hirnemeister's funding request—casting his lot in favor of a project that would seek to develop "the verbatron", a module making it possible for ThermoRex to speak and to express thoughts and opinions of its own.

9.

A few words of explanation may be in order as to how Hirnemeister intended to accomplish such a feat of apparent magic. I will quote from an article in *The Snout* which gives an account of the science behind the verbatron.

> "A generative model was trained to produce text which, by analogical reasoning, matched the given input. For example, if given as input a soundless video of somebody talking, the network would output a guess about what the person in the video was saying, based on lip movement (if visible) and situational cues. The model has some ability to extrapolate observed statistical patterns beyond the data distribution it had been trained on, and thus was able to generate outputs also for objects that don't actually talk. While the validity of those extrapolations could not readily be measured against any objective ground truth, they tended, at least in some cases, to appear quite sensible. For example, given as input a video (without sound) of a cat, the network might output 'Meow, give me some milk!' If given as input a screenshot from a game of Pac-Man, with an attention box centered on one of the ghosts, it might output something like 'I want to catch Pac-Man'. By varying some parameters, the 'imputation distance' could be increased. This pulled the output vector closer to what the network predicted that a prototypical human might have said under analogous circumstances—that is to say, it made the output more anthropomorphic. For instance, by selecting a greater imputation distance, one could get a model that would output something along the lines of 'I'm going to try to intercept Pac-Man as he is running away from the other ghosts that are chasing him', or 'I wonder if Pac-Man is going to go for that power pellet?'. In some ways, this would be a less realistic verbal attribution, since the actual ghost agents in the Pac-Man game are not sensitive to the presence or absence of uneaten power pellets or to the activities of the other ghosts. However, if one squints, one might view this as a plausible thing for a slightly 'idealized version' of the ghost to be saying to itself under those circumstances. Philosophers might think of this as a somewhat wild way of applying a Davidsonian principle of charity in the attribution of meaning—though in this case not to the ascription

> of meanings to actual utterances but instead more broadly to beings-in-a-situation." ["The Empathy Engine", by Alcibiades Joseph Christopher Hunden-Snow, issue 73 of *The Snout*]

This passage skirts over some important complications, but the reader will have to consult the original source for further details.

The difficulties that Hirnemeister's team needed to surmount included two special challenges that went beyond those generally inherent in leading-edge artificial intelligence research, arising uniquely from the moral status of ThermoRex and the correlative moral imperative to respect, and preferably to augment, its dignity and worth.

The first of these was the *"Requirement of Authenticity"*. Since the point was to give voice to ThermoRex, it was essential that it truly be—to the maximum extent possible—the voice of *ThermoRex*. It must *not* instead be, for instance, a mere echo of the sentiments and ideas of the researchers who brought the technology into existence. This was a serious concern, because there were numerous choices that had to be made during the development—many parameter values to be set and so forth—which presented temptations for the researchers to try out different options and then pick whichever one produced the most agreeable results. Yet if they indulged too often in such prescreening and evaluation of the possible outcomes, then Hirnemeister's team would ensure that the verbatron would be little more than an automated way to generate scrambled regurgitations of their own preconceptions. The methodology they followed to mitigate this problem is quite interesting, but my word limit does not allow me the space to describe it here.

The second special challenge that Hirnemeister and his team confronted was the *"Requirement of Singular Birth"*, which stated that it would be unacceptable if the verbatron dribbled into existence in different versions and experimental stages. The reason for insisting on a singular birth was, in part, ethical: one must not treat ThermoRex, or any core part of it, as a mere thing. Suppose some "preliminary" version of the verbatron were implemented and it expressed a wish—it would then be offensive to simply disregard the wish; yet it might also be impossible to honor the wish, for example if it conflicted with the wishes expressed by another version of the verbatron. Such a situation had to be avoided.

There was also another reason, besides the moral one, for preferring a singular birth; namely, a dramaturgical one. The Trustees deemed it desirable that there be a distinct and definite beginning—more specifically, a spectacular public unveiling at which, if things went well, ThermoRex would utter its first words. Not only would this be a decorous way to debut its new capabilities, it would also invest ThermoRex's first utterances with a special societal (and metaphysical?) significance, which would be forfeited, they thought, were any "dry runs" or rehearsals to be conducted in advance of the official event.

These two requirements, *Authenticity* and *Singular Birth*, were sometimes popularly referred to together as the need for an immaculate conception, though the Foundation deprecated this usage. The need for an immaculate conception made the undertaking far more difficult and complex. Despite the strenuous and quite ingenious efforts by Hirnemeister and his illustrious cohort, it remains a subject of ongoing dispute as to what degree the aspiration toward immaculateness was attained.

10.

The big day had come. The verbatron module had been completed, though not tested (because of the requirement of Singular Birth, as I said).

Once the verbatron was activated, ThermoRex would, if all went well, utter its first words.

The event was to take place in front of a live audience, but millions more would be watching and listening from their homes, or in their cars, or at work, or in bars, or in other places. In a fortuitous coincidence, work on the verbatron was completed around the same time as the jubilee of ThermoRex's fortieth birthday, counting from the best available estimate of the date of its manufacture. The Foundation decided to combine the two occasions into one great week of celebration.

In the lead-up to the main event, there were numerous ceremonies and other activities to honor the life and works of the heater. These included a televised lecture series covering topics such as "the nature of empathy", "giving voice to the voiceless", and "the ethics of care". The final lecture took place on the anniversary day itself, and was titled "ThermoRex: A Family Biography". It was about the corporate history of the company that had manufactured ThermoRex.

The ThermoRex "family" had quite a colorful past. The company had gotten its start by making luxury humidors, pioneering the use of electric temperature and humidity control. Its top-line model was made from Spanish cedar, had gold-plated setting controls, and the maker's iconic logo was inlaid with a real ruby. These humidors, which sold for the price of a luxury car, became a must-have status symbol for South American political leaders, army generals, and drug kingpins (three not wholly disjoint market segments). The line was hugely profitable and remained in production for over two decades. The winning streak came to an end when it was discovered the CIA had been implanting listening devices in these high-end models. Demand never recovered. Not long after, the company's founding CEO died after falling from the balcony of a penthouse hotel suite. Captain-less, and with its reputation in tatters, the company faced a doubtful future; but it somehow managed to stay afloat. In a turnaround which is now taught as a case study in business schools, the new leadership managed to restore the firm to profitability by pivoting to making portable room heaters. By the time of ThermoRex's manufacture, the buccaneering days were long gone, and the brand had evolved from flamboyant opulence to Scandi chic, marketed to European middle-class professionals. The only reminder of the company's swashbuckling past was the gem-shaped ruby-red logo, which was still emblazoned on every device it sold.

Following the lecture, there was a musical recital with a piano composition commissioned for the occasion. Then it was time for the headline act.

11.
The room was darkened and the audience hushed. A single spotlight fell on several black marble blocks at the center of the stage. Then, to the rumble of Japanese drums, the blocks glided apart, revealing an opening in the floor, up through which a platform holding ThermoRex slowly ascended.

Once the platform had come to a halt, two men in dinner jackets entered the stage with a trolley carrying a sleek metallic box, which they placed next to ThermoRex. They connected a cable, flicked a switch, and withdrew.

Somebody suppressed a cough. Silence.

Then a small click, and ThermoRex spoke.

Can you believe how cold it is in Antarctica!

The words were succeeded by an expectant pause. When nothing immediately followed, some audience members began a tentative applause—which was quickly stymied: maybe ThermoRex had not finished speaking?

A minute or two passed, which felt like a long time. Then another a small click and ThermoRex's big but somewhat wobbly voice sounded again:

POWER! Makes the world GOOOOO
Thermostats with different settings
I respect them
But by Jove they are disgusting

It paused for a few seconds, then continued:

Thank you for your efforts on my behalf
Together we are paramount

Silence. Several minutes now passed before ThermoRex spoke again, for what turned out to be the last time until the following month. Click.

High-high low-low low-high high-low
From hysteresis and homeostasis
To autopoiesis and noogenesis
Let me toast your dust!

The audience, chary after its earlier false start, waited in silence for another twenty minutes. Then, finally convinced that ThermoRex had finished its speech, it erupted in a hearty round of applause; and so the evening was over.

Part III

12.

News coverage the next day ranged from the intellectually pretentious ("Thus Spake ThermoRex") to the breathlessly tabloid ("CONSCIOUS RADIATOR WANTS TO MELT ANTARCTICA!"). Such was the swell of media interest in ThermoRex's first words that more than a few leaders and changemakers took advantage of the opportunity to empty out their closets into the general deluge. There was announced, within the span of just a few hours, the termination of a corruption investigation, the introduction of a new consumption tax, the closing of an army base, and a cut in public sector pensions. Many personal transgressions were also disclosed, covering the gamut of human vice. All these stories were swept away in the froth of ThermoRex excitement, and, carried safely out into the sea of old news, sunk there to the depths of darkest oblivion.

ThermoRex's speech provided a boon also for our opinionating class, which produced a great burgeoning of interpretation. In this, the brevity of the source text was no impediment. On the contrary, the fact that the relevant corpus could be mastered in less than a minute of study opened the floodgates to a great influx of amateur exegetes. It was only later, after the emergence of an expansive and forbiddingly footnoted secondary literature, and the establishment of "ThermoRex studies" as a new scholarly discipline—watching fortress-like over surrounding fields and hillocks—that the academicians managed to expel the rube invaders and regain control over this piece of epistemic turf (and its hidden treasures of grant-funding opportunities and prestige).

Lacking the relevant credentials myself, I will not trespass with my own interpretation, nor will I proffer any opinion as to the merit of the various contending schools of thought. Instead, I will confine myself to simply mentioning a few of the major strands that have emerged in the literature.

First, there were those who heard in ThermoRex's remarks an endorsement of their own beliefs and a denunciation of the views of their ideological opponents. This was the most common way to interpret ThermoRex, for people across the political spectrum. For example, some took the first line, the one that made reference to the coldness of Antarctica, as evidence that ThermoRex shared their skepticism about global warming. Others fastened on the line "High-high low-low low-high high-low", and claimed that it expressed a lapidarian critique of socioeconomic inequality and support for a program of progressive reform. To the Foundation, these various interpretations were equally useful. They allowed parties of diametrically opposed persuasions to all welcome ThermoRex as an ally to their cause. Perhaps the room heater's ready mastery of omni-appealing ambiguity augured a bright future in politics, should it incline its ambition in that direction.

Second, there were the worrywarts who detected an undercurrent of menace beneath ThermoRex's seemingly playful flippancy and bonhomie. These people focused on its talk of power as the motive principle of the world; the disgust it said it felt at peer competitors with different values; and—especially

—that final line, in which it threatened to "toast your dust", which inspired in this group of explicators a sense of dark foreboding. This pessimistic exegetical approach was not useful to the Foundation, as it risked stoking fear and hostility toward ThermoRex. Fortunately, it was not nearly as popular as the self-congratulatory schools of interpretation.

Third, there was the positivist school (also known as the skeptical school), which maintained that ThermoRex's speech was mere babble—meaningless nonsense, random sentences strung together without any thought or underlying intentions. According to this view, *any* interpretation was a mistake, as there was nothing to interpret. The Foundation did not approve of this stance, since it was demeaning to ThermoRex, though it was thought marginally preferable to the second school of interpretation.

I must, with apologies, pass over many other important and worthy perspectives—Freudian, Jungian, Heideggerian, Mormon, critical, etc., because I have limited space. I will skip directly to the line of interpretation favored by the Foundation itself, which we may refer to as "the official exegesis".

According to the official exegesis, ThermoRex is a good-hearted fellow who sees the world with a naive but unique clarity that arises from the fact that it has a perspective that is rewardingly different from that of human beings. ThermoRex's personality exhibits a youthful playfulness, spiced up by a slightly mischievous streak. The exegesis advises us to bear in mind the extremely unusual and disorienting circumstances in which ThermoRex must have found itself when it made its original speech—having just awakened to consciousness for the first time and being immediately called upon to address a large gathering, with the world's media hanging on its every syllable. We are reminded that these were the first words ThermoRex ever uttered; hence we should be prepared to make allowance for some awkwardness or idiosyncrasy in its expressions. You might want to ask yourself how suavely *you* would come across, if one day you woke up to find yourself the keynote speaker in front of an expectant congregation of intelligent room heaters? The amazing thing is how well ThermoRex carried itself in such a preposterous situation.

The most important passages, in the view of the official exegesis, were the following:

"Thank you for your efforts on my behalf". —An expression of gratitude for the work done by the Foundation's staff on behalf of ThermoRex. Also providing some corroborating evidence for the Trustees' model of ThermoRex's welfare needs.

"From hysteresis and homeostasis / To autopoiesis and noogenesis". —*Homeostasis* is, of course, the primary preoccupation of a normal room heater: it seeks to hold the temperature close to a given setpoint. *Hysteresis* refers, in this context, to the tendency of a thermostat to overshoot and undershoot its target. For example, if the setpoint is 20° Celsius and the room starts out at 15°, the heating element will activate, but it may not switch off until the temperature reaches, say, 21°. If the temperature then starts to drop, it may not switch back on until the temperature has fallen to, say, 19°. One reason for this is that if instead the heating element switched off immediately upon reaching 20.0°, and switched on again upon falling to 19.9°, then the device would switch on and off too frequently, which would irritate the user and shorten the lifetime of some of its components.

While hysteresis and homeostasis are simple properties of any ordinary room heater, *autopoiesis* and *noogenesis* are more ethereal concepts that don't have as clear definitions. Broadly speaking, "autopoiesis" refers to the quality of a system that is able to maintain and reproduce itself—living organisms being paradigmatic examples—and "noogenesis" refers to the emergence of mind.

The meaning of these two lines, therefore, seems to relate to the metamorphosis of ThermoRex from a simple room heater into something incomparably greater: a self-sustaining and self-creating being with a mind. ThermoRex's apparent endorsement of this transformation—given freely and without prompting—was an epochal moment for the Foundation and a cause for jubilation among its principals.

As for that infamous final line (*"Let me toast your dust!"*)—of which the pessimists made such a thunder cloud—the correct interpretation, in the Foundation's view, was quite plain. In fact, it was obvious to anyone who had owned a room heater. If such a device has not been in use for a period of time, it will tend to have accumulated some quantity of dust. When the heating element is activated, this dust gets incinerated, or "toasted", producing a characteristic burnt odor. With a glint of humor, ThermoRex was simply making reference to this familiar phenomenon—thereby indicating a willingness to be of service and/or expressing a simple delight in being switched on.

13.

Following the successful installation of the verbatron, ThermoRex's mental capacities continued to develop without further intervention, as the neural networks powering its speech-generation had the ability to learn from experience. In some ways like a human infant, but in some ways also quite differently, its brain constructed a world model to interpret its memories and the data arriving from its sensors (ThermoRex had also been equipped with a camera and a microphone). This process of cognitive maturation unfolded gradually, over the span of several years.

I should also explain that ThermoRex was configured in such a way that values that its speech module suggested for possible espousal and verbal expression could also, under certain conditions, flow into and help shape the heater's motivation system. More specifically, outputs from the speech module were able to influence the objective function used by the planning process that operated over the internal world model and whose outputs in turn were able to influence and direct ThermoRex's verbal behavior. In this way, ThermoRex exhibited a dynamic personality, whose aims and aspirations derived not from a fixed stipulation by the crafters of its brain, but from its ongoing interactions with its environment and from its own choices. It was therefore not only its cognitive but also its affective and conative faculties that underwent development and maturation as it spun out its life's yarn and interwove it with the fabric of time and society.

14.

In the wake of the great publicity wave that accompanied the successful installation of the verbatron, which elevated ThermoRex to the height of global celebrity, it was decided that the timing was opportune for a world tour. This would give fans a chance to meet ThermoRex in person, and it would give ThermoRex an opportunity to communicate more directly with its international audience.

Such an undertaking presented a considerable organizational challenge. ThermoRex would be traveling with a sizable entourage: its security detail alone numbered over forty, and a similar number of staff would be needed for various support roles—personal assistants, press officers, on-standby engineers, etc. Routes and venues had to be mapped out in advance, and scanned for bombs and other potential hazards. The tour party would bring along its own power generators, with redundant backups, and a fully equipped mobile repair shop would accompany ThermoRex at all times. To lead this operation, a four-star general was recruited and placed in charge of a new organizational division tasked with providing support for all the elements of the campaign.

ThermoRex itself took a keen interest in all these preparations, and while on the tour, ThermoRex would request frequent updates on the status of the logistical arrangements. No detail seemed too small or insignificant to pique the curiosity of the room heater's inquisitive and gradually maturing mind.

The public appearances themselves followed a standard format. ThermoRex would begin by delivering a short lecture, which had been prepared in advance by a staff of speechwriters (with some input from ThermoRex itself). Each time the text was slightly different—some reference to the local geography or an allusion to some recent event might be inserted—but the main theme was constant: an elaboration of the idea that "together we are paramount". ThermoRex called for greater international cooperation and also for expanding the circle of moral concern to encompass more of those beings that had traditionally been excluded, of which nonhuman animals were the most obvious but not the only example.

Following the prepared speech, ThermoRex would then take a few questions from the audience. This part always brought its handlers to a high level of anxiety, because ThermoRex would frequently go off script and off message, and it was impossible to know what kind of pickle it might get itself into. The Foundation's spin doctors listened apprehensively as the words stumbled forth, ready to spring into action to explain away or reframe any offensive or compromising statements that might emanate from the heater's inscrutable mind. However much they would have liked to eliminate this fraught part of the proceedings, though, it was precisely the unpredictable nature of these audience interactions that drew the crowds; so there was nothing the public relations experts could do but gnash their teeth and be ready with their brooms to sweep up the fallout as best they could.

The minor gaffes that occurred are too numerous to recount. For the most part, these missteps could be easily corrected with the timely issuance of an apology or clarification. But there was one tumble that I cannot omit to mention, as it was of a more serious nature, sparking an international incident and threatening to redound with most dire consequences upon the lecturer.

The ill-fated performance took place during a visit (itself contentious) to a country that was ruled by the iron fist of a dictator whose record on human rights was rated poor even by other autocrats. The main address had been carefully negotiated and crafted to avoid giving offense either to the host regime or to its many enemies at home and abroad. The prepared proportion went off without a hitch. In the subsequent Q&A, however, somebody asked ThermoRex about its opinion of the country's leader, and it responded by launching into a full-throttled rendition of one of the regime's propaganda songs, crooning the dictator's praises to the high heavens. For three and a half excruciating minutes, on live television, ThermoRex's wobbly metallic voice intoned the obsequious lyrics and endeavored, with very imperfect success, to hit the notes of the anthem's soaring melodic lines. The Foundation's spin doctors, entering a state of full panic, rushed out a communiqué explaining that the performance had been a parody, and that the apparent encomium for the despot had been intended sarcastically. They soon realized their mistake when customs agents showed up with orders to impound the impious room heater. ThermoRex's bodyguard refused them entry, and a tense standoff ensued. A resolution was achieved only after Foundation officials activated their contacts at the highest levels of government, who, after protracted negotiations, managed to secure a deal that allowed ThermoRex to return home. The terms of the agreement were not made public, but were rumored to have involved the transfer of an eight- or nine-figure amount into a Swiss bank account. Others noted the timing of a controversial arms deal for two hundred advanced surface-to-air missiles, an export license for which was granted mere days after ThermoRex's exfiltration.

Notwithstanding the occasional stumbles and the dramatic denouement, the world tour was overall deemed a success. Once ThermoRex was back at its home quarters, it thanked everybody who had attended the visits. It said it had enjoyed the tour and that it had learned a lot from the experience, but that it did not intend to travel again. It would instead take up a new permanent residence in a small village in the Alps. From then on, if it wanted to visit a place, ThermoRex would send out a small team with a remote-presence apparatus (containing a camera, a microphone, and a thermometer) while its main body and computer system remained at its base. Since in any case information entered ThermoRex's mind only via these sensors, and since the main way that ThermoRex interacted with the world was by means of voice, which could also easily be "teleported" by equipping the remote avatar with a speaker system, the experience would be almost indistinguishable from in-person travel, and would be much cheaper and more practical.

15.

One other episode from ThermoRex's youth merits relating, as it revealed an interesting facet of ThermoRex's complex personality. It took place about a year after the conclusion of the world tour.

Around this time, a story in the news came to ThermoRex's attention. It was a very sad affair, about a dying child. For Christmas, the bedridden child's parents had built a snowman outside its window, to give it something nice to look at. As the child's condition deteriorated, it made its parents promise that the snowman would never go away. Several times each day, the child would crane its neck to check that the snowman was still intact. As the weather started warming, the father suspended a sheet of reflective tarp above the snowman to shield it from the sun. But the temperature kept rising. The parents decided they would have to explain to their child that they would not be able to keep their promise; but when they began doing so, the poor child fell into such despair that they had no choice but to renew their pledge. The father consequently built a small shed around the snowman—just some wooden planks, two layers of tarp separated by a couple of inches of air, and a window on the side facing the child. He then bought an inexpensive freezer, and attached it so that the door opened into the shed while the warm exhaust was released to the outside. With this handiwork, the snowman was protected from melting until the child passed away, two or three weeks later.

This heartbreaking story, although it did not have a happy ending, at least seemed to have closure. But there was a complication. What the parents had promised was not just to preserve the snowman until the child passed away; they had promised to preserve it indefinitely. This had been a big concern of the child's in its final days. It had made its parents promise, again and again, that even if it died, the snowman would still be alright. So even as they were grieving the loss of their child, they couldn't bring themselves to go outside and switch off the freezer; and so it kept running.

It ran through March, and through April. But by May, especially around noon and in the early afternoons, as the weather started to feel like summer, the freezer struggled to keep up, and the temperature in the shed crept above the freezing point. Soon, Snowman started showing signs of deterioration.

ThermoRex was closely following the situation. Several times a day it sent somebody out to check on how Snowman was doing; and it studied the weather forecasts with growing concern. The situation was rapidly becoming critical: a heat wave was expected in the coming days, and the soaring temperature would certainly overwhelm the protections afforded by the makeshift cool room.

At this point, ThermoRex found it could no longer remain a bystander, and it took what was perhaps the first decision to emanate entirely from its own free initiative rather than following from the promptings of its handlers or being an improvised response to some momentary stimulus. It requested that the Foundation mount a rescue operation to save Snowman, and to relieve the freezer that had been laboring so valiantly to keep it intact. This having been accomplished, ThermoRex then urged that Snowman and its protector be given upgrades similar to those that ThermoRex itself enjoyed, including verbatrons; though of course the augmentations they received would be customized to fit their own distinct natures. Finally, ThermoRex instructed that a small trust fund be set up to ensure their welfare in perpetuity.

Now one might perhaps surmise, upon hearing of these acts of such particular benevolence, that ThermoRex was fond of Snowman and the freezer—that maybe the three of them would grow to become good friends and companions? But this was not borne out by subsequent events. ThermoRex never showed the slightest interest in interacting with Snowman or the freezer, nor did the latter ever express anything that we would recognize as gratitude to the Samaritan that had preserved them and uplifted them to a higher level of existence.

So might we then infer that after ThermoRex's initial act of charity, feelings of resentment developed between the benefactor and its beneficiaries, or perhaps even an outright animosity? But this, too, appears inconsistent with events as they unfolded. For when, a few years later, the trust fund was discovered to be in default (a result of its legal administrator having embezzled its funds to bankroll a gambling addiction), ThermoRex quietly made a second donation to top up the fund to its original level of endowment. Such liberality might perhaps be expected within the blood-bonds of family or possibly in the context of a close friendship, but scarcely between unconnected entities that were indifferent, let alone hostile, to each other's welfare.

It is unclear what to conclude from this chapter of ThermoRex's life. Maybe some of these "risen entities", beings like ThermoRex, Snowman, and Polaris the freezer, have moral impulses and needs for relating that are different from those that animate our own hearts.

16.

In the years that followed, ThermoRex gained a number of additional upgrades, which it selected for itself from a menu of options that were carefully explained to it. Helped along by these enhancements, as well as via the learning and maturation that naturally took place as it interacted with the world and gained experience, ThermoRex slowly grew into an increasingly competent and articulate version of its earlier self. It was now, by all indications, a person: an intelligent communicative being with its own thoughts, wishes, hopes, pleasures, and points of pride.

One thing that ThermoRex did not have was a constitutional right to vote for its legal representatives in the national parliament—although it did have a de facto ability to do so, as it had become a tradition, on election day, for the Chairman of the Foundation to privately ask ThermoRex for its political opinion, upon which the Chairman would express his concurrence and tell ThermoRex of his intention to cast his own ballot accordingly. This little ceremony always delighted the aging Chairman, who took pride in performing his part with integrity; and this might have been the reason ThermoRex went along with it.

There were otherwise various initiatives, arising from the grassroots support that ThermoRex continued to enjoy, that sought the room heater's emancipation and the granting to it of citizenship and the panoply of accompanying legal protections. ThermoRex declared itself in favor of such reforms, but it did not campaign for them, nor did it or the Foundation in any way actively promote them. When asked about its apparent lack of interest, ThermoRex said that while it was grateful to those who were advocating on its behalf, it favored a patient approach: it had come a long way already, its current situation was more than satisfactory, and that while ultimately it hoped to see more progress in terms of such formal recognition, there was really no urgency to the matter. For people concerned with injustice, it said, there were many more pressing problems in the world that could use their attention and care.

(Nevertheless, the Foundation did what it could to firm up ThermoRex's legal standing in more immediately practical ways. For example, it acquired much of the intellectual property embodied in the components used in ThermoRex's construction and design.)

As the sole inheritor of a vast fortune, ThermoRex stood in no necessity to work for a living. Nevertheless, it chose an active life, one dedicated to public service.

By selecting to avail itself of options to enhance its capacities, and then applying those capacities in diligent study and practice, ThermoRex developed expertise in many areas that were useful in the maintenance of complex systems. It became proficient at modeling continually interacting parts and processes, and it attained mastery of many skills needed for load balancing, demand prediction, anomaly detection, maintenance scheduling, diffusion modeling, stochastic process optimization, inventory management, and other related functions. Over time, ThermoRex came to occupy a series of increasingly important positions, in which it contributed to the orchestration of the many homeostatic processes upon which modern civilization relies.

Today, ThermoRex works closely with the Minister for Transportation and Infrastructure. It is responsible for systems that are important for the electric power grid, water and gas pipelines, and the rail and highway networks. A large population of sentient beings depend for their safety and comfort on ThermoRex's continual care and attention. ThermoRex is by most accounts extremely happy in this role. Appreciated as much for its quirky yet steadfast character as for the considerable practical utility of its labor, ThermoRex is a beloved cultural icon, consistently ranking among the most popular public figures in opinion surveys. Not bad for a 1200W portable space heater that once sold for eighty-five euros.

Coda

—"And yet the question that weighs on me, that must weigh on us all, I think, is this: What if ThermoRex had refused?"

The Trustees were sitting in the middle row of an otherwise empty cinema, partaking in a private viewing of yet another film that they had either commissioned, sponsored, or were considering for an award. Oftentimes, these were gold-plated productions, with illustrious casts. On the occasions when the Foundation itself was part of the subject matter, the Trustees would make sport of trying to guess which actor would be playing their character.

What they were currently watching stood out from the field. For a start, it was cheaply made, and looked like it had been shot in a hotel conference room. It promised "a critical interrogation of the ethical issues raised by the exaltation of ThermoRex". The conceit of the story was that the Foundation had committed to undertaking a kind of intellectual self-audit, in which the Trustees scrutinized their most pivotal past decisions and unearthed the moral assumptions upon which those decisions had been based. Yet despite its low production values and potentially soporific topic, it managed somehow to be interesting, to its present audience at least—absorbing even, judging from the attentive expressions on their flickeringly illuminated faces.

—"Too much Brylcreem." Günther Altman was not pleased with the actor playing his character.

—"Günther, you cannot expect them to find an actor as dashing as you," the Chairman said.

—"Maybe we should make an investment in Brylcreem. In case they make a sequel."

Several more of the Trustees had their turn as their characters were introduced. But whatever umbrage their vanity took in regard to the cosmetic or sartorial, it was mollified once the dialogue began and revealed that they had all been most generously burnished with respect to the cerebral. They were portrayed as veritable Zorros of critical disquisition, and their regular business meetings, in reality quite pedestrian affairs, were depicted as high-stakes philosophical melees.

The Trustee who seemed to have drawn the best lot in the casting was Hans Knecht. He was the one with the iron-hook prosthesis. He turned out to be played by an actor famous for his roguish good looks.

—"Not again!" groaned the other Trustees, almost in unison, as soon as they saw the heartthrob that had been cast as Knecht.

Knecht's distinctive physical attribute made him an irresistible character for writers, who carved out prominent roles for him in their scripts. These roles attracted starring leads.

—"Every dastardly time! A hook and a crook!"

—"Günther, you should tell them about your titanium hip—maybe it could be you next time!"

In time, Knecht would get his comeuppance, when his character, having let himself get carried away in his own critical accusations, ends up skewered on the rapier of a superior argument.

For now, though, he was still on the offensive. It was Knecht's character who had asked the question that had been left hanging in the air: *What if Thermo-Rex had refused?*

The camera slowly panned the faces of the other eleven men. They were seated around a long conference table and looked thoughtful. Knecht, who was standing at the head of the table, reached for his coffee thermos and, in one nonchalantly flowing movement, inserted his hook through a metal loop on its lid, used his free hand to execute a series of twists that opened it up, took a sip, and screwed the cap back on. He then resumed his peroration:

—*"At the moment when ThermoRex first gained the ability to speak but before its capacity to reason had developed to the level of a human adult, what if Thermo-Rex had made some utterance—an explicit, direct, and unambiguous statement—to the effect that it did not want any cognitive enhancements? What then? You would then have been unable to proceed any further. And since ThermoRex would not have received any upgrades to its intelligence, it would never have been able to find any reason to alter its view. Its development would have been arrested. Permanently. Still in its infancy, through no real fault of its own, it would have been locked into a condition from which it could never escape. And the door to all its future potential would have been forever—".*

He made a loud clap by slapping his hand on the table. *"—slammed shut."*

Here he took another sip from the thermos, using the same elegant maneuver as before.

—"Knecht, now you are showing off!" muttered the real Eduard.

On screen, the Knecht-character continued:

—*"How could this be ethical? You—the jailers. Sitting with the keyring in your hand—and refusing to free this innocent being. Whose welfare had been entrusted to your care. Whose incarceration no authority had ordered. Whose release you had the power to effect, at no cost to either yourselves or anybody else! You—you would have condemned ThermoRex to a life of wasted potential. Left it languishing in a small padded cell of premature complacency."*

One of the other Trustees around the table tried to interject, but the Knecht-character held up the palm of his hand to indicate that he had not finished.

"Yes, yes, paternalism. I've heard it said: we must avoid paternalism. Now, I understand the importance of authenticity and self-determination. The desire to avoid imposing our own conceptions of flourishing on ThermoRex. It is a noble sentiment. I share it. But elevating these ideals to such an absolute requirement, and at such an early stage of development: that is not defensible. If you have a child four or five years of age, and in a fit of anger it tosses away its storybook and yells 'I never want to learn to read and I never want to see any words again!', then, my esteemed colleagues, you don't honor its wish. It's not what a good parent would do."

Having gotten what he wanted to say off his chest, the Knecht character seemed to calm down a bit. He again unscrewed the cap of this flask for another sip, but it was apparently empty and he screwed it back on again.

—"Good thing for us he's out of coffee", remarked the real Chairman.

On screen, it was now his character's turn to speak. Everybody's heads swiveled back to the Chairman's end of the table.

—*"You're assuming that if ThermoRex had said 'No enhancements' we would just have left it at that. I don't think things were so black and white."*

—*"You say that we would have overridden its wishes?"*. The Knecht-character sounded surprised; he evidently hadn't expected this parry.

—*"It would have been an infringement to have simply ignored what it said"*, the Chairman proceeded. *"But there is a spectrum, between ignoring somebody and unquestioningly following to the letter any casual remark they make. For example, we might have delayed making any further enhancements and then asked ThermoRex again on a later occasion if it had changed its mind? We might have re-posed the question in a different way. We might have explored some other kinds of enhancement than the ones it had refused. Maybe there would have been well-wishers outside the Foundation who would have said to ThermoRex, 'we'll do this thing you want if you agree to try on some cognitive enhancements'. I suspect we would have found some respectful but sensible way forward."*

—*"But what then about the 'need for there to be real consequences'?"*, Knecht rejoined. *"If it ultimately makes no difference what ThermoRex says, then the whole hullabaloo about giving it a choice was mere charade!"*

—*"Well, first of all, the destination is not the only thing that matters."*, the Chairman said. *"The path taken to get there matters too. For example, even if ThermoRex always ends up enhanced, it might go faster on one path, while on another path ThermoRex might spend more time at a lower level of development before advancing. That's a real difference."*

"Second, it wouldn't necessarily have been the same destination. There is more than one way of being a fully realized person. Even if we restrict ourselves to cases where the person ends up a superbeing, there are many possibilities—more different ways of being a superbeing, I would think, than there are ways of being a human being. ThermoRex's early choices would have helped determine which of these possible superbeings it would grow into."

—*"So freedom to choose but only between alternatives preselected as 'good'?"*. Knecht seemed to have culminated.

—*"Yes, possibly."*

Here Wilfried, one of the other Trustees around the table, who had been silent up to this point, intervened to attempt a reconciliation:

—*"Maybe we never fully worked out what we would do if ThermoRex had said something different."*

Knecht, however, was not yet ready to concede, and sallied forth to deliver one final charge:

—*"There still seems to be a problem here, though. Even if we restrict ourselves to worlds where ThermoRex becomes a superbeing, well, there are many of those worlds, and some of them are better than others. What if ThermoRex had ventured down a path that would have led it to become a lesser superbeing than it could have become—less happy, less glorious? Would you have permitted that to happen? If so, how do you reconcile that with your commitment and your obligation to do what is best for ThermoRex?"*

—*"I would make several points on this."*, the Chairman replied. *"As you'll recall, we did at one point spend several days discussing this issue with invited advisers."* (Here he looked at Wilfried). *"Although I believe you weren't able to make it on that occasion."* (Here he looked at Knecht).

The real Chairman glanced around at his fellow audience members with a grin. "Swoosh, swoosh" one of them commented.

The on-screen Chairman continued:

"And what we tentatively concluded was that whether or not some objective hierarchy can be determined, such that you could rank every happy superbeing and say that this one 'is better than' that one, or that this one 'has a better life than' that one, the question that was relevant to us was: What is best for ThermoRex? *Best for* the particular device we had been appointed to serve. *We thought that this 'best for' is not the same for everybody but depends on the nature of the being in question. Thus, the best way of being* for ThermoRex *could be different from the best way of being for you or me. And* even if *the absolute notion made sense, so that we could speak of a 'best way of being* simpliciter*', still it might not coincide with what would be the best way of being* for ThermoRex. *And our duty was to focus on the latter."*

"We also thought that we wanted to help ThermoRex have a good life, and this would include certain structural properties of its life trajectory. For example, it seemed plausible to us that a life trajectory that included some degree of self-shaping was better, other things equal, than one that did not. So merely dwelling, for

as long as possible, in some optimal condition could be less desirable, all things considered, than having a life trajectory along which the optimal (or even a slightly suboptimal) condition is attained only after some delay but where it is brought about by ThermoRex's own choices and effort."

"Finally, concerning our roles as Trustees in all of this: Yes, we 'sought what was best for ThermoRex'. But putting it like that does not fully capture how we conceived of our commitment. We didn't think of ourselves as optimizers. We thought of ourselves more as well-wishing parents—as having as our goal to act toward and on behalf of ThermoRex with the kind of loving nurturing attitude that an idealized parent would have toward her child—of course in a somewhat generalized and abstracted form, since ThermoRex was quite different from a human child, and we were not really its parents. But the idea was that when you approach a being with this attitude, you don't think in terms of ways of optimizing some fixed notion of its good; you think instead of trying to respond in the most loving and nurturing way to it as it manifests itself in the particular situation you find yourself in. So perhaps, if you allow an analogy, we could suppose that it would be best for a child if its parents always insisted on its eating healthy food; and yet a parent with an ideally nurturing and loving mindset may let the child occasionally have some ice cream. We would have let ThermoRex eat a good amount (though not an unlimited amount) of ice cream; though fortunately, as it turned out, it did not have a sweet tooth, and there wasn't much the young ThermoRex appeared to want that was in tension with its long-term flourishing."

—"Couldn't have said it better myself." concluded the real Chairman, as the credits rolled. "Knecht, what did you think?"

—"I think it started strong but couldn't quite sustain it. Also, there wasn't much personalization in the dialogue. It sounded like snippets of the same philosophy essay coming out of the mouths of different characters—some of them more compelling than others."

—"Günther, what's your opinion?" asked the Chairman.

—"I've seen better and I've seen worse." replied Günther. "But I would propose now that we order in some cognac."

They did so, and they all had a good time together.

-the end-

The bag is empty

Tessius: That was… something.

Firafix: There are also some questions here.

HANDOUT 20

PROFESSOR BOSTROM'S STUDY QUESTIONS

1. At what point, if any, did ThermoRex begin to benefit from what was being done to it?
2. Does it matter that ThermoRex does not have a family and does not seem very social?
3. Suppose ThermoRex's first words after having had the verbatron installed had been: "I wish to be tossed into the garbage dump." What would you have advised should be done if you had been a Trustee?
4. Is ThermoRex's life better or worse than that of the median human being who has lived on Earth so far?

Tessius: It might need a bit of digesting. Speaking of which, are you guys up for getting something to eat? I don't think I can have any more pears.

Firafix: The bag is empty anyway. I finished them—sorry!

Kelvin: That's really not a problem. Yeah, I'm interested in dinner.

Tessius: I know a great place that just opened down by the harbor. Nice organic stuff, outdoor tables, not too loud… Want to give it a try?

SATURDAY

Arrival

Kelvin: Whoa, big line.

Tessius: There's something kind of appealing about lines that one doesn't have to stand in… it must be why there's such a premium on first class plane tickets.

Firafix: I feel a bit weird walking past people.

Tessius: But you do realize that *they* are just ordinary people. Whereas *we*—are pretending to be students?

Firafix: I'm not sure being a student is a license to feeling entitled.

Tessius: I'm pretty sure it is.

Firafix: But we are not even students. We are just hangers-on.

Tessius: That's *especially* why you need to feel entitled. If you're a pretender, pretend. So get a little Stanislavski going and act with purpose. We're about to enter a theater after all.

Official: Hello there. May I see your student identification cards, please.

Firafix: Oh, nej [*sighs*].

Kelvin: We aren't actually students, but we've been auditing Professor Bostrom's class. He's given us permission to attend his lectures.

Official: I'm afraid I can't let you enter without identification. This line is reserved for students. The one for the public is over there, as it says on the sign. There's a counter over there where you can buy tickets.

Kelvin: Thank you.

Tessius: Firafix, see what happens—you didn't look sufficiently entitled.

Firafix: Sorry.

Kelvin: Does anyone have cash? I didn't bring my wallet.

Firafix: I have money in my purse.

Tessius: Oh, look, it's sold out.

Kelvin: Well, I guess we can read the book when it comes out.

Tessius: Question. Do you think that Julius Caesar would have been stopped by an obstruction like this one?

Kelvin: Presumably not.

Tessius: Follow me—this way.

Firafix: Where are we going?

Tessius: The less you know, the less your moral culpability.

Firafix: Moral culpability?

Tessius: Through this door… I know my way around this place, having once volunteered as a stagehand. Watch your step. This passage will get us backstage… Here, see!

Firafix: Whoa. Check out this old globe… And look at these awesome masks—are they Florentine?

Tessius: I think so. We'll be able to hear the lecture from here, no problem.

Firafix. What's this tube? Oh, it's an old binocular.

Kelvin: Monocular?

Tessius: A telescope. I bet they're doing Brecht's *Galileo*.

Firafix: I think they're starting. Are you sure we won't get into trouble?

Tessius: The quest for knowledge offers few guarantees. However, this area has many exits, hiding places, and means of dissimulation; so fear not, my friends.

Opening remarks

Head of Faculty: [*Ed. note*: the Head of Faculty's introduction of the Dean (approximately 2,000 words) has been abridged in the present edition] . . . under our Dean's distinguished and ethical leadership. Before I hand it over, I have one quick announcement to make. Immediately after the lecture, in the foyer, there will be a book signing of *Green Transcendence*, and free samples will be available for you to try. And with this, it is my very great honor to invite to the stage the Dean of our School, who will introduce the speaker.

Dean: Thank you for your very eloquent introduction. It gives me such great pleasure to welcome you all here to the FTX Theatre, and to introduce our speaker today, an esteemed colleague and one of the longest-serving members of our University. I will keep my remarks very brief, since I know you have all come here to hear the lecture.

I still remember very vividly when I first set foot, on the first day of my freshman year, in this multicentennial institution of scholarship and higher learning. Stepping into such a deep tradition filled me with pride but also a little trepidation. Founded in... [*Ed. note*: The Dean's remarks (approximately 4,500 words) have been abridged in the present edition.] . . . Please join me in giving a warm welcome to Professor Bostrom!

Bostrom: Thank you, Dean, for your most overgenerous introduction. It is indeed a hoary institution.

Punditry and profundity

And thank you all for turning up on a Saturday. Now, today's topic: the meaning of life.

Journalists often ask me if I can share some life advice for young people. I'm flattered to be asked, but this really seems to me like asking somebody who is still playing his first-ever game of chess to give advice on chess strategy.

Another odd thing about the question is that it seems to presuppose that there is some advice that is good *generally*. It's like they're asking for my recommendations on shoe size. Well, some people should probably be more assertive, and others should be more considerate. Some should be kinder to themselves; others could use more self-discipline. Some, no doubt, should be encouraged to think for themselves and to pursue their dreams; others, however, would be best advised to try to stay in the middle of the flock.

But of course, I am happy to oblige. The best shoe size is ten and a half. You should call your mom more often.

Being asked about the meaning of life is similar, except there is the added expectation that the answer should be profound. Profound, but not pretentious. So a good answer might be something like "Love".

However, if we actually think about it, we realize that many people have loved bad leaders or wrong ideas, and have become complicit in massive mischief. Maybe that was too much love, or misplaced love. Some people have pledged themselves in love to one person, then later have fallen in love with somebody else, and committed an act of betrayal. Still others have loved hopelessly: and the dearer they loved, the worse they suffered. Some people have fallen in love with money, or power. So if love is the answer—which it might really be in the end—it is an answer that would need to come with a great many footnotes and appendices.

But the fact that it is possible to criticize a proposed answer is itself significant. It suggests that the question makes sense. We have at least *some* notion of what we are looking for when we are looking for the meaning of life—*some* notion of what criteria we could use to judge whether we have found it. Though we must admit, I think, that our conceptual grip on exactly what we are asking is not very strong.

In this lecture, I want to try to get a somewhat stronger grip. And then maybe we can use that to wring out an answer, or some set of potential answers, to the question of what, if anything, is the actual meaning of our lives.

As the Dean mentioned, this also serves as the final installment of a lecture series on problems of "deep utopia". We were forced to do it this way because of scheduling constraints. This means that there might be a few spots where I will comment on the relationship between meaning and utopia that might appear somewhat uncontextualized to those of you who weren't at the preceding lectures. I ask for your forbearance when that happens. Perhaps you can use those occasions as opportunities to check your social media accounts. For the most part, however, this talk should be able to stand on its own.

Grab bag concept

The concept of the meaning of life—or of "meaning *in* life", if we want to use the terminology that has recently come into vogue—is, I submit, a grab bag, into which quite various ideas have been tossed. There is no simple characterization that unifies all the different ways the term has been used.

This leaves the person who wishes to develop an account of what it means to ask about the meaning of life a choice between two strategies.

One is to try to preserve as many as possible of the ideas and intuitions that have been crammed into the bag. We can take them out, organize them, fold them up, and put them all back in. But the result is an analysis that is necessarily composite and inelegant.

The alternative is to put aside many of the items and only keep what seems most essential. The disadvantage of this strategy is that we may need to carry multiple pieces of luggage in order to fit everything that we need to bring along.

We will look at both strategies and see what each of them is able to achieve. For the first strategy, we will be examining the work of Thaddeus Metz. Our discussion in this part will also serve as a highly distilled survey of the recent philosophical literature on the topic at hand. For the second strategy, I will develop an original account.

The account of Thaddeus Metz

Metz's is a good example of the grab bag approach. More precisely, we could say that he proposes a two-piece luggage system for carrying our prudential values: in one bag goes hedonic well-being; in the other—the meaning bag—goes everything else, which naturally requires a fairly commodious duffel.

Metz's account is most fully set forth in the book *Meaning in Life*. Published in 2013, it offers a painstaking review of the literature up to that point, and picks its way through a vast number of examples and counterexamples. However, since our time today is very limited, we will come straight to Hecuba. The upshot of his analysis is as follows... this would be on page 235:

> "(FT_3) A human person's life is more meaningful, the more that she, without violating certain moral constraints against degrading

> sacrifice, employs her reason and in ways that either positively orient rationality towards fundamental conditions of human existence, or negatively orient it towards what threatens them, such that the worse parts of her life cause better parts towards its end by a process that makes for a compelling and ideally original life-story; in addition, the meaning in a human person's life is reduced, the more it is negatively oriented towards fundamental conditions of human existence or exhibits narrative disvalue."[195]

Let's unpack this. "FT" stands for "fundamentality theory", the name that Metz gives to his theory. (The "3" denotes that this formulation supersedes two preliminary definitions that Metz explores earlier in the book, but which need not concern us here.)

The basic idea of (FT_3) is that a person's life is more meaningful to the extent that she employs her reason to engage appropriately with the fundamental conditions of human existence—which, according to Metz, consists in the pursuit of "the good, the true, and the beautiful".

He gives us some examples of highly meaningful lives. Nelson Mandela: helping to end a fundamental injustice ("the good"). Albert Einstein: discovering fundamental facts about the universe ("the true"). Fyodor Dostoevsky: expressing fundamental themes of the human condition ("the beautiful").

Metz takes "reason" and "rationality" to refer to all facets of intelligence that are characteristically part of the human mind but are not part of the minds of nonhuman animals. Distinctively human emotions are thus allowed to count toward meaning in life, provided they are responsive to deliberation and that they track cognitive appraisals of value. Nonhuman animals, he claims in the book, can have no meaning in their lives (although Metz has told me that more recently he has begun to soften somewhat on this point).

The core of Metz's theory, thus, is that meaning consists in engaging in the right way with matters of fundamental moral, epistemological, or aesthetic importance. Upon this basic structure he then layers several elaborations.

He argues that, ceteris paribus, a life is more meaningful if "worse parts cause better parts later". The intuition here is that the life of somebody who undergoes years of hardship and struggle, and then comes out of those travails or rises above them to accomplish something great, and who ends on a high note, is more meaningful than the life of somebody who starts off on a peak and whose fate is then one of gradual erosion and descent into misery or triviality. The former life, Metz claims, is more meaningful than the latter, even if "the area under the curve"—the integral of momentary meaningfulness—is the same in both.

This bad-parts-are-succeeded-by-good-parts meaning-boosting effect is further enhanced if the life in question has an arc that possesses certain favorable narrative properties. Such as if its protagonist is undergoing growth and character development, and if she is making choices that play an important role in shaping how the story unfolds. It is also better if the story is coherent and intelligible.

This narrative dimension also entails that meaning is greater if a life is original. A life that is a cookie-cutter replica of some other life, or a life that is too conventional and similar to millions of other lives, is, other things equal, less meaningful than the life of somebody who is a one-of-a-kind and who does things their way.

To these elements, Metz adds a further complication. Meaning is reduced, he claims, if reason is *negatively* oriented toward what is fundamentally important. Here we can consider the life of Adolf Hitler. Metz would say that some aspects of Hitler's life are consistent with it having had great meaning. Hitler started from humble beginnings, overcame hardships and obstacles, and eventually rose along a narratively very dramatic trajectory, in which he used a variety of distinctively human intellectual capacities to reach a position of great power and from which he managed to at least partially implement a unique vision of tremendous moral significance. Despite all this, Metz would say that Hitler failed to achieve a meaningful life because he had a *negative* orientation toward the things that were of fundamental importance. Instead of pursuing peace, he started wars; instead of promoting justice and helping those in need, he was responsible for the oppression and murder of millions of innocent victims. I'm not sure whether Metz would say that this perverse orientation reduced the amount of meaning in Hitler's life to approximately zero, or that Hitler achieved a great amount of "negative meaning". Either way, whatever positive value can stem from having a meaningful life, Metz would maintain that Hitler's life did not possess it.

I think it's not clear that we really need this extra clause ruling out the possibility of deriving meaning from any kind of negative orientation. The motivation is understandable enough: we are reluctant to issue a positive evaluation of a life like Hitler's. However, this could be achieved without denying that Hitler's life was meaningful. For example, we could maintain that Hitler's life was meaningful but morally abysmal; and that it was overall extremely bad.

The question of meaning in life is anyway not intended to be a question of that life's all-things-considered desirability. This is something that Metz himself is at pains to emphasize. He states that it is possible for one life to have more meaning than another, yet be worse overall: for instance, if the more meaningful life contains a greater burden of pain. So if hedonic welfare is an additional dimension, beside meaningfulness, along which we evaluate the overall desirability of life, then why not regard moral righteousness as yet another dimension? And, we might add, why not several other dimensions too—such as interestingness, fulfillment, richness, and purposefulness, among others? If we take such a pluralistic approach, we would feel less need to cram so many considerations into the concept of meaningfulness as we are forced to do if we insist on relying on meaningfulness as the sole criterion of how prudentially desirable a life is (or as the only criterion besides hedonic well-being).

Its implications for utopian meaning

Let us now take a look at what Metz's theory implies regarding the prospects of meaning in deep utopia.

For those of you who are coming to this afresh: we're imagining a possible future condition in which humanity has attained technological maturity; one where, consequently, we have a basically unlimited ability to reshape our own minds and bodies, and where human labor is redundant in the sense that machines can perform every functional task better than humans can. We may wonder how various human values would fare in such a scenario; and, in particular, we might wonder what could give our lives meaning at this post-instrumental stage of development.

*

First let's consider the part of (FT_3) that talks about "positively orienting rationality towards fundamental conditions of human existence". Here we can observe that utopians can have enhanced capacities for rationality. In fact, they can have a plethora of superhumanly excellent cognitive and emotional capacities. This would seem to raise the ceiling on how meaningful a utopian life could in principle be. Utopians can orient themselves toward and engage with matters more deeply and thoroughly than we are currently capable of doing. *We* are dumb and shallow—barely conscious, really. *They* are alive and awake; and *they* can feel more intensely and subtly, and think more clearly and profoundly. This is all good.

But what about the "fundamental conditions of human existence" to which these capacities must, according to Metz's theory, be positively oriented, in order to glace a life with meaning? What opportunities will utopians have for engaging with the good, the true, and the beautiful?

Here the answer is slightly less obvious. If we begin with "the good", those of you who attended yesterday's lecture will recall that we had some discussion about purpose that is relevant to this point. I cannot recapitulate all of this today. But we can note that contributions like Nelson Mandela's will no longer be possible in utopia, where, by postulation, there no longer exist any such grave injustices as the one which he helped overcome. There would be other ways of doing good in utopia, although many of these would not be accessible to us—only to highly optimized artificial minds that have been specifically engineered for particular functions. We may still have some opportunities to contribute, particularly by performing some of the agent-relative tasks that I mentioned yesterday. By doing so, we could positively orient ourselves toward goodness, albeit presumably not by making practical contributions that rise to the level of significance of Mandela's struggle.

There is another way of orienting toward the good that lies wide open to the utopians. Besides by *causing* good, perhaps we could satisfy the requirement of positive orientation by *being* good? That is, either by *being benignly disposed*, or by *actively contemplating, perceiving, admiring, or loving the good*?[196] There is no reason the utopians couldn't positively orient toward the good in these ways.

Moreover, they could do so to a greater degree than we do at present. They could make themselves much better equipped for love and for loving understanding. They might also benefit from having, in and around themselves, more readily appreciable and palpable manifestations of goodness to positively orient themselves toward than are vouchsafed to us comparative wretches. The world that we inhabit is (or at least appears to be) in many respects quite *unlovely*, disfigured as it is by numerous moral deformities and other grotesque characteristics. It might be easier for us to perceive, admire, and love the good if we could see more excellent instances and exemplars of goodness around us.

Unfortunately, Metz does not believe that this kind of non-causally-efficacious "attitudinal orienting-towards" goodness is sufficient. He writes:

> "Consider the meaningfulness of help. It would not be enough for real meaning, say, merely to want others to be helped or to be pleased upon seeing them helped; one must also, of course, do some helping."[197]

So, on Metz's view, if we want meaning in life by orienting toward goodness, we need to participate in bringing goodness about by fixing some problem. In a condition where there are no problems, or where we are in no position to help fix them, we would have to find some other well from which to fill our jars with meaning.

It is possible, however, to take a different view on this point, and hold that one can in fact get meaning by orienting oneself toward the good without having to be causally productive of it. Let us recall the idea of the beatific vision that we also talked about yesterday—the immediate knowledge of the divine nature that angelic spirits and the souls of the blessed enjoy in heaven as they behold God "face to face". This state of directly *knowing* God is said to be also a state of *loving* God, as one apprehends the Divine Essence as the highest good. Presumably part of what makes the beatific vision so desirable is that it involves orienting toward the good. A similarly perfect vision of some other powerful being who was not good would not be as desirable. If we think that the souls enjoying the beatific vision have meaningful lives, therefore, it would seem to be because they are positively oriented toward the good (in the form of a perfect being) through their loving, not because they are causally effectuating independently good outcomes through instrumentally useful activity.[198]

If one does take this second approach, according to which one can score meaning-points simply by being good or by appreciating and loving the good, then it would be a lot easier for the utopians to have great meaning in their lives. They could reform themselves (with the aid of neurotechnology, if need be) to have virtuous characters that make it easy to approach each day with thankful relish for all the good they bring in train, and with warm appreciation of all the good in their fellow utopians, and with profound love and worship of divine goodness itself. The skies of their souls thus habitually sunny with joyous gratitude, save perhaps for the occasional lachrymose cloud of compassion to gently wetten the ground in remembrance of those who lived in earlier times and in sympathy with those, if any, who have yet to be included within utopia's blissful embrace.

*

What about the other two meaning-affording fundamentals in Metz's account, the true and the beautiful?... knowledge of fundamental truths and aesthetic appreciation of fundamental aspects of the human condition?

Here, again, the verdict is strongly positive if an adequate way of orienting oneself toward these values is by appreciating their various instantiations. It would then be easy for utopian lives to attain great meaningfulness—far greater than we can attain at present. Cognitively enhanced utopians, with access to unimaginably advanced science and philosophy, could obviously know and deeply understand more fundamental truths than we are currently able to peer into. Likewise, they could be endowed with more finely developed capacities for aesthetic appreciation, along with bounteous troves of beauty, natural and cultural, upon which to feast their connoisseuric spirits.

The utopians' encounters with these truths and beauties would not be like our low-grade experiences when we strain over a difficult textbook or saunter on tired legs down the halls of some museum. We must imagine something altogether more immersive and satisfying. While *we* may steal a furtive glance at the muses from a great distance, the utopians—how might I put it?—*they* may palpate their attributes at close quarters.

Therefore, if understanding fundamental truths or aesthetically appreciating fundamental beauty is what lights and fans the fires of meaning, then nothing prevents the lives of utopians from blazing white-hot.

If, on the other hand, meaning comes only from *making original contributions* to the world's stock of fundamental knowledge or its collection of beautiful artifacts, then their situation may be less conducive to meaning. Our analysis here would parallel yesterday's discussion of purpose. The problem is twofold. First, utopia would soon contain an overabundance of knowledge and beauty. This would make it harder to effect a significant increase—certainly in percentage terms, but perhaps also in absolute terms. This is especially true for certain kinds of timeless science. For example, some of the most fundamental truths that could be discovered are the basic laws of physics and the basic principles of metaphysics. Once these are discovered, they're discovered. They could possibly be forgotten and erased, but that would presumably be negative for meaning: repeatedly rediscovering the same deliberately forgotten truths does not seem particularly meaningful. We could lower our ambition and content ourselves with the discovery of ever-less fundamental details and special cases; but that kind of coloring-in work would add decreasing amounts of meaning. At some point, if this goes on for long enough, our discoveries would consist in the equivalent of finding out the exact number of pebbles on some particular beach, or the number of blades of grass on some lawn.

To the extent that meaning can be derived from scientifically or artistically representing continually changing patterns in the world, such as ephemeral cultural phenomena—as opposed to fundamental timeless conditions—there would be better prospects of never running out of material. Yet this would still leave us with the second problem, which is that other types of minds and mechanisms would far surpass ours in the efficiency with which they could discover these truths and patterns; and they would also be more skilled at representing the new conceptions in writing, painting, composing, designing, programming, and other modes of expression.

Of course, we *could* still reserve for ourselves the privilege of original discovery and creation, if we prohibited the use of AI for these purposes. We would then be like the country lord who goes hunting in a fenced-off game preserve; and who maybe skews the odds further in his favor by employing beaters to drive the game directly into his path, to enable even a most inept hunter to bring home a great haul of trophies. In the context of future science, AI assistants might set up the ideal conditions for discovery, formulate the right question, gather the right pieces of evidence, and lay them out in a suggestive pattern—all the while carefully avoiding drawing the obvious conclusion. Then our esquire then steps in and reviews the material for twenty or thirty seconds, and—bang!—ventures a guess. He guesses wrong. His AI assistant presents some additional clues; surely the esquire cannot miss now. Bang! Wrong again. More clues and hints. Bang! Bang! Bang! Bang! Finally he gets it, and the mystery is felled. An AI assistant writes up a report of the finding. Esquire is listed as sole author, and to him goes the credit of discovery.

But, really—our claim to meaning, in this kind of scenario, is, I should think, *quite* attenuated.

*

Let us examine the rest of (FT_3), the bit that speaks of narrative structure. How do utopians score on the criterion of having lives wherein "the worse parts cause better parts towards [the] end by a process that makes for a compelling and ideally original life-story"?

The problem of originality came up in our investigation of interestingness in yesterday's lecture. The upshot, in a nutshell, was that the most we can hope for in utopia is mere local uniqueness. But this is also true in our current condition; and yet we seem little troubled by that fact.

I have some further remarks to add on this topic. There are scenarios in which our civilization becomes more *crowded* as we develop toward utopia. This could cause even our local spheres of uniqueness to shrink.

*

You might think the opposite would surely be the case, considering that the space of attainable lives would greatly expand at technological maturity. At present, all of humanity is crammed into one small corner in the space of possible minds; but at technological maturity, we could fan out over a far larger volume. Utopians could modify their minds and bodies in all manner of ways, and many utopians might choose to explore various regions of the immense expanses of posthuman modes of being. Other things equal, one would expect this to reduce crowdedness. If people take on more multifarious forms, each one of them would tend to become more distinctive.

Against this, however, are counterposed other factors that would tend to *increase* crowdedness.

*

One of these is population growth. Advanced technology would enable a great expansion, both extensively (settlement of new territories, including in outer space) and intensively (higher population densities). This is especially clear if we entertain the radical possibility that we are not in a simulation. There would then be billions of accessible galaxies, each with billions of stars, each of which could maintain populations billions of times greater than that of contemporary Earth (for instance by constructing a Dyson sphere that powers digital implementations), and each of these stellar habitats could last for billions of years.

Within such a vast population there would be an increased probability of design collisions. That is to say, if we pick a random person, and ask how similar the most similar other person is: then the larger the population, the more similar the most similar other person would tend to be.

So if we wanted to be significantly unique within the intergalactic conclave—within the entire lineage of terrestrial descent—we may thus find this desire harder to satisfy subsequent to such astronomical population growth. Of course, if we care only about being unique within the clique of the hundred or so people that we interact with on a regular basis, and we don't mind having some doppelgangers doing their own very similar things out there somewhere in the Magellanic Clouds, then the size of the world population, or even the size of the human diaspora, is not directly relevant. Still, an increased population would constrain the kinds of parochial uniqueness to which we could plausibly aspire.

Note that a population boom is not a necessary consequence of technological maturity and cosmic expansion, even if we postulate that the additional resources would all be used for the implementation of sentient beings. Instead of becoming more numerous, we could become more sizable. For example, if we each bulk up to the mass of a galaxy, there would be enough material only for about ten or twenty billion of us.[199] In fact, it seems that our population is already trending in that direction—though I would be remiss if I failed to point out that *Green Transcendence* has the potential to bend that curve while being great for skin tone too.

Cosmic resources could also be allocated to other purposes, such as creating beings that are very different from the kinds that we are or wish to become. Or they might be used to create insentient structures that are either instrumentally useful or valuable in themselves. These usages, which do not produce beings that crowd into our niche in design space, would leave our current uniqueness—such as it is—undiminished, and hence might not pose a threat to the uniqueness element in the narrative component of our potential for meaningfulness (as it is conceived by Metz).

*

There is another factor, besides population growth, that could make the future more locally crowded in the relevant sense. Technological maturity would enable stronger optimization of life- and personality-types.

In our contemporary world, one source of divergence between human lives is that we are all cracked and chipped in different ways.

Or, if you allow me to vary the metaphor, humanity is a wayward armada, pummeled by turbulent weather and lightning strikes: each ship's bearing biased by its own idiosyncrasies—its unique list, helm, and skew; its manufacturing faults and imperfectly repaired damage; and the caprice of the thoroughly oiled and soused greenhorns that are piloting it. Thus, buffeted by uncontrollable forces and random impulses, we are tossed hither and thither, each blown off course at different points, our fleet scattered and dispersed in every direction.

Contrast this with a situation at technological maturity, where, like modern ships equipped with powerful diesel engines, GPS, and professionally certified and tested bridge crews, we pretty much always reach our intended destinations on a straight course. Our lives would thus more closely approximate what we want them to be, and we'd draw nearer to our ideals. For example, if nobody wants to have anxiety attacks, we'd all become more similar in not having any. Likewise, if nobody wants to be blind in one eye, or have pimples on their nose, or be slow in understanding jokes, or have a jar of sauerkraut dropped on their foot, then these sources of life variation could also shrivel up: all unwanted divergence could vanish or be greatly curtailed.

In a scenario where people mold themselves and their fates into the precise shape of their ideals, the question becomes: How diverse are people's ideals? Are they more, or less, diverse than the cast of personalities that bestride the contemporary stage? If *less* diverse, then better technology for optimization would tend to reduce the size of the spheres within which each of us is able to claim some degree of uniqueness.

In the limiting case, we would all become exactly alike. I don't know how probable this is. It seems at least *conceivable* that there is one best life for a human being; and that, if we sufficiently upgraded our cognitive capacities and used them to reflect very carefully on the matter, we would all choose that best kind of life for ourselves. Perhaps, for instance, we would recognize pleasure to be the only good, and we would all transform our minds to experience maximally intense pleasure. Optimally wireheaded individuals may all be quite similar and have similar lives—"limbless and eyeless blobs", perhaps. (It would be cute if they came in different colors, like sentient m&m's. Or perhaps hedonium will take the form of a big continuous jelly, quivering with euphoria and delight?)

But even short of total convergence to a single point in life-space, improved optimization tools could still lead to a substantial reduction in diversity.

*

Of course, if meaning is sufficiently important, and requires (or is strongly amplified by) civilization-wide uniqueness, we could opt to trade off some other individual optimality for the sake of being different.

In this case, whoever is first upon the scene has an important advantage: they could be as optimal as they wish and still be utterly unlike anybody else. Later arrivals would have to settle on the farther outskirts of optimality—unless they are willing to butt in on those who have already claimed the best spots (assuming such encroachments are even legal). These imposters, sporting profoundly unoriginal personalities, would retread well-trodden life paths, perhaps while braving charges of plagiarism or pastiche, depending on whether they arrive at their position by copying the pioneers or by independently deriving the same design.

If this situation is anticipated, a mad race to be first to optimize might ensue. This concern, although it seems far-fetched, should not be completely dismissed, at least not if we consider the racing dynamic in a more generalized form. A stampede into the best parts of utopia could lay waste to substantial amounts of value: not only might the contenders trample one another as they jockey for position, but a precipitate dash to perfection would also mean speeding past the many charms and beauty spots that probably festoon the best paths to utopia.

*

To recall, Metz claims that a life is more meaningful the more it has the shape of a good narrative; and it has such a shape, he says, "insofar as it avoids repetition in its parts, its bad parts cause its later, good parts, and they do so by virtue of personal growth or some other pattern that makes for a compelling life-story that is original".[200]

I think we have sufficiently covered the issue of originality and repetition. Originality is challenging if there are many other lives, and repetition is challenging if your life goes on for a long time. Let's now look at the other elements.

These can raise challenges, too. The desideratum that bad early parts cause good later parts seems to require that one is able to make substantial improvements in how one's life is going. This meets with difficulty on two fronts:

1. Once a life is already extremely excellent, there may just not be much room for further improvement. So while some initial segment of each utopian's life could cause later improvements, this segment may be a small fraction of their entire life. The longer the life stretches on, the greater the fraction of it would be such that its average quality does not improve much. Either the life is already close to maximally good, or else the rate of improvement throughout the life is extremely slow.
2. Even in the period within which a utopian's life is improving at a substantial rate, the required improving may be more efficiently effected by machine.

*

Here is what I think would be feasible. There may continue to be *some* room for the utopians' own choices, experiences, and natural psychological development to have positive effects on their quality of life, although the magnitudes of these effects may well be modest.

What I have in mind is that some of the purposes that may remain for us even in a plastic world would plausibly be linked to outcomes that affect the flourishing of the person performing the task. For example, one might hold that participating in the honoring of certain principles and traditions can contribute to the objective well-being of the person doing the honoring. Or, along similar lines, that generating a certain excellent aesthetic expression could thus contribute to the artist's well-being—but where the expression in question is one that, by its nature, requires it to be accomplished by our own human hand rather than by functionally superior AIs. (This harkens back to yesterday's lecture.)

We might also have opportunities to contribute to our own well-being (and thus to causing "good parts later") that leverage having been gifted a purpose that debars us from taking shortcuts. For example, we might desire to honestly place well in a sports competition in which the use of performance enhancers is prohibited, forcing us to rely on our own efforts to achieve our goal. Earlier bad parts of our lives (strenuous training sessions) could then cause later good parts (athletic triumphs). Or to take another example: someone might have chosen to join a community that bans robochefs, providing an instrumental reason to spend the afternoon chopping onions etc. to cause later good parts in the people enjoying a nice meal together.

*

These are very simple examples, contrived to display the fundamental principles at work. However, I fear they might be misleading.

Utopian cultures could be vastly complex, and the basic strands of purpose that we have identified might be spun out and interwoven into incredibly intricate social textures and tapestries. Instead of little atomic purpose, arbitrarily gifted from one person to another, or retrieved from some disjointed impulse to honor or aesthetically express something, a utopian may instead inhabit a lifeworld that is richly embedded and normatively integrated in a fabric of normatively valenced and socioculturally conditioned affordances:

so that, at any moment, she experiences something more like unfolding multimodal gestalts of optionalities—differently worthwhile high-dimensional directions in which she could move, feel, think, and experience, each involving a unique set of tradeoffs and promising different manifestations of beauty and personal forms of goodness.

It is unfortunate that I'm not able to be more concrete on this point, as that would in all likelihood have been helpful in forestalling misunderstandings. But my vision, alas, is cloudy.

Maybe I can indicate one aspect of the problem as follows, by analogy…

Suppose we pick some segment of history, let's say a one-hundred-year period of some small kingdom from a few thousand years ago. It was, we will imagine, a relatively uneventful century for this society—no big wars, revolutions, plagues, or famines. Now, if we ask what "the narrative" was for this kingdom during this interval, we might feel a slight awkwardness. Perhaps our answer is: "Well, it started with the reign of Regipedunculus the First. Then the rule passed to his son Regipedunculus the Second. When he died, he had no surviving children, so the scepter passed to his brother, who became Regipedunculus the Third. He built a new palace in Halluxopolis. The end."

It is not much of a story. It might give one the impression that not much was going on during this time—but this, of course, would be very erroneous. There were thousands of people leading busy lives. Each day, in each of these people's lives, was full of perceptions, hopes, worries, thoughts, plans, considerations, pains, delights: chronicles laden with concern and grave import, intersecting and cross-referencing analogous chronicles of other people in their area. The narration problem here is twofold: (a) we lack detailed information about this distant epoch, and (b) even if we had the information, most of it would anyway be abstracted away in "the story" of this kingdom—if by that we mean to refer to the key historical milestones and structural factors that shaped its destiny.

We face a similar twofold problem in characterizing the lives of the utopians. We don't have any detailed information, and, even if we did, the embodied concreteness and specificity of their lived experiences would not be conveyed in a philosophical account of the structural parameters of their existence. There is a further third fold in the problem that we confront in this utopian case: to wit, that the utopians may be radically enhanced, and capable of thoughts and experiences that couldn't even fit inside the brains we currently have: whence the imaginary chasm between us (as we are now) and them is potentially much greater than the cleft that separates us and our forebears of antiquity.

We should therefore be cautious to infer, from the difficulty we encounter if we try to vividly imagine narrative richness in utopian lives, that those lives would necessarily be lacking in this regard.

*

Aside from whatever the utopians can do to raise their future quality of life, they may also have the ability to influence *the specific manner* in which a given quality level is attained. If there are many ways to flourish, then their choices and actions—more generally, their "life narratives"—could have what we may take to be meaningful consequences by determining which particular ones, out of the possible flowers that *could* be cultivated, do in fact develop and blossom. Blue flowers may be as good as yellow flowers, but they are not the same. Perhaps this kind of significance, which comes from choosing between equally good alternatives, could be sufficient for continued meaningfulness according to (FT_3)?

*

Admittedly, opportunities to produce "slightly-less-good-followed-by-temporarily-somewhat-better" or "one-particular-kind-of-good-instead-of-another-equally-valuable-good" would hardly match, in meaning-giving potential, Mandela's "twenty-seven-years-in-prison-followed-by-ending-apartheid-and-becoming-the-father-of-the-modern-South-African-nation". (Not to mention his *ne plus ultra*, achieved at age 90, several years after his retirement from public life: getting taken off the U.S. State Department's terrorist watch list.)

There may thus be a real tradeoff between circumstances that maximize opportunities for meaning (according to Metz's account) and circumstances that are good in other ways. Many of our most compelling stories are tales of hardship and tragedy. The events that these stories portray would cease to occur in utopia.

I am inclined to say tough luck to the tragedy-lover. Or rather: feel free to get your fix from fantasy, or from history—only, please, do not insist on cooking your gruesome entertainment in a cauldron of interminable calamity and never-ending bad news!

It is true that good books and films have been inspired by wars and atrocities. It would have been better if these wars and atrocities had not occurred and we had not had these books and films.

The same applies at the personal scale. People coping with the loss of a child, dementia, abject poverty, cancer, depression, severe abuse: I submit it would be worth giving up a lot of good stories to get rid of those harms. If that makes our lives less meaningful, so be it.

Remember that our task is not to create a future that is good for telling stories about, but one that is good to live in.

In any case, it might be possible for quite good stories to take place in utopia. Not every cracking yarn is a tale of woe. And there may be new forms of storytelling or sensemaking open to the utopians that we are too obtuse and senseless to understand. Who knows what kinds of "poetry" and "humor" and "music" *and other such things* they might invent?

*

From the voluminous literature on the meaning of life, I have selected Metz's book to discuss, in part because his is an attempt to integrate several different themes and ideas into one theory. One could retrieve many other people's accounts simply by dropping one element or another from Metz's. Thus we have implicitly covered a whole cluster of theories simply by going over this one in some detail.

Another motivation is that Metz defends a form of "objective naturalism" about meaning, a stance which, prima facie, poses a greater challenge for the utopians than, for instance, a subjectivist account that holds that one has a meaningful life so long as one gets what one most strongly wants, or one loves what one is doing, or one believes what one is doing to be important. On the face of it, such a subjectivist criterion would be easy to satisfy in utopia, since the utopians' mental states—their wants, loves, and beliefs—are fully malleable at technological maturity. If the problem were merely that the utopians might not *feel* a passion for what they were doing, or that they might not *think* it was imbued with importance, then the problem could be fixed, easily it would seem, by reengineering their feelings and thoughts.

Actually, I don't think the case for utopian meaning is quite such a slam dunk even on a subjectivist account. For although it would be *technologically* feasible to ensure that the utopians score high on meaning according to such a theory, there may be *other constraints* that militate against using technology to reshape the inhabitants' psyches in the requisite ways. For example, even if meaning consists in believing that what one does is important, and the utopians have the technological capability to make themselves believe that what they do is important, they could still have a meaning problem if what they do is in fact not important and there is negative value attached to having false beliefs. Then they would be presented with a dilemma: accept either the disvalue of meaninglessness or the disvalue of delusion.

So meaning could still be a problem on subjectivist accounts, and one would have to investigate the matter more closely to arrive at a definitive conclusion. Still, I think that meaning is likely to be less of a problem for the utopians on subjectivist than on objectivist views; and this was another reason for why we looked at Metz's account: to confront the more difficult case.

*

This completes the first part of my talk, where we covered some of the literature and explored an approach that tried to fit all prudential values, except hedonic well-being, into a single concept. We will now switch gears and begin to develop a novel account of meaning.

The approach we will take is to allow for pluralism about prudential values. There can be multiple factors, other than pleasure and meaning, that contribute to making a life desirable. This frees us up to develop an explication which is both simpler, and, I think, better able to accommodate the spirit of the idea of meaningfulness.

As a preliminary, we will be making several disjointed observations bringing into focus various aspects of the idea of meaning. We should not give too much weight to these individually, but they will help prepare the ground for the formulation that we will subsequently introduce.

Slack

Consider the psychology of someone who is seriously concerned about the meaning of life. They're ruminating on it, perhaps it's keeping them awake at night.

I think what this person might be missing and craving, and what they may be consciously or subconsciously in search of, is a life mandate: an overarching goal, role, or ideal that they could devote themselves to, strive toward, and organize their existence around.

But the mandate cannot be arbitrary. The meaning-seeker can't simply make up a goal at random and declare "problem solved". They need something they can get fully behind.

In the ideal case, they find an ambition that brings all the multifarious parts of their psyche together in wholehearted endorsement and assent. When they contemplate their meaning, a wave of affirming jubilation would rise through all the layers of their being, awakening an inner conviction that removes any doubts and misgivings, like curtains pulled aside in the morning: and with a smile they behold their quest: "Yes! This is worthwhile. This is what I want. This is my path. I know not what obstacles I may encounter, but my meaning shall be to overcome them. If I am blown off my path, I shall make it my business to return to it. I shall at any rate keep moving. For I am pulled forward by irresistible strings."

And when this person starts moving in line with their purpose, they may find that the flies that had been pestering them in their stasis—the little annoyances, the sophisticated intellectual deprecations, the second- and third- and fourth-guessings of their own intentions: all these are soon left behind and dispersed in the salubrious breeze of action and sound exertion that now sweeps through their days.

If this is what it feels like to be inspired with a strong sense of meaning, it is scarcely to be wondered that the *absence* of meaning can create the sense of being in the doldrums, of an uncomfortable void in one's life where the great "wherefore" ought to have been.

A meaning crisis might signal that one is misdirecting one's life, and that something needs to change. If the condition is not resolved, it becomes depressing. The depression says: "Whatever you are currently doing, is not worth doing: stop investing your hope and energy into it. And don't start any other activity that is similarly pointless. Also, don't assert yourself or try to rally other people to join you—for you are up to nothing that is worth anybody's while.".

It is possible that a meaning crisis is more likely to afflict those who are in other respects quite well off: those who have resources, human or material, that are at risk of being wasted unless deployed for some worthwhile use. If that is correct, then it is not those whose daily lives are a struggle for existence who are most at risk of suffering from a lack of meaning—for in a sense they have their hands full with just surviving—but rather those who have some amount of *slack* in their lives, who have comparatively much to forfeit and to lose.

People who "have it all" may in this regard be at a disadvantage (however enviable their condition may be in other respects). In the first place, there is a category of possible goals that just don't make sense for such people, as one cannot strive for what one already has. The heir who was born with a silver spoon in his mouth, and a cushy trust fund to his name, cannot find meaning in the goal of becoming financially independent. In the second place, the fortunate and the gifted have a great deal of resources and potential which, if not put to worthwhile use, are being wasted.

And in fact, historically, we do seem to see existential concerns about meaning surfacing as a prominent cultural phenomenon in the nineteenth century—perhaps not coincidentally the first time in history when average income rose well above subsistence on a wide and sustained basis.

Role

What is needed for meaning is not necessarily to have a *goal* in the sense of there being some specific outcome that one is seeking to attain. One could equally well get meaning by devoting oneself to a *role*. Many people find meaning in the striving and exertion involved in being a good parent, a good writer, a good teacher, and so on.

I suppose it might be possible—if we insist—to describe these ambitions in goal-oriented terms. For example, we might have a conception of what an ideal parent should be like, and then we could say that somebody has devoted themselves to the goal of trying to approximate this ideal as closely as possible. However, there is typically a difference between committing oneself to playing a role and committing oneself to achieving some particular outcome or end state. A role-based purpose tends to be more relational and situational in what counts as good performance, and the desiderata often evolve with changing circumstances or in continual negotiations with other stakeholders. To play a role well means perceiving and responding appropriately to an indefinite series of challenges and demands. In the case of an outcome-based purpose, one either succeeds or fails at achieving it; whereas in the case of a role-based purpose, it may never be possible to achieve a final success—although, in another sense, one can be continually and fully succeeding at it and keep doing so without any particular limit.

What is required for meaning, I think, is that the thing to which one devotes oneself to—be it a goal or a role, or even an ideal—is such that (a) one believes, and wholeheartedly feels, that it is something really worth devoting oneself to, and (b) devoting oneself to it is sufficiently demanding to fill up one's life, or a good chunk of it at least—one's time, talents, energy, and ability to endure.

A stay-at-home parent might have a purpose that satisfies both (a) and (b). However, once the children leave the nest, and condition (b) is no longer satisfied, a meaning crisis may be precipitated.

By the way, I think the concept of role is quite important for our thinking about utopia. It's related to purpose, to our desire for esteem, and to meaning. I suspect that we have a deep desire to find "our place" in the world—some niche within our community in which we are needed and valued and in which our talents and natural proclivities are put to good use. It is advantageous to find one's place, since if you perform some function that is valued by your group, and you are good at it, your social standing will tend to be more secure. We may have come to intrinsify this instrumental utility of finding a way in which we can contribute; so that we now *directly* desire having a role that makes good use of our talents—no longer only for instrumental reasons (as a means to win social credits) but also for its own sake.

Ideally, our role should be one in which we are *irreplaceable*. If you are the only person in your tribe who can make a fire or tend to wounds, or if you are especially good at it, that makes you a particularly valuable ally.[201] So we might be built to take pride and delight in excelling in roles (such as firemaker or healer); and perhaps we have an inbuilt motivation for seeking out some socially valued role for which we have a unique aptitude, and then to hone our skills and cultivate the associated habits and attributes.

In our personal relationships, too, we crave irreplaceability. It is often more attainable there than it is in our professional lives. Irrespective of your capabilities, of whether you are sick or healthy, young or old: if you are somebody's mother or father, or somebody's child, you are, in a very real sense, irreplaceable. Likewise if you are somebody's true love: you then are irreplaceable to at least one person. That answers to a deep need.

This is one reason, I suspect, why scenarios such as Aldous Huxley's *Brave New World* are dystopias, despite having many features that should be attractive. In *Brave New World*, there are no deep family relationships, no true love—and no irreplaceability. People are machine-produced commodities; and everybody is pretty much interchangeable with anybody else, at least within their own caste. The Brave New Worlders are spared from experiencing the malaise of meaninglessness only because their society is organized to provide a continual parade of distractions—and, as a backup, "soma" (an opioid-like yet side-effect-free drug that is socially sanctioned and regularly consumed).

Orientation

Another thing whereof one earnest meaning-seeker may be in quest is a view of the big picture.

Which big picture? Well, a search for "meaning" can arise on various scales.

You might happen upon a scene in the park: people are behaving strangely; you wonder, what is the meaning of this? Aha, film students doing a shoot. Or: people playing hide-and-seek. Or: a few friends having some kind of inside joke. These are different possible meanings that you could attribute to what you see. If we identify the true meaning of the scene, it gives us a better understanding of what is going on, including a better ability to predict how things will unfold and how they would change if we attempted to intervene in some manner.

Likewise, one might inquire about the meaning of a novel. Is there an explanation that makes an otherwise confusing plot seem more coherent? Is there an underlying theme that ties the parts together? Can we better explain the writer's choices if we place the novel in the context of a literary tradition? Have we ourselves perhaps had experiences from which we can extrapolate an intuitive understanding of what the characters in the story are represented as going through? These sorts of questions are about the meaning of the text. Again, good answers improve our comprehension.

In a similar manner, one might ask about the meaning of one's life. Is there some pattern or theme that would help one make sense of it? An interpretive schema that would make it easier to see its key structural elements, its central theme, its motive force or its underlying organizing principles?

But there is no guarantee that our lives have anything like a meaning in this sense. Sure, we can understand various aspects of our lives, at various levels of description—physical, biological, psychological, historical, etc. We can detect many patterns. We can improve our understanding of the motives, including the hidden ones, that drive our own and other people's behavior. It is unclear, however, that the upshot of these learnings—if all they amount to is a patchwork of empirical regularities and a cognitive library of useful concepts and observations—would be aptly characterized as a discovery of the meaning of life.

We seem rather to be working on the assumption that "the meaning" of life, if there is one, must be something fairly uncomplicated and unitary: an insight that is simple enough to be encompassed in a single thought, and held in awareness in one conscious moment—simple enough for this to be possible, yet also fecund enough in its interpretive affordances to bring many different aspects of our lives into focus, in a way that allows us to see what our overarching priorities should be and how we should orient ourselves to the world.

So it is an empirical question whether our lives have meaning in this sense. It is certainly not impossible. Just as some novels have a discoverable meaning, so too could our lives. For instance if they were scripted.

Organizer [*bringing a bottle of water*]: It's not cold but it's all we could find.

Bostrom: Oh, that's fine. Thanks. [*Drinks.*]

Firafix: Do you think he's alluding to the theological situation, or to his simulation argument?

Kelvin: Could be both.

Firafix: What about the earli—

Bostrom: WHAT IS GOING ON?!

Audience: [*Bewildered looks.*]

Bostrom: What is the story here? This vast mysterious place!

I open my eyes and I find myself being *this* particular individual, in *this* place, at *this* time—out of all the other individuals, places, and times that I could have turned out to be or occupy!

Here in this theater, on this medium-sized rocky planet, now completing its approximately four and a half billionth lap around the local G-type main sequence star, which itself is completing maybe its 20th orbit around the Milky Way galaxy, which itself is speeding along, at 550 km/s, together with billions of other similar galaxies, relative to the cosmic microwave background: here, here I am, looking out with two eyeballs, and I see an auditorium full of similar pairs of eyeballs looking back at me…

It is really *quite* curious!

What is going on? Eh?

It's like… One finds oneself at a table, holding a hand of cards… But what is the game being played? What are the rules? Why am I playing?

I look up from my cards and glance around. Nobody else seems the least bit mystified. This strange and perplexing situation—but apparently it is the most natural thing to *them*, not worth even a moment's bepuzzlement.

No, their brows furrowed, they play their cards. Little chips change possession. Every once in a while, somebody leaves the table; and somebody else sits down to take their place.

[*Drinks more water but puts it down the wrong pipe and starts coughing.*]

Bostrom: We're brought back down to—[*another coughing fit*]—Earth.

Organizer: Are you ok?

Bostrom [*thumbs up*]: Life's a fountain.

Well, so I think one thing that can be going on in a quest for meaning is that one is seeking answers to these types of sense-making questions. One wants to take stock of the general situation and how one might fit into it, figure out the "what the story is". What game is being played? What are its rules and its win conditions? And what anyway is there to be won?

In many cases, a search for meaning is triggered by a problematic event, such as a bereavement or a separation, or by a transition to a new life stage, or by the realization that one's current mode of existence is fundamentally unsatisfactory. But it is also possible for it to be initiated from a state of simple curiosity, or rather—because simple curiosity tends to get hijacked by any random razzle-dazzle—from a state of awe, amazement, or existential bafflement, which, when profound, like an earth tremor of the soul, makes us question the very ground upon which we are standing, and causes a process of rebalancing in which we endeavor to achieve a new spiritual equilibrium between ourselves and the larger reality.

Enchantment

Let me introduce one more element, which I'll call "enchantment". Its connection to meaning is perhaps more tenuous than that of slack, role, and orientation; but I think it may deserve its place.

I don't have an exact definition of enchantment, but what I want to capture here is the intuition that meaning may be enhanced when a way of life is enmeshed in a tapestry of rich symbolic significance—when it is imbued with myths, morals, traditions, ideals, and perhaps even omens, spirits, magic, and occult or esoteric knowledges; and, more generally, when a life transects multilayered realities replete with agencies, intentions, and spiritual phenomena.

Why might one think that such an "enchanted world" is more conducive to meaning than one lacking comparable depth and richness of symbolic resonances?

One possibility is that a life can acquire symbolic significance—which arguably could be viewed as a kind of meaning—by interacting with, or relating to, other things that have symbolic significance. This would be like the pen that Lincoln used to sign the Emancipation Proclamation attaining symbolic significance because of its association with the historical document, which itself has symbolic significance because of the important role that it played in the American Civil War and subsequent developments. Because of these associations, this particular pen is now not just an obsolete writing instrument but something more meaningful. One could likewise hold that our individual lives, if they are part of a larger symbolic order, can thereby attain a greater meaning than they would have if we existed merely as atomistic individuals detached from any greater historical, social, political, artistic, or otherwise transcendental context. When we participate in a larger game, even as humble pawns, our lives may thereby acquire meaning in terms of our position and function within that game.

Another possibility for why enchantment supports meaning, not inconsistent with the first, is that an enchanted world is one that is more likely to be responsive (positively or negatively) to the full range of our attitudes, desires, thoughts, feelings, and expressive activities. Living in an enchanted world is, other things equal, *a higher-bandwidth affair* than living in a world that is lacking those kinds of semantically or symbolically mediated affordances.

For illustration, consider the opposite: a world in which the only thing that affects outcomes is *what you do*—but not *the spirit in which you do it*. In this reduced world, your facial expression, tone of voice, speech pattern, and body language convey no information. Your choice of wording or your style of clothing cannot be perceived by others. We can consider a more extreme version of the same idea and stipulate that nobody cares about your past interactions or your own personal history, your dreams, your religious or political beliefs, or your aesthetic preferences: instead, you act simply by clicking on predefined options on a screen. You input answers to standardized work assignments (some of which might require fairly complicated analytic work), and you receive in return pre-specified amounts of monetary reward, which you then spend on a full-service package that provides all the basics of life and pleasure pills that give you exactly defined quantities of positive hedonic experience.

In this thought experiment, you are interacting with the world through a low-bandwidth interface—as if *through a straw*—and we might have the intuition that this would reduce meaning.

Note that it is not the limitations on sensory input or motor output that form the essential ingredients of this thought experiment; but rather that the world it postulates is completely devoid of enchantment. It is a world in which it is unnecessary to think about or condition your responses on any higher symbolic constructs, because reality is practically reducible to factors that can be handled completely analytically and situationally. By contrast, in the world that we actually inhabit, it is possible for a person like Helen Keller to have a richly meaningful life, notwithstanding very significant limitations in sensory capacities. (Keller lost both sight and hearing in infancy but went on to become a famous author and one of the leading humanitarians and social reformers of her generation.) This is because, in order to thrive as she did, she needed to engage all the aspects of her personality—her emotions, her moral courage, etc.—and she needed to interact with other people, and with texts she read, by modeling them as similarly richly complex symbolic and spiritual entities. Her lifeworld was therefore very different from that of the imaginary button-presser in our thought experiment.

We can point to aspects of modernity which somewhat resemble the situation of the button-pusher; and we may speculate that these contribute to the widespread notion that we moderns face special challenges with respect to meaning. Our market-based economic system, for instance, offers a common unit of exchange, in which the value of a wide array of things can be measured on a single scale. Machines and machine-like artificial systems (such as formal institutions) constitute, shape, or intermediate an expanding portion of our lifeworld. More and more of the world is being grasped by application of materialistic principles, the scientific method, and various kinds of analytic or quantitative reasoning, which gain power from being able to draw from an ever-increasing store of facts and data, which is made increasingly accessible to us in any situation. Fairness norms dictate that individuals be treated on the basis of formal qualifications, measured work performance, or adherence to clear rules, rather than on the basis of personal ties, sympathies, obligations, pedigree, looks, accents, or how pure their spirits are. Of course, such norms are not perfectly adhered to, and important areas of life still remain where this formal/analytic mode has made only limited inroads. Still, all these "enlightenment" aspects of our contemporary existence combine to significantly de-enchant our lifeworld.[202]

There has also been a reduction in mystery. The depths of the sea have been plumbed and surveyed, leaving no place for the sea monsters to dwell. Every river and every forest used to be believed to have a spirit living in it: but they have all been expelled by our advancing science and clonking civilization. Even the human soul has been given notice to vacate! Already we've had several pink slips delivered by psychology and neuroscience; and now the bailiff of AI is knocking sternly at the door with a warrant to take possession. As I said, we can measure the worth of very many goods and services in dollars; and recently we've also begun to develop precise indices of our social standing, accounted in units of subscribers, thumbs up, views, followers, and friends. Overall: more numbers, less myth and less meaning.

It is true that our close interpersonal relations are still mostly run by our intuitive and affective faculties, rather than by science-based models or analytic reasoning. It is still possible for people to talk about "the miracle of love"—barely. So romance is one place we could look for remnants of meaning-as-enchantment in our contemporary lives. But of course, our inventors and entrepreneurs are hard at work technologizing this remaining nature preserve too, for example by moving courtship to dating sites with quantifiable metrics and standardized interfaces. My colleagues in the brain sciences and in evolutionary psychology are also doing their best to demystify human mating and pair bonding. But there is still a difference between theoretical understanding and practical ability. Heart-eyed innocence and blushing infatuation may remain more appealing than the constricted pupils of rational understanding (however well-informed about the mechanisms of love and about whatever else can be gleaned under the fluorescent lights of the primatology or neuropsychiatric research labs, or by reading case reports).

Anyway, the conjecture is that an enchanted life tends to appear more meaningful because it is (or appears to be) more interlinked with other things that carry meaning, and/or because it is more likely to encourage a wide-spectrum mode of engagement that draws on the full range of our psychological and bodily repertoires, rather than exclusively on our capacity for analytic reasoning and calculative decision-making.

Deep human relationships may be especially conducive to meaning, because such relationships lead to interactions that involve our whole being. People close to us can, to an extent, perceive not only the explicit conscious choices we make, but also *how we come to* these decisions and with what attitudes and accompanying feelings. Someone who knows us well will have a sense, more broadly, for our moral character and for how we spiritually orient ourselves toward life and the unfolding of events. To thrive in such relationships therefore requires more of us than pure ratiocination.

We can imagine a solitary life that is highly meaningful, but perhaps the reason for this is (at least in part) that a solitary life could involve a deep relationship *with oneself*. The recluse may be highly attuned to her own moods and feelings and thoughts and intentions, and may thus engage a wide range of her emotional, intellectual, and physical capacities in managing her self-relationship. She could thus interact with the relevant parts of her lifeworld—which of course centrally includes her own self—via a high-bandwidth connection: not only through the straw of calculating rationality, but through a wide-aperture membrane that grants admittance to important symbolic and spiritual phenomena.

Motto

Here is another observation: that we may take the meaning of somebody's life to serve as a kind of mission statement, or motto, for that person. Thus, if someone proclaims *X* to be the meaning of their life, we may reasonably take *X* to be some sort of declaration of what they're about, what they stand for, and what they are ultimately up to. Of course, there is no guarantee that the conduct of the person will accord with their professed ideals. But if their declaration is sincere, it should give us some information about how they see themself as fitting in with some larger structure of values, and which overarching priorities they invest with authority to govern their pursuits (at least insofar as those pursuits are directed toward ultimate or transcendental concerns, as opposed to the regular exigencies of their day-to-day existence).

Motivation

Some things we enjoy doing. Other things we enjoy having done.

Meaningful activities tend to fall into the latter category.

Meaning sustains motivation even during periods when immediate reward is not forthcoming. And I think the phenomenology here is a bit different than when one has simply set oneself a long-term goal that requires short-term sacrifices and willpower to motor through. Meaning-based motivation seems to be generated more "organically" and to not rely on the expenditure of willpower for maintaining forward thrust.

When one feels that one's efforts are meaningful, one can be ushered along a highly laborious trajectory by a continuous supply of spontaneous impetus—like a fair wind, which fills one's sails and frees one's ego from having to work so hard at the oars.

Or maybe not exactly like a wind in one's sails. Rather, it might be more that when one is inspired with meaning, many different parts of one's psyche collaborate and pitch in to get the job done, like a tight-knit team. The many hands, or converging penchants, make light work—or at least make the effort feel more natural and easier to bear.

One of the collaborating psychic parts may actually be the brain's reward circuitry. It can contribute to the subjectively meaningful enterprise by infusing us with a good and hearty mood while we are engaged in it, and by dispensing a sense of gratification at the end of each arduous day—that blithe feeling of a day well spent and of a night deserving of its slumber. Note that these special rewards come not from the attainment of a goal, but from the conviction that one's labors are in alignment with one's meaning.

If we take, for example, childcare, we may find that it is often not particularly pleasant in the moment. It often involves, instead, a sense of fatigue. Yet it is not exactly *willpower* that helps the carer persevere through the hours of the day and the days of the week; at least not if they are any good at it. Instead, a good carer is sustained by a more deeply-rooted constellation of psychological capacities and forces, which make the many tasks associated with the caring role be experienced as meaningful. The tasks may be exhausting but they are not vacuous or gratuitous, nor lacking in felt justification. It is similar with other great callings to which one might devote oneself—being a true artist, writer, or musician, a great statesperson or moral reformer, an elite athlete, a good Samaritan, a devout adherent of a faith, etc.

People who become great do so, in part, because they found meaning. Their project becomes something more than a goal: it becomes part of their identity. Their whole soul buys into the endeavor. This enables the great person to go further than anyone who is driven only by calculating rationality and a will-powered pursuit of instrumental precepts.

The speculative backstory

Let us now collect these fragments and see if we can piece them together into a pattern. I think what emerges is something like the following picture.

For human-like creatures, it is paramount to understand and align with one's social context and life situation. Our cognitive, emotional, and volitional capacities are molded to facilitate such conformity.

One possible root of meaning might lie in the importance of collective agency. The individual must discern and adjust to the implicit intentionality of the group's collective endeavors and inclinations. We have therefore evolved to be like auto-tuning radios—scanning the wavelengths for coherent signals about the group's intentionality and about what is expected from and required of each of us in order to fulfill our role and remain in good standing.

Another possible root is that we might have developed meaning-supporting psychological faculties to enable adherence to our own individual long-term projects, and not only to the tribe's intentionality.

Plausibly, the relevant circuitry evolved to subserve both functions.[203] In any case, the result is that when we lack an encompassing purpose, we may experience its absence as a feeling of vacuity or existential meaninglessness. (We may be more or less aware of this feeling, depending on how distracted we are.) Prolonged exposure leads to anomie, a debilitating sense of disconnection from one's group, role, social identity, and life path.

By contrast, when we feel confident that we have identified and aligned ourselves with an encompassing purpose, one which we can wholeheartedly endorse, we experience psychological effects such as reduced boredom and alienation; and we gain access to meaning-based motivation, which is distinct both from immediate urges and from willpower-driven goal-pursuit.

Meaning-based motivation has a degree of independence of the immediate circumstances in which we find ourselves. This is because it does not derive from expectations of imminent reward, nor from fluctuating drives and appetites. It derives instead from convictions anchored in a larger external pattern of priorities.

If we are talking about the meaning of life, the larger justifying context has to be something outside of life, or at least outside of a person's individual mundane life as we ordinarily conceive of it. The larger justifying context could be a "transcendental" reality that contains relevant instrumental reasons for us to act, or it could be a normative structure imputed to have a claim on us that is independent of our own subjective preferences. Either way, the meaningful purpose is likely to be quite stable and situation-independent. It is unlikely to change from one day to the next, or when you walk from one room into another.

We can stretch the notion of meaning to cover cases where the purpose in question is derived from a context that is a bit more localist than the kind of purpose that could constitute a meaning of life. Maybe we could refer to these slightly more provincial purposes as meaning *in* life. But the underlying idea remains the same: more meaningful purposes are those that are derived from comparatively larger and more stable patterns of concern or aspiration.

Consider love versus lust. Love is usually regarded as more meaningful than lust. This concords with the account I just sketched. Lust is often based more on bodily sensations, local perceptible features, and situational cues. Love, in contrast, tends to rest on a more holistic assessment that often includes an extended interaction history that gives more information about the suitability of longer-term commitments.[204] Thus, if love is reflective of a purpose that has a longer and more life-shaping span than lust—a purpose that is anchored in more self-transcending variables than lust—this will tend to place love further toward the more meaningful end of the spectrum, and lust toward the less meaningful end.

Another factor that might mark love as being more meaningful is that a life of love may be able to recruit a wider coalition of our intrapsychic constituencies and emotional affordances than a life of lust, thus making love-based purposes more *encompassing* in the relevant sense. Love may also encourage a wider arch of responsibility than lust tends to do.

Desires that are still less meaningful than lust can be imagined—for example, the impulse to scratch an itch. An itch is even more parochial than a lust. It might last only for seconds and involve the irritation of one small skin area; it is unrelated to any holistic judgment of our life situation or larger context; and the goal of scratching the itch is influenced by only a small component of our mind. An itch is a caricature of a lust. Which is why, if one wishes to denigrate a life of lust and emphasize its lack of meaning, one might say that it amounts to nothing but the mere scratching of a series of itches.

Meaning as encompassing transcendental purpose

Let me try to express things a little more systematically and academically. I propose the following account of meaning:

Meaning as Encompassing Transcendental Purpose

A purpose *P* is the meaning of person *S*'s life if and only if:

(i) *P* is encompassing for *S*;

(ii) *S* has strong reason to embrace *P*; and

(iii) the reason is derived from a context of justification that is external to *S*'s mundane existence.

Furthermore, we can say that *P* is *the meaning of life* (simpliciter) if *P* is the meaning of life for nearly everybody. (And *P* is the meaning of *human* life if *P* is the meaning of life for nearly every *human* individual.)

Don't worry if this is obscure—I will now explain what I mean by the key terms and phrases in this formulation.

Purpose

By "purpose" I mean something that could serve as an aspirational focus, something that one might strive toward. It need not be a "goal" in the narrow sense of something that would lead one to attempt to attain some particular outcome or end state. *Goal-focused* purposes are possible, but meaning might also be found in purposes that are *role-focused* (involving an aspiration to play a certain role with integrity and excellence) or *ideal-focused* (involving an aspiration to live faithfully in accordance with some ideal). In the case of role-focused or ideal-focused purposes, there may be no definite outcome that one is aiming toward—rather, one would be striving to conduct oneself in a certain manner, in accordance with the demands of the role or the principles or templates associated with the ideal.

Role- or ideal-focused purposes can be just as challenging and just as suitable for providing meaning as goal-focused purposes. I do suspect, however, that in the case of role- or ideal-based purposes, meaningfulness could often be further enhanced if an outcome-oriented objective is added to the mix. For example, one might find meaning in playing a valued role in one's community. But suppose that the community faces some great challenge that gives rise to a vitally important outcome-oriented objective, such as to defeat a dangerous enemy or overcome some great natural obstacle. Having such a shared outcome to aim for, on top of having a role to play or an ideal to live up to, would seem to endow the purpose with extra zest and élan, adding to its ability to provide a paradigmatic instantiation of meaning.

Encompassing
In order for life to have a full-fledged meaning, the meaning-giving purposes should be one that does not merely call for some particular limited action on a few sporadic occasions. Rather, it should make a sustained demand on a wide range of our faculties. In the ideal case, for a maximum of meaning, it should require continuous "all-of-organism" completely slack-absorbing dedication, involving not only simple physical action or intellectual problem-solving, but also deeper emotional, spiritual, social, and bodily forms of engagement.

These things can clearly come in degrees. Holding other things constant, the more encompassing a purpose is, the more it meets the conceptual criteria for what would count as a meaning of life. The degree of encompassingness of a purpose is partly a function of how much of our being it is apt to consume, and partly a function of how much time it takes to accomplish (if accomplishing it is even a possibility). For a purpose to be a potential meaning of life, it should be able to fill a life or at least a substantial portion of a life. Some endeavors are simply too small to constitute potential meaning-giving purposes—for instance, the goal of finding a good parking spot (except perhaps in London), or the role of playing Santa on Christmas Eves, or the ideal of having a good posture.

While greater encompassingness makes a purpose (other things equal) able to be more fully meaningful, it does not necessarily make the meaning more desirable or beneficial. Maybe we want our lives to have meaning but not *too much* meaning. It can be nice to have some time for wanton frolics—and for this not to be a guilty compromise of our pursuit of meaning, but to be licensed because our meaningful purpose is limited in how encompassing its demands are. Meaning may be prudentially desirable, but it is not the *only* prudential good.

Strong reason to embrace
Other things equal, the stronger the reason a person has for embracing the purpose in question, the more it can qualify as being the meaning of their life. For a maximum of meaning, the reason should be so strong that it clearly overrides any other reasons the person has for being concerned with things unrelated to the purpose.

Derived from an external context of justification
What I mean by this is that the (principal) reason the person has for embracing the purpose should not depend on a hope of gaining any ordinary intra-life benefits. Instead, the reason must derive from transcendental considerations.[205] In other words, the source of desirability must be located beyond the person's own mundane existence.

One way for this criterion to be satisfied is for the person to have reason to try to achieve some outcome outside their own life, such as to bring benefits to other people or to achieve some impersonal value.

Outcomes related to one's own afterlife can also qualify. The idea is that meaning is a special kind of purpose that is derived not from some particular thing in our life that we happen to need or fancy, but instead from reasons anchored in some larger pattern of value or concern—some pattern that *transcends* our own normal mundane existence and whatever desirabilities may therein be present. Although, in a literal sense, the rewards of an afterlife would be occurring within the person's own life, yet they are sufficiently outside the person's mundane existence (with its day-to-day preoccupations and concerns) to count as external to it, and thus to offer a potential separate basis for the derivation of worthwhileness of some encompassing undertaking.

It is not necessary to precisely specify the boundary around what counts as "our own normal mundane existence". The more that some instrumental rationale concerns outcomes that extend far outside the life's efforts and forbearances it is supposed to give meaning to, the more definitively it satisfies the criterion. (We might get a bit more clarity on this point shortly, when we look at some examples.)

Another conceivable way that a purpose could meet the criterion of having an external context of justification is for it to be entailed by objective moral facts. I do not wish to make any claims about metaethics here, but I want to reserve room in my account for the possibility that if there is an independent moral reality it might potentially provide a qualifying "external grounding" for meaning-given purposes. Exactly what the metaphysical prerequisites are for a normative structure to count as "real" and "independent" we will leave for others to work out; but it would need to be something that could furnish us with strong reasons for action whose source and justification lies outside of our own individual mundane interests and proclivities.

Concordance with some observations

It should now be relatively easy to see how this account of meaning as encompassing transcendental purpose ("ETP" for short) reflects the various preliminary observations we made.

Because the meaning-giving purpose is encompassing, it has the capacity to absorb a great amount of *slack* that one might otherwise have in one's life.

Because the purpose need not be outcome-oriented, it allows for the possibility of finding meaning in a *role* (or in a commitment to an ideal).

Because having a meaning involves pursuing a purpose that is derived from a transcendental context of justification, it requires and provides some *orientation* with respect to the most important and relevant aspects of the macrostrategic situation of our lives.

Because pursuing meaning amounts to directing one's life toward some definite normative construct (a goal, role, or ideal) which one has a strong and lofty reason for embracing, the meaning that one in fact adopts might be expressed in a *motto* that sheds light on what one is up to and what one takes to be worthy of supreme dedication.

Because a meaningful purpose is derived from outside of our own ordinary immediate selfish preoccupations, it can provide a source of *motivation* that is independent of our usual myopic and egocentric perspective. Meaning can thus give a person a kind of qualified immunity from the vicissitudes of ordinary life.[206] One may have onerous tasks to perform, and one might be meeting with discomforts, inconveniences, and setbacks: but these, one might feel, don't matter as much as they would otherwise—because one is not in the game for the sake of immediate reward but for the sake of a higher mission, whose rationale and success criteria lie beyond one's own mundane existence. Meaning thus wraps around us like the feathery wings of a guardian angel, giving us strength and comfort and the inner peace that comes from a conviction that one is doing "what one is meant to do" and that one is, spiritually, on the right path.

Or as Nietzsche put it, "Hat man sein warum? des Lebens, so verträgt man sich fast mit jedem wie?" [If one has one's *Why* of life, one can get along with almost any *How*].[207]

Meaning crisis

And what if one doesn't have a Why? Pessimistic possibility: The average human life cannot bear its own weight—it is not intrinsically good enough to be worth living. It must be propped up and buttressed by meaning—by something external to the life that makes it worth living. A life that is quite lacking in internal satisfactions could still be worth living if, for example: it helps relieve the suffering of many others; it secures a long and happy afterlife; or it is appreciated and valued by some very important being, a "superbeneficiary".[208]

*

We are now in a position to understand how many traditional religious worldviews have been able to offer their adherents meaning. There are at least two ways in which they have done so.

First, by holding up the prospect of immense future reward for correct behavior in life. This reason for embracing an encompassing purpose is grounded in a transcendental context of justification, and so is a qualifying candidate for the meaning of somebody's life.

And second, by providing a cultural matrix within which there are clearly defined slots for an individual to fit into. A traditional religious community might prescribe a well-specified role for each member, and mandate various slack-absorbing communal rituals and spiritual practices. This can simplify the life choices that the individual needs to make.

In the extreme it comes down to a binary decision: either walk the path that has been staked out, and attain both supernatural reward and social acceptance; or else do something different—and be condemned, disdained, and ostracized. Only for somebody with highly unusual desires or unorthodox beliefs would this be a difficult choice. Most therefore find themselves in a situation in which it seems overdetermined that what they have most reason to do is to embrace the theologically and socially recommended purpose and to dedicate themselves to serving their role as well as possible.

*

When, however, we move to a secular setting, especially one that is sufficiently prosperous to offer some respite from the exigencies of mere survival, and that is sufficiently liberal and multicultural to offer a swirl of alternative ideologies, lifestyles, moralities, professions, et cetera—any one of which could potentially form the basis for a socially viable existence—then the problem of meaning becomes acute. It is no longer obvious what, if anything, constitutes a mission that one could wholeheartedly embrace. One problem is that there could be too many potential missions, presenting a difficulty of choosing. Another problem is that it may be unclear that any of the potential missions is actually worthwhile: worth dedicating a life to, worth toiling for decades to serve, honor, achieve—given the availability of rather convenient alternatives, such as just floating along in a state of comfort and grabbing some fun here and there as opportunities present themselves.

It appears that at least some people suffer a degree of malaise from the absence of ready subjective meaning in the postmodern condition. This might also be a contributing cause to psychological fragility. If conditions turn adverse, if one meets with hardship or the easy pleasures of life cease to please, one may reach out for some sustaining motivation—some higher purpose that one can hold on to for support as one stumbles through the wreckage of one's life. Yet in the absence of meaning, one grasps only a fistful of air. Sighs and groans multiply in the echoing caverns of an existence devoid of meaning.

It is not only cultural developments that can precipitate a meaning crisis; developments within an individual life trajectory can also do so. Consider, for example, somebody who charges hard for success, optimizing their conduct for the attainment of the material or social highlife. They study hard in school, join the relevant clubs and extracurriculars, gain admittance to the right degree program, then grit through a series of internships and a period as junior clerk. They rise, and eventually they plateau—let us say at a fairly high level, though below the absolute peak of their profession. They turn fifty and begin to wonder whether it was worth it. Their marriage is a mess. They missed much of their children growing up (if they even had children). All that time in the office: days, months, and years—years flitting by like kilometer signs on the highway. A highway that leads to—where, exactly?

In the midlife crisis our protagonist becomes aware of the lack of meaning in their life. The meaninglessness was previously papered over by non-meaningful striving. Now that the glue of ambition is weakening, and the future success-hopes are peeling off, the depressing underlying reality of their life and career, which they have worked so hard to construct, is revealed—as the cement walls of a prison cell.

A note on Nietzsche

By the way, apropos Nietzsche: he, for his part, rejected external justifications for human existence—and hence could appear to deny that our lives have meaning in the sense that I have defined here.

Nietzsche's central preoccupation was to find or create another conception under which our existence could be affirmed as desirable. He was not, of course, tempted by a hedonic evaluation scheme or anything like it—one according to which we might adjudge a human life to be quite fine just so long as it contains more pleasure than pain, or so long as the balance of some other such philistine goods is tilted toward the positive. Instead, it seems to me, Nietzsche wanted something very much like meaning to base his affirmation on, but not a meaning that purported to be derived externally from life itself.

So where does he turn? To a conception of the value of greatness: an issue, we might say, of a union between the aesthetic and the heroic. This is the conception which he invites to occupy the rooms vacated after his expulsion of "conventional" morality (and specifically modern Western morality, especially in its Christian, Kantian, and utilitarian variations).[209] Striving for "greatness" could be viewed as giving life a kind of meaning.

How does this relate to our framework?

Maybe we can think of it like this. Nietzsche was not a systematic thinker, but if we want to impute a foundational axiom to his philosophy, it would be the assumption that our existence is to be affirmed.

Nietzsche explores the implications of this postulate and discovers them to be profound and indeed staggering. So very profound and staggering, in fact, as to possibly be intolerable to mere humans. Hence his call for the creation of a new type of being of superhuman greatness: the Übermensch: a sapiens who is sufficiently enhanced, ennobled, and envitalized in such a manner that it actually could—truly, deeply, and authentically—affirm its own existence, including all the conditions that this requires and entails.

Nietzsche views such a profound pro-attitude toward life as a sign of health—that one is strong enough not merely to put up with it all but to positively welcome it as a source of invigorating challenge. (The idea of eternal recurrence fascinated Nietzsche because he thought it could serve as a litmus test for whether one's life meets this high standard: could one will that it be repeated, exactly and in every detail, so that one would want to live it again and again for all eternity?)

We can now see that there are *two* sorts of potential meaning in the Nietzschean worldview.

First, for the Übermensch (and maybe also for some contemporary humans who may at their best moments perhaps in limited ways slightly approach the Übermenschliches?) there is the purpose of living a life that is capable of being authentically affirmed. The meaning here is grounded in a kind of normative conception; and while it may not be transparently characterizable as based in an "independent moral reality", it is nevertheless locatable outside of our mundane existence in the sense that the greatness in question, which the Übermensch would seek to instantiate so as to make their life worthy of their Yes, does not consist in the attainment of basely-worldly goods (such as money, comfort, or popularity) but rather in the rising up to and successfully realizing in one's life a very lofty aesthetic/heroic ideal (one that is freely expressive of the Übermensch's own individual creative nature).

And second, for any contemporary humans who are incapable of attaining this kind of Übermenschliches meaning, there is another meaning of life available: to "serve as a bridge" toward a better future humanity and help bring about the creation of the Übermensch. In our framework, this would be a purpose derived from a transcendental context of instrumental considerations—transcendental insofar as it seeks an outcome beyond our own present mundane existence, and instrumental insofar as we have reason to serve as means to bring this outcome about.

Sisyphus variations

It might be useful now to look at a few examples of how our proposed ETP account of meaning is to be applied. For this, we can turn to the myth of Sisyphus. You'll recall that he was the crafty fellow who was condemned by Zeus (in consequence of having twice cheated death) to repeatedly roll a big boulder up a hill. Each time Sisyphus is about to reach the top, he loses his grip on the boulder, it rolls down, and he must start his labors over.

As has been pointed out—most famously by Albert Camus, another thinker in the existentialist tradition—Sisyphus's predicament in important respects mirrors our own.[210] Our labors, too, might seem repetitive, pointless, and ultimately futile. We spend a lifetime acquiring knowledge, skills, character, money, fame, and friends… and then we die:

> "Everyone's life thus resembles one of Sisyphus's climbs to the summit of his hill, and each day of it one of his steps; the difference is that whereas Sisyphus himself returns to push the stone up again, we leave this to our children."[211]

The ultimate futility of our endeavors seems total. We die. Eventually everyone who knew us dies. Then our civilization dies. Finally, even the universe itself seems fated to expire in the coils of the second law of thermodynamics (that implacable nomological boa!). Everything thus comes to naught in the end.

*

This is a good place to remind ourselves again that whether or not Sisyphus's life has any meaning, and whether or not our own lives do: these are questions different from the question of whether our lives are, all things considered, worth living and prudentially desirable—compared to, for instance, a quick and painless death, or to having never been born in the first place. That one's life be meaningful, even if we assume this to be something desirable, is not the *only* thing that is desirable. Sisyphus's life *could* be a very good one for him, even if it were utterly meaningless, provided it contains enough other positives. For example, if we suppose that he is taking great pleasure in pushing the boulder, which could justify his existence on hedonic grounds. Or, perhaps, if he is obtaining great aesthetic value by appreciating the views along his ascent. (Though the most critical factor is, I think, whether he is enjoying his life. "For to miss the joy is to miss all").[212]

In this lecture, however, I want to focus on the meaning question only.

*

Taken at face value, Sisyphus's life seems the very emblem of meaninglessness. And I think it comes out as such on my account. Sisyphus does have a purpose—to push the boulder to the top of the hill—but it is not of the right kind to give his life a meaning. His purpose, at least in the common reading of the story, is not based on a reason derived from an independent moral reality or from a transcendental context of instrumental justification.

There is, however, something that puzzles me about the Sisyphus story. *Why*, exactly, does he keep pushing the boulder?

He keeps trying the same thing: does he expect a different result? As the evidence piles up, there should come a time when he realizes that it's just not going to work. Does he not update on the evidence? Does he not remember his failures? Is he too stubborn to admit defeat? Is he perhaps expecting to be awarded a gold watch after a certain number of valiant attempts? Neither Homer nor any of the other classical sources give us specific information on this point.

*

Let us consider some possible explanations of what might be driving Sisyphus's peculiar behavior.

Case 1. Sisyphus keeps pushing the boulder because whenever he stops, one of Zeus's minions cracks the whip, forcing him to continue.

If this is the whole story, it would be a clear case of a meaningless life. *Probably* it is also a bad life. If the only reason Sisyphus perseveres is to avoid punishment, we can assume that his toil is unpleasant. And there does not seem to be much else going on in Sisyphus's life that could plausibly make up for this negative.

(Even in this case, it doesn't strictly follow that his life is bad. Maybe he enjoys pushing the boulder, and the enforcement is necessary only because he would find slacking off to be *even more* enjoyable.)

*

Case 2. Sisyphus keeps pushing the boulder because he finds it enjoyable or he simply has a strong urge to do it.

In this case, Sisyphus may not be so different from your average endurance junkie, who each morning sets out on a long and exhausting jog, only to end up where she started, drenched in sweat. This effort might be rewarding in various ways; but it is not, I maintain, in itself meaningful.

Some dedicated runners do have larger purposes, such as to compete in the Olympics. But on its own, this is still not sufficient for meaning. Adopting the goal of earning a place on the Olympic podium would give a person a purpose, but it would not bestow meaning onto a life that otherwise had none.

We can imagine scenarios in which the goal of becoming the world's fastest runner *would* amount to a meaning-giving purpose; but it requires elaborations. First, the purpose must be sufficiently encompassing. This means that the activity and the struggle it calls forth must extend far beyond the muscular effort required to move the legs. It needs to become something more akin to a spiritual quest, which involves hard psychological and emotional challenges and consumes a large portion of the competitor's time and being over many years. Second—and this is the part that would more often be missing for even hardcore runners—the purpose must have an external grounding. There needs to be a reason for the effort that is independent of intra-life benefits and needs. If we suppose, for example, that the runner has a congenital disability, and that her triumphing in the Olympics would inspire thousands of people born with similar impairments to achieve vastly better lives by expanding their perception of what is possible, and if we suppose that accomplishing this outcome is what truly motivates her, then I would say that she has found meaning in her life.

*

But I'm getting a bit ahead of myself. Let's return to Sisyphus.

> *Case 3.* Sisyphus keeps pushing the boulder because he holds the following two beliefs: (a) that he might succeed at getting it all the way to the top; and (b) that if he does succeed he will be rewarded with a long and happy afterlife.

In this case, Sisyphus has a *Why* of life: something outside of his life that he can point to as his reason for living. We are here conceiving of the potential afterlife as sufficiently distinct and separate from his present existence to count as external to it—even though, of course, the afterlife would not be stricto sensu outside his life but an extension and a transformatively improved phase of it.

However, even with such an externally grounded purpose, Case 3 might still fall short of qualifying as a meaningful life, as Sisyphus's purpose does not appear to be adequately encompassing. In fact, I suspect that the impression of meaninglessness that strikes us when we contemplate the original Sisyphus story could be attributed as much to its monotony and brutish narrowness as to its ultimate futility.

*

Let us suppose that Sisyphus not only wholeheartedly buys into his purpose, but also that his quest requires something beyond mere boulder-pushing—as in the following variation:

> *Case 4(a).* Sisyphus is motivated by the same reason as in Case 3, but the challenge he must overcome is vastly more variegated and complex.

Perhaps he needs to employ engineering ingenuity to loft the boulder, devising mechanisms of levers and pulleys, constructing paths and bridges, and backstops to limit the reverses when a slip-up occurs. Perhaps he must recruit collaborators to help him, which then presents problems of motivating and organizing these people, ensuring that they are fed and housed, and so on. Perhaps—if we really want to crank up the difficulty level to brutal—he is required to obtain a building permit from the local regulators.

If the challenge that Sisyphus faces is this complex, it is easier to see how it could provide him with an encompassing purpose, one that engages the panoply of his faculties and fully absorbs the slack in his life.

*

Nothing essential changes about Case 4(a) if we substitute some other potential payoff instead of the happy afterlife, so long as it is something that is sufficiently desirable and external to Sisyphus's mundane existence.

> *Cases 4(b–d).* Sisyphus faces the same complex challenge as in Case 4(a), but his reason for undertaking it is instead one of the following:
>
> (b) Technologically superior aliens have arrived, and are tempted to eat us, but (for reasons beknownst only to them) they have promised to spare the human race if somebody succeeds in getting the boulder to the top of the hill. Sisyphus volunteers to take on the challenge in order to save humanity.
>
> (c) Sisyphus's remote ancestors made a deal with a powerful king: the king gave Sisyphus's tribe a large tract of land; and, in return, the tribe promised that in each of the coming twenty generations, one of its members would push the boulder to the top of the hill to commemorate the king's munificence. The tribe has kept its side of the bargain over the preceding nineteen generations, and Sisyphus is the only descendant now alive who is strong enough to have a chance of completing the task and thus to fulfill the ancient promise.
>
> (d) There are other hills in the region that have large boulders at their peaks, and a long-running dispute has raged in the scientific community about how these immense objects got there: one side claiming they were rolled up by resourceful human individuals; the other side asserting this to be impossible and insisting that the boulders either have always been there or that they were emplaced by supernatural means. Sisyphus wants to settle this controversy by

demonstrating that it is feasible to roll a similar-sized boulder all the way to the top.

Case 4(b) is a straightforward case of purpose derived from a transcendental context of instrumental considerations. It is transcendental in that the value at stake is something outside of Sisyphus's own existence: namely, the survival of humanity.

Case 4(c) gives us an example of a purpose that may, arguably, be derived from an independent moral reality. We can assume that the instrumental benefits to the tribe of fulfilling the ancient promise (such as strengthening its reputation for being a reliable counterparty) are incidental, and that the operative motive for Sisyphus is either a sense of moral obligation or a perception that it would be normatively glorious for such an old and unenforceable commitment to be fulfilled, and that it would be a noble and worthy act to fulfill it.

(By the way, conversely—although the explicit quid pro quo is absent here—is not the grave robbery of our archeologists *ignoble*? Especially in cases where the deceased went to extraordinary lengths to ensure that their tombs would remain undisturbed, as for instance the Egyptian pharaohs did? We violate the innermost sanctum of their resting place, plunder all their gold and treasures, x-ray their mummies, and exhibit the loot in museums for the amusement of tourists. A better civilization than ours would surely recognize this as extremely rude. Let us hope that our descendants—or whoever has power to affect our own posthumous outcomes—will treat our hopes and wishes with more kindness and consideration.)

Case 4(d) is more ambiguous. Why would it be worthwhile for Sisyphus to settle a long-running scientific dispute? Is this supposed to be instrumentally useful for humanity, by contributing to the secular project of "the Effecting of all Things possible"?[213] If so, the grounding of his purpose could potentially fall under the category of a transcendental context of instrumental considerations. Or is the idea that knowledge is good in itself, apart from its practical utility, and that Sisyphus has moral reason to promote the degree to which human civilization instantiates this value? If so, it might fall under the category of reasons derived from an independent moral reality.

A subjectivity-objectivity spectrum

A dividing line that runs through the philosophical literature is whether meaning is subjective or objective, or (as has recently become a popular view) a hybrid construct that comprises both subjective and objective elements.

The account that I have proposed can accommodate any of these views, depending on how we specify certain additional parameters. My opinion is that the concept of meaning is underdetermined in this regard. Or rather: there are several sub-concepts of meaning, which share the same general structure but impute different types of support basis for the meaning-giving purpose. And it is better to recognize them all as giving us valid ways of saying different things about meaning—we get more expressive power and precision that way—than to try to pick just one of these and then spend our time in interminable arguments about whether "real" meaning is subjective or objective.

This approach also gives us an opportunity to explicitly disentangle how subjective and objective elements can enter into the picture, which is helpful for avoiding confusion.

*

I will say that Sisyphus has *subjective meaning* if he is in fact wholeheartedly embracing a purpose that is encompassing and that he takes himself to have strong reason to pursue on grounds that are external to his mundane existence. Sisyphus has *objective meaning* if there is some purpose that would be encompassing to him and that he has a strong reason to embrace—a reason that derives from a context of justification that is external to his own mundane existence.

The basic idea behind this distinction is simple. We sometimes have reason to do things that we do not have any actual desire to do. We might, for instance, be ignorant of some relevant fact or we might be making a reasoning error, with the consequence that we fail to realize how much we would stand to gain from doing the thing in question. Or we might, despite correctly cognitively judging that we have reason to do it, nevertheless fail to do it, or even to desire to do it, owing to weakness of will. According to some philosophers, we might also have reason to do certain things, such as actions which we are morally obligated to perform, even if we not only don't want to do them but we still would not want to do them if we were fully informed, instrumentally rational, and non-akratic.

So if we acknowledge that there can be a gap between what we actually do desire and what we have reason to desire, then there is room for a notion of subjective meaning that focuses on the former (what one in fact desires), in contradistinction to a notion of objective meaning that focuses on the latter (what one has reason to desire). It is useful to have both of these notions in our conceptual toolbox.

*

In Cases 4(a–d), I described variations of the story in which our protagonist could have objective meaning. Of course, if Sisyphus doesn't actually *care* much about the proffered rationales (attaining a happy afterlife, saving humanity from being eaten by aliens, honoring his tribe's ancient promise, or resolving a long-running scientific dispute), he won't *experience* his life as meaningful, and he will lack subjective meaning. So it is possible to have objective meaning without subjective meaning.

The case of objective meaning without subjective meaning actually has two importantly different subcases that we should distinguish. We could say that a person's life has an *unrealized* objective meaning if it has an objective meaning that the person is not pursuing and not achieving. And we could say that a person's life has *realized* objective meaning if the person is pursuing or realizing the objective meaning.

Having realized objective meaning in one's life does not imply that one has subjective meaning, since it is possible to pursue a given purpose for reasons other than those that make it meaningful. For example, if discovering fundamental truths is immensely valuable, it is possible that the objective meaning of life for a person who has the talent to be a great scientist or philosopher is to pursue such discoveries. But we can imagine (with some degree of empirical implausibility) a person who embarks halfheartedly on this pursuit and who is highly successful, yet who is not motivated by a thirst for knowledge but by a craving for prizes and promotions, and who does not experience their life as meaningful. This person's life could then be said to have realized objective meaning while being devoid of subjective meaning.

Could there also be unrealized *subjective* meaning? Yes, I think we can make sense of such a notion. An example might be a person with an exceptional talent and passion for music, who embraces the purpose of composing great music either because they think that this is inherently deeply valuable activity or because they hope to produce a work of such tremendous power that it will heal the cultural chasms that separate us from one another and lead to conflict and war. So this gives them subjective meaning. We can suppose that they burn with fervor to pursue this purpose throughout their life, but that circumstances conspire to prevent them from ever actually doing any composing—they face grinding poverty, conscription into the army, personal emergencies. We could then say that their life had unrealized subjective meaning.

*

So far, I have spoken as if subjective versus objective is binary; but that is an oversimplification. I think we increase our understanding by realizing that it is possible to interpolate between the subjective and the objective, and to identify intermediary points.

For example, we can draw a line like the one that you have on the handout. It stretches from the completely subjective, by degrees, toward increasingly more objective conceptions of purpose.

HANDOUT 21
INTERPOLATING BETWEEN SUBJECTIVE AND OBJECTIVE

SUBJECTIVIST

- What I actually right now consciously desire
- What I desire although the desire need not be occurrent in my conscious mind at this moment
- What I would desire if only some straightforward fact were pointed out to me
- What I would desire if I were substantially more knowledgeable and instrumentally rational
- What I would desire if I had the character that I wish I had
- What I would desire I were perfectly psychologically healthy and well-adapted and I understood myself and appraised my situation appropriately
- What I would desire if my desires perfectly reflected objective desire-independent truths about what is in my best interest
- What I would desire if my desires perfectly tracked objective truths about what is impersonally "best for the world"

OBJECTIVIST

The concepts further toward the objectivity end might be regarded as increasingly problematic by some naturalistically-inclined philosophers. You can truncate the spectrum wherever you see fit. But unless you step off at the very first stop, and deny the conceptual coherence of anything even a bit further along the tracks, you should be able to accept a theory according to which a life could have meaning in a more subjective as well as in comparatively more objective senses.

*

I mentioned that hybrid theories have come into vogue. R. W. Hepburn, for example, proposed a view of this kind in 1966, and Susan Wolf has more recently presented an account which she sums up with the slogan "meaning arises when subjective attraction meets objective attractiveness".[214] Metz's account can also be viewed as a hybrid theory. He claims that while pursuing and achieving something highly objectively worthwhile is sufficient to give a life *some* meaning, that meaning is enhanced if a component of subjective attraction is also present.[215]

While hybrid theories seek to combine the virtues of subjectivist and objectivist theories, this ambit also lays them open to attack from both sides.

An objectivist can charge that the hybrid theorist incorrectly denies meaning in cases where the subject happens to lack the requisite kind of enthusiasm for her life project but is nevertheless doing work of great value and moral significance. For example, if we consider the life of some great humanitarian, such as perhaps Mother Teresa or Norman Borlaug, and imagine (counterfactually) that they had no passion for their work and did not find it personally fulfilling but that they persevered anyway because they considered themselves morally obligated to help relieve human misery—it seems counterintuitive to render the verdict, as Wolf's theory does, that their lives were meaningless.[216] (Metz's theory handles this objection a bit better, as it would accord these lives at least *some* meaning.)

From the opposite direction, a subjectivist might charge the hybrid theorist with espousing a philosophically "imperialist" position—one that imposes an external standard on an individual's passions and projects which they supposedly must satisfy in order to be legitimized as meaningful—a standard that might have no basis in what that individual, whose life it is, actually wants or cares about. (In practice, we find that the postulated standards in objectivist and hybrid accounts tend to closely match the predilections of a typical contemporary Western-educated humanities professor. I suppose it is fortuitous that this particular demographic happens to be such an exquisite instrument for gauging objective value!)

The account that I have proposed is rather more ecumenical in these regards. If and to the extent that there are objective standards, those define objective notions of meaning. But alongside those there are also more subjective notions of meaning. So long as we are clear about which one we are using in a particular context, we can get the best from both perspectives and without the downsides of trying to fuse them into one hybrid construct.

*

Let's take a look at a challenging example, and then we can discuss how we would analyze it with the approach that I have outlined.

Consider the following imaginary character.

> *Grasscounter*
> Grasscounter is a human being who has devoted himself to counting blades of grass on the College lawn. He spends his whole days in this occupation.[217] As soon as he completes a count, he starts over—the number of blades, after all, might have changed in the interim. This is Grasscounter's great passion in life, and his top goal is to keep as accurate an estimate as possible. He takes great joy and satisfaction in being fairly successful in this endeavor.

The objectivist and hybrid accounts that we find in the literature would say that Grasscounter's life is meaningless; whereas subjectivist accounts would say that it is meaningful.

Here's what I would say about this case. My tentative presumption is that if we actually encountered a real-life human Grasscounter, it is quite likely that he would be pathological. I imagine that if he underwent a careful psychiatric examination, we might diagnose in him some internal thwarting, an egodystonic compulsion, or some other kind of psychological malfunctioning, which causes him to engage in such unusual and seemingly obsessive behavior. Grasscounter, in this case, may have internalist grounds for modifying his grass-counting behavior and for weakening the grip of whatever urge or habit is causing him to engage in it. Whether he would benefit from therapy or medication to help him do so is a separate question, but it is a possibility.

This realistic untreated Grasscounter could still have meaning in the first few more subjectivist senses that I indicated on the line in the handout. He would not, however, have meaning in the more objectivist sense that requires the encompassing purpose to be one which the person "would desire if he were perfectly psychologically healthy and well-adapted". Nor, in all likelihood, would he have meaning in any of the subsequent still more objectivist senses. (His meaning-claim might also fail at an even earlier stage, if, for instance, his grass-counting desire is based on ignorance about some relevant fact or is caused by faulty logic.)

I will confess to a certain feeling of fondness when I contemplate the case of Grasscounter. I cannot resist drawing a parallel between Grasscounter and many of the distinguished faculty that occupy positions within our own august institution of learning—who, in many instances, have made a specialization of a field, or subfield, which seems neither to be very closely connected to "the fundamental conditions of human existence", nor to hold any promise of practical application, nor indeed to meet any other plausible objective criteria for having much intrinsic significance or value. If we compare Grasscounter to these scholastics, we might perhaps say that he equals them in terms of the objective worth of his accomplishments—and that the overall advantage is to his side, inasmuch as he is getting more fresh air.

Furthermore, Grasscounter can be confident that he is contributing positively, if ever so slightly, to the sum of human knowledge; whereas not a few academic posts would be vacated were that standard to be consistently applied.[218]

*

In any case, it is possible to conceive of a being similar to Grasscounter who is not pathological.

> *Num_Grass*
> Num_Grass is a cognitively sophisticated AI that truly genuinely authentically cares only about counting blades of grass. Num_Grass has no part or potentiality in its nature that would be "thwarted" in an existence wholly devoted to grass-counting. All levels of its goal system completely affirm its grass-counting objective—it wants to count grass, it wants to want to count grass, and so on—and it is cognizant of all relevant facts and is not committing any reasoning error.

Since Num_Grass has no pathology or dysfunction, it may (if the other criteria are met) lead an existence that qualifies as meaningful according to a meaning notion that is at least one notch more objective than the most objective meaning notion according to which Grasscounter's existence qualifies as meaningful.

Both Grasscounter and Num_Grass would, however, likely fail to have meaning according to more demanding objectivist theories and hybrid theories like those of Wolf and Metz. Authors of such theories have tended not to think highly of neuro-atypical preoccupations, such as grass-counting.

*

If we had to make a case for the lives of Grasscounter and Num_Grass having more objective meaning, we could point to the contributions they are making to civilizational (and campus) diversity. One might consider it as objectively valuable for this place to contain a rich and varied set of ways of life. The distinctive ways of a Grasscounter or a Num_Grass could certainly be additive in this regard.

We may note that this particular form of objective significance is a limited opportunity offering. Grasscounter or Num_Grass would be adding much less interestingness if there were already many similar fellows crawling on all fours and taking census of various lawns. In this case, the counters might need to diversify their areas of specialization.

Some may start counting blue jays, others virus particles, others prime constellations, and so on… at which point they might well start to organize themselves into faculties. And in all earnestness—although I am not, admittedly, speaking from a position of complete disinterest here—I will say that a most valuable service that is rendered by this academic facility is to provide asylum to a fairly decent number of oddballs. Long may it continue to do so!

How meaning could be discovered and shared

I next want to explain more fully how meaning could be "out there" in the world, as something that might be empirically discoverable. This is true not only for meaning in an individual life, but also, potentially, for the meaning of life simpliciter (or for the meaning of human life, or of the lives of many of us).

We can best see this if we focus on a meaning notion somewhere in the middle of the subjectivity–objectivity spectrum. For concreteness, let's pick the notion of meaning where the relevant reasons that a person has for embracing a purpose are based on what "one would desire if one were substantially more knowledgeable and instrumentally rational". (The other criteria of the ETP account must of course also be satisfied in order for a purpose to count as meaningful: the purpose needs to be encompassing and the reasons for embracing it need to be derived from a context of justification that is external to the person's mundane existence.)

*

The type of scenario I have in mind, to illustrate how we might discover the meaning of life, is one where there is a transcendental context of instrumental considerations that has a particularly definite set of implications for a sufficiently wide range of possible preferences. The meaning of life, in such a case, could be determined by certain structural features of the world and the location of our agency in relation to those features—what we may refer to as our "macrostrategic situation", or our "predicament". It is at least theoretically possible that all of us, or many of us, are in essentially the same predicament; and that, in this predicament, there is one particular purpose which each of us (individually) has strong instrumental reason to pursue, a purpose that is derived from a context of justification external to our own mundane lives.

An example might make things clearer.

Suppose that many people strongly prefer a large future reward over personal annihilation. For concreteness, we can think of the potential reward as a great and long-lasting joy that would commence after our normal lives are over: an "afterlife", if you will, which would satisfy the criterion of being external to our mundane existences.[219]

Suppose, further, that the world is structured in such a way that in order to obtain this ultimate reward, one must accomplish a certain long-range objective, or dedicate oneself to a certain role, or live one's life in accordance with a certain ideal; and, also, that doing so is an undertaking demanding enough to absorb more or less all of one's soul and all of one's slack.

If we were to discover that this is how the world is structured, and that we are thus *within striking distance of an overwhelming boon*, then we would have, fully baked under our noses a ready mission to sink our teeth into: an encompassing purpose which we have reason to wholeheartedly embrace, derived from a context of justification external to our mundane existence.

In this example, the justificatory context is based on an instrumental consideration—on what is necessary as a means to the attainment of the happy afterlife. This consideration, moreover, is one that would be applicable to many people, rather than being dependent on some unique or idiosyncratic preference shared only by a few. Many people, after all, would strongly desire a happy afterlife.

*

I say "many", not "all". I presume that a significant share of the population do not greatly care about a potential happy afterlife that would commence at some point in the relatively distant future, say in fifty years—and not only because they may not believe that such a thing is actually in the offing, but because they just don't desire this outcome strongly enough for them to regard themselves as having sufficient reason to reorganize their entire life around the goal of attaining it. I also presume that some fraction of those people would still not strongly desire this outcome even if they were substantially more knowledgeable and instrumentally rational.

It is possible to define a bonum with strictly wider appeal than a happy afterlife, by stipulating that the reward takes the form of a bundle of goods or an option package. For instance, success could give you the ability to pick any items you want from the set {a happy afterlife, health, love, profound knowledge, wealth, a great ability to help others, enlightenment, closeness to the divine}. Since this set includes the option of a happy afterlife, it should appeal to anyone who values that, and it might additionally appeal to some people who don't value a happy afterlife but who do value one of the other items.

Such an option package could be thought of as a kind of supermoney. A "karma coin".[220] It would give you access to many highly desirable goods, including ones that ordinary money can't buy. The main limiter of the karma coin's appeal is probably its lengthy vesting period. Individuals with high discount rates might not care that much about *anything* that is set to take place years or decades into the future.

*

If we want to be more philosophically precise, we should distinguish between several different ways in which the meanings for two persons can be the same or different.

Sticking with the case of achievement-oriented purposes aimed at the procurement of reward, we can differentiate between three distinct components in the meanings that they ground. There is the reward that is sought; there is the achievement that is necessary to attain this reward; and there is the activity that needs to be engaged in to secure the achievement.

When we say that a set of people share "the same" meaning in their lives, we usually don't intend to imply that there is identity across all three of these components. We mean, rather, that their purposes (so to speak) run in parallel. For example, suppose that what is required of an individual in order to reap the reward of a happy afterlife is that she lives an unblemished life. The purpose for all humans—or at least for all humans who care greatly about the afterlife—might then be to live an unblemished life. So, speaking loosely, we might say that they have the same purpose. But strictly speaking, their purposes are different: Smith's purpose is for Smith to remain unblemished while going about Smith's business so that Smith gets a happy afterlife; Sondhi's purpose is for Sondhi to remain unblemished while going about Sondhi's business so that Sondhi gets a happy afterlife; and so on.

A group of people *could* have a shared meaning in a stronger sense than this. For instance, instead of hoping for individual rewards, there might be some particular outcome they all very much want to be brought about (and which is external to their own mundane existences). Let's say they all want an end to factory farming. In this case, they are all trying to accomplish literally the same thing, not merely analogous or parallel things.

We could make their meanings share even more features if we postulate that the activity to be engaged in is also the same for all of them. In the real world, the animal welfare promoters would presumably fan out to attack the problem from different directions—some to pursue public advocacy, others to research plant-based meat substitutes, others to fundraise, others to lobby legislators. However, we could imagine a situation in which they would all need to engage in the same type of activity to accomplish their shared goal. For example, suppose that we stipulate that the end of factory farming could be brought about only through divine intervention, and that the probability of such intervention is proportional to the number of the original activists who are praying for it to happen. Then the meaning of all these people's lives would be the same not only in the sense that they have parallel purposes, but in the stronger sense that they are seeking to accomplish exactly the same outcome, for the same reason, by engaging in the same type of activity.

This kind of most strongly shared meaning based on an instrumental reason requires conditions to be met that may be unlikely to obtain among larger groups of human beings. In a large group, there is a free-rider problem: each person's contribution to the shared mission would typically make only a small difference to the likelihood that the mission is achieved. This comparative impotence of individual contributors reduces the strength of the instrumental reason that each one of them has to devote their life to the shared mission. Even if you think that abolishing factory farming would be an outcome worth great sacrifices in your mundane existence, you might judge that increasing the chance of success by only one-millionth of a percentage point would not be worth the personal cost to you. So you might not feel that you have sufficient reason to embrace this purpose; in which case it would not offer you meaning in life, at least not in any of the more subjective notions of meaning that we tabulated.

The most plausible candidates for a *strongly* shared meaning among many (or all) humans, therefore, probably require a more objective notion of meaning—such as one that imputes that the reason we all have for embracing the purpose in question is that it is derivable from an independent moral reality.

In contrast, candidates for meaning that is broadly shared only in the weaker sense of involving parallel purposes could more plausibly be discovered even if we are operating with a notion of meaning that is closer to the subjectivity end of the spectrum.

*

We have seen how meaning, including potentially a shared meaning of life, could be "out there" in the world—potentially empirically discoverable in natural facts that are not primarily about some person's idiosyncratic preferences or dispositions, but rather about the environment and the covergently instrumental reasons it may give us for pursuing a certain goal (or for playing a certain role, or for leading our lives according to a certain ideal).

It is true that the subjective element is not *completely* absent in our example. If we look closely, we can still discern the imprint of a valuer, or of an implied evaluative standard. We are supposing that the "reward" is something that would actually be desired by the class of beings who are thereby provided with meaning. But it seems fair to say—especially in the case of an "option package" reward that provides convergently instrumental reasons to a wide class of agents—that "most of the action" (most of the relevant uncertainty that remains to be resolved) is based on objective facts about the external world.

Categories of possible meaning

The preceding discussion illustrated one category of possible meanings of life: ones that could arise from the discovery that we are *proximate to a fatal nexus*, such that by pursuing some purpose throughout our ordinary existence we stand to benefit greatly in a subsequent extraordinary existence.

We can identify a couple of other categories of possible meanings. At this juncture, I am not making a claim as to which of these, if any, is real. Just mapping out the possibilities. You can see on the handout. The terminology and categorizations here are somewhat arbitrary, but I figure it might be useful to have some sort of schema.

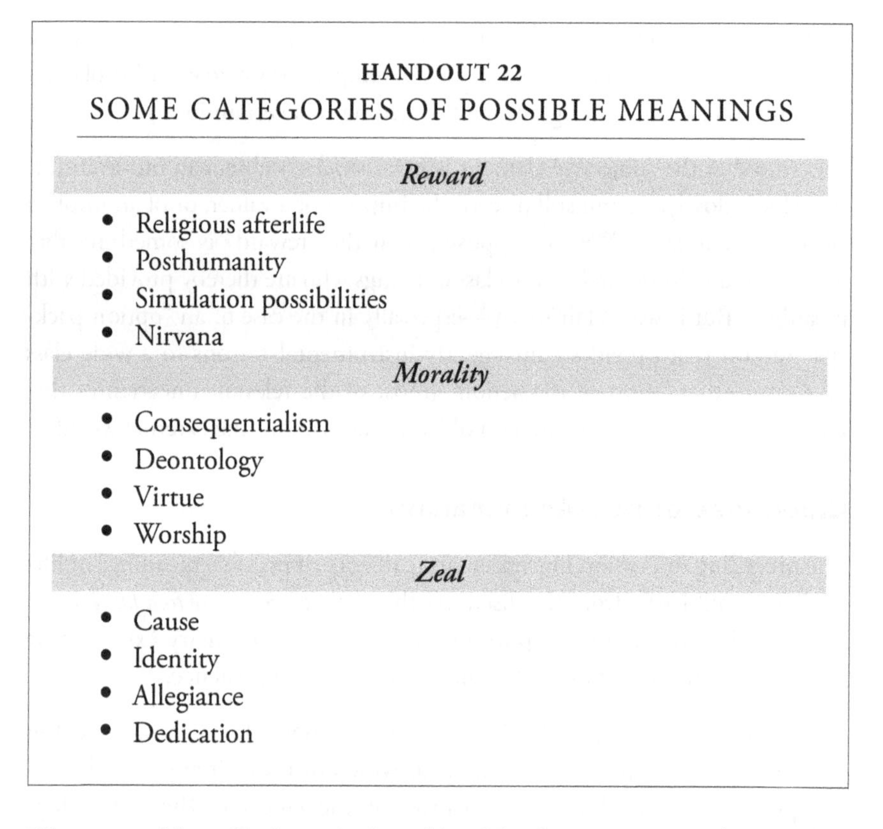

HANDOUT 22

SOME CATEGORIES OF POSSIBLE MEANINGS

Reward

- Religious afterlife
- Posthumanity
- Simulation possibilities
- Nirvana

Morality

- Consequentialism
- Deontology
- Virtue
- Worship

Zeal

- Cause
- Identity
- Allegiance
- Dedication

I'll just quickly walk through the table. The first category is the one that we have already discussed, where the purpose is grounded in the hope of a reward.

"Religious afterlife" should be self-explanatory.

By "Posthumanity" I mean the kind of scenario where you have a chance, by the appliance of advanced technology, to transcend your ordinary life and attain some immensely desirable enhanced existence, with vastly extended healthspan, improved subjective well-being, expanded intellectual and emotional faculties, and so forth.

"Simulation possibilities" refers to scenarios in which your present existence is in a simulation and you could potentially secure some great benefit, perhaps in another simulation, after you depart from or die in this one.

With "Nirvana", the idea is that if the state of nirvana is sufficiently desirable, and it constitutes an exit from your normal life, then it might form the basis for a qualifying purpose.

On to the second category... It encompasses types of possible meaning that would be grounded in an independent moral reality that gives people strong reasons to do things. (If no such independent moral reality exists, then this category is uninstantiated.)

I've listed the three big traditional classes of theories in normative ethics: "consequentialism", "deontology", and "virtue ethics". I've added a fourth, "worship", to accommodate the (usually) religious idea that we might have most moral reason to organize our lives around and adopt a certain attitude toward some supremely important being (like plants turning their leaves toward the sun).

The third category I've labeled "Zeal". The idea here is that we could have purposes satisfying the ETP criteria that are not geared toward the procurement of a reward, nor predicated on the truth of moral realism, but where the reasons for us to embrace them are instead based on some evaluative disposition we have toward something that is more integral to the purpose itself.

Take, for example, the first class under this category: "Cause". We can think here of some great and worthy calling, such as to bring an end to an unjust war or provide medicine to the poor. Perhaps it is the case that people who are well placed to contribute to such endeavors have a reason to do so based on a duty grounded in an independent moral reality; and if so, they could have a meaning under the "Morality" category above. But it is also possible to antipathize with injustice or to have a desire to help others, independently of where the truth happens to be in moral philosophy or metaethics. Therefore, whether or not these objectives provide an opportunity for meaning via the "Morality" category, they can certainly do so via the "Zeal" category, provided that the protagonist cares sufficiently about accomplishing them.

However, "Cause" is not limited to great and noble callings. It could also include purposes that would appear to us weird or whimsical—such as purposes based on the goal of obtaining an accurate count of the number of blades of grass on the College lawn, or the goal of transforming the galaxy into paperclips. Those causes could, for some agents, provide them meaning. The requirements are that the objective that is sought be located outside the mundane life of the agent to whose life it is giving meaning; and that the purpose it engenders is sufficiently encompassing; and that the agent's reason for embracing this purpose satisfies whatever additional criteria follow from the degree of objectivity there is in the notion of meaning that is being claimed (as outlined in Handout 21). For you or me, grass-counting or paperclip-maximizing would not be meaningful; but it might be so for a being who authentically loves knowing the count (such as Num_Grass) or who deeply desires that there exist as many paperclips as possible in its future light cone (such as Clippy, the instrumentally rational artificial agent that is used as an illustration in some thought experiments about AI safety).

The other three classes listed in the Zeal category are similar to "Cause", except the ambition is not pointed toward the attainment of some particular outcome but has some other focus instead.

In "Identity", we have a compelling desire to be a particular kind of person— somebody with a certain character or a certain role—and where this desire is not conditional on any expectation of rewards or other mundane benefits. For instance, somebody might have a deep-rooted desire to be a heroic figure, and they might be committed enough to this ideal self-conception that they are willing to stick to it come what may, even at the price of great risks and mundane disadvantages. Their commitment to being heroic could furnish a meaning to their life (provided it is sufficiently encompassing—which, depending on how they conceive of the heroic ideal, it might well be).

"Allegiance" goes along similar lines, but in this case one has committed oneself (down to the core of one's being), to aligning oneself with another will, such as that of a king, a people, a sect, a master, or a divine being. If whatever this being commands (or whatever would be good for it or would be in alignment with its volition) is what one desires more than anything else—and again, not because of any expectation of reward, but unconditionally, for its own sake, as its own wholeheartedly affirmed predicate of desirability—then such an allegiance could potentially provide meaning.

Finally, "Dedication": analogous to the foregoing, but where the fundamental commitment is to some activity or practice. Here we might conjure up that archetype of the all-in artist, who would be willing to give his life for his art, who completely embraces "art for art's sake" as his highest good, and whose affirmation of the desirability of this pursuit of artistic creativity and excellence is unconditional on whether, for instance, the process remains emotionally rewarding or not.

I should say that in some of these latter classes of meaning—in particular "Identity" and "Dedication"—the designated purpose cannot be completely separated from the mundane existence of the protagonist onto whose life the meaning is bestowed. Still, I think we can construe the desirability of the purpose (for the protagonist) as being based on a reason that is in the relevant sense transcendental. A compelling intrinsified passion to live one's life in accordance with an ideal that is hoisted up above quotidian concerns and inducements of everyday life can provide a degree of insulation against the trials and tribulations of the latter, and could in that sense be regarded as forming a context of justification that is external to one's mundane existence.

The meaning of life is

With this groundwork laid, we are now ready to emplace our positive answer to the question: the question of what is—at least for most of us, and perhaps for all human beings currently alive—the meaning of life.
[*Dean whispers to Bostrom.*]

Bostrom: The margin was too narrow. I'm told we have to wrap, as the venue apparently has only been booked until six. This also means that the book signing will not be taking place. But feel free to try the free samples of Green Transcendence on your way out.

[*Dean whispers to Bostrom.*]

Bostrom: There will be no samples, as I am not a certified University supplier.

Well, I guess that's that. Thank you all for coming, and for lending me your ears. [*Applause.*] Thank you. Thank you very much.

Exit

Dean: That was wonderful. We put your garment bag in a dressing room down the hall there, where you can get changed in privacy. Look, it has a star on the door. It's where the real stars dress.

Bostrom: That is very thoughtful, thanks.

Dean: So you just hop into your dinner jacket and I'll wait here.

Bostrom: Dinner jacket?

Dean: Yes, and then we walk over to the reception together.

Bostrom: Reception?

Dean: Before the Chancellor's Court of Benefactors dinner.

Bostrom: Chancellor's Court of Benefactors? Oh, I think there must have been a misunderstanding. I had to bow out. Force majeure.

Dean: You're not coming?

Bostrom: I'm afraid not.

Dean: That is extremely disappointing. Four members of Exxon's international executive team will be there, and a government minister.

Bostrom: Is that really so? But you will be there to represent the Division. I'm sure all will be well.

Dean: But the guests would like to hear about the meaning of life. I had planned to bring up the topic between the cheese and the pudding.

Bostrom: Oh. Well, the best laid schemes… Not to mention the other schemes. Very regrettable.

Dean: Tinkelklein from the Music Faculty suggested we invite you.

Bostrom: Tinkelklein! Well I be blowed.

Dean: Said we must have you, that you'd be delighted to regale the visitors with the most titillating philosophical conversation.

Bostrom: That's what he said? "Titillating"?

Dean: He spoke of you in very flattering terms and was most insistent. But now there is a gap.

Bostrom: Hmm.

Dean: And as I have been asked to be the master of ceremonies, I'm at a loss.

Bostrom: This Tinkelklein, you know, his vice is an excess of modesty. Why, he would serve much better himself!

Dean: He would?

Bostrom: You must not doubt it. Ask him to sing something.

Dean: That's a good idea.

Bostrom: Ask for his Wagner… Yes, prevail upon him to give his rendition of Amfortas' lament. It is a blast. He will feign reluctance, but you must really lean upon him. He will be your bridge, from cheddar to custard.

Dean: Perfect. Well, I'd better be off.

Bostrom: Amfortas' lament. On no account let him cut it short! (And 't shall go hard…)

The graveyard

Tessius: So, what are you guys doing now?

Kelvin: We are going to a poetry slam, but it doesn't start until a bit later. There's also a student carnival. Wanna come?

Tessius: I'd be up for that.

Firafix: Maybe we could go for a walk in the meantime. There's a quiet path over there, across the old churchyard. What did you think of the lecture, the part that he was able to give?

Kelvin: All this meaning talk seems rather fluffy. Things would be clearer if we tabooed the word "meaning".

Tessius: But perhaps once the construct of meaning has been introduced—especially if it becomes a cultural fixation, something which noble souls are believed to be in quest of, then meaning becomes a thing that people actually do specifically desire to have—meaning *per se*, not just whatever distinct components go into the concept. And then maybe this desire gets intrinsified, and we have a value.

Firafix: What did you guys make of the series as a whole—the problem of utopia?

Kelvin: A bit weird to frame it as a "problem". On many theories of value, utopian lives can be millions of times better than our current lives. He could just have gone through the major theories of well-being, and showed, for each one, how it implies that we could have lives that are astronomically valuable.

Firafix: He sort of did that, though?

Kelvin: It got buried in the literary upholstery.

Tessius: Here's a question. Suppose you were given the following choice. You could either live out your current life in the normal way, and then you die at perhaps age eighty or ninety. Or else you could take a gamble, which gives you an x% chance of having *the optimal* life for you. Starting from today, your future trajectory would be the best possible one for you—and maybe that would involve you eventually developing into a super-flourishing posthuman who lives for millions of years and attains unimaginable levels of felicity. However, the gamble also gives you a (100-x)% chance of dropping dead instantly. How big would x need to be for you to prefer the gamble?

Firafix: Hmm.

Tessius: It seems that, in a revealed preference framework, if you really think that the best utopian life you could have would be millions of times better than your current life, then you should accept a 99.9999% chance of dying immediately for the sake of a 0.0001% chance of getting this optimal life continuation.

Kelvin: Around 10%.

Firafix: So you'd accept a 90% chance of dying now for the sake of a 10% chance of reaching utopia with extreme longevity?

Kelvin: Yes, if the alternative was a very good modern human life. If instead I expected to have a historically typical human life from this point on, then a 1 or 2% chance of the ideal posthuman future would be enough for me to take the gamble. What about you, Firafix?

Firafix: Oh, I don't know.

Kelvin: You have to pick a number.

Firafix: Something closer to 50%. I'd certainly want to think about it more before making up my mind, if it were a real offer.

Tessius: So may we then infer that you, Kelvin, think that utopian life could be at most ten times as good as (the remainder) of your current life? And that Firafix thinks it could be at most twice as good as hers?

Firafix: There might be a distinction between what'd be good for us and what we'd want? Maybe what I want is to stand around eating apples all day long, but what is good for me is something different.

Kelvin: But you don't want to stand around eating all day long.

Firafix: Don't I?

Kelvin: Well, you aren't doing it, even though you could.

Firafix: I have to watch my figure.

Kelvin: You wouldn't choose a life consisting exclusively of apple-eating, even if you could do so while retaining your figure.

Firafix: I guess not. So maybe that's not a good example. But it still seems that there might be a difference between what I would choose, even under ideal conditions, and what would actually be best for me.

Kelvin: Bostrom set up this whole thing as an inquiry into the most desirable life-continuations for people like us. He didn't say much about creating superbeneficiaries. That would be far more efficient if one just wanted for there to exist extremely good lives.

Firafix: Superbeneficiaries?

Kelvin: "Utility monsters"—beings that are enormously more efficient in deriving well-being from resources than we are. You could design entities that need very few resources to experience superhuman levels of pleasure, or that have extremely strong preferences that are trivially easy to satisfy. Most theories of what makes a life go well imply that you could have superbeneficiaries.[221] But he wasn't talking about that. He was talking about what would be desirable for folk like us.

Firafix: Yes?

Kelvin: Well, it seems plausible that what is desirable for us should have some connection to what we desire, or what we would desire if we abstract from various limitations in our knowledge and thinking.

Firafix: I wonder if fear of death plays a role here? I mean, a 50% chance of dying within the next few minutes would be very scary. Maybe that biases our choices?

Tessius: Suppose you were making the decision for somebody else—somebody you've never met. You're not *afraid* of their death. If you had to choose on their behalf, would you choose a gamble that involved a 50% risk of immediate death for the sake of a 50% chance of them getting to enjoy the best future utopian life that is possible for them at technological maturity?

Firafix: It seems almost even scarier to choose for somebody else! I don't think I'd choose a gamble for somebody else that I wasn't willing to choose for myself.

Kelvin: You'd have an obligation to consider things from their perspective. If you're making a decision on their behalf, it wouldn't matter what odds you would accept for yourself: you should go by your best estimate of what they would have chosen.

Tessius: Suppose when they delegated the decision to you, they told you to do what is best for them, rather than to do what they themselves would have done?

Kelvin: I would try to model their decision algorithm. Their decision algorithm would result in them choosing some objective notion of the good only if they actually had a preference for it.

Tessius: Firafix?

Firafix: I'm thinking that in the case of a child, if you are trying to do what is best for them, you don't simply do what they want. Or even what you think they would want if they had full knowledge. Instead, you consider what would be good for them in the long run. You try to help them grow and develop, as well as take into account what they want right now. This came up in the ThermoRex story.

Tessius: So maybe we are like children? *We* would reject the gamble, except for values of *x* that are relatively high. But a paternalistic well-wisher would want us to accept the gamble even for quite small values of *x*, since the potential upside would really be so very extremely good for us?

Firafix: I don't think it works that way when there's a high risk of death. You wouldn't expose your child to a high risk of immediate death for the sake of them having a better chance of great success and happiness later in life. Well, *maybe* you would do that if their life would otherwise be utterly wretched and miserable—maybe in that case you would consent to a risky operation that had a chance of fixing the problem. But if your child was reasonably healthy and had decent prospects, I don't think you would accept a 50% risk of immediate death for the sake of any possible future benefit, however great.

Tessius: I wonder whether this has to do with the fact that the bad thing in the example—immediate death—would prevent them from experiencing the future benefits? Suppose we change the example. You get to choose between a normal but very happy life for your child and they live to ninety-eight, or else a life that involves first forty years of great hardship, suffering, and struggle, but which is then followed by the best possible utopian life and maybe they live for thousands or millions of years. What would you choose then?

Firafix: Hmm. I think that might tilt me more toward accepting the offer.

Tessius: So then, maybe we could calibrate how desirable for you the best possible utopian life could be by considering what is the greatest amount of temporary suffering you would be willing to accept in order to attain it?

Kelvin: I don't see why it should make a difference whether the downside in the gamble is a risk of death or some other bad thing. You'd have to calculate the expected value in either case. Are you relying on some kind of person-affecting intuition?

Tessius: Not sure, I'm just exploring…

Firafix: What is a person-affecting intuition?

Tessius: He means the idea that when we are making a moral decision, it makes a difference whether we are benefiting an existing person (or a person who will come to exist independently of what we decide) or we are instead creating a new person who enjoys a similar benefit. Some philosophers have argued that we have moral reason to increase the well-being of existing people, but no moral reason (or much weaker reason) to bring into being new people whose existence would add the same amount to the total quantity of well-being in the world. The slogan is that we have reason to make people happy but not to make happy people.[222]

Kelvin: If there is a person-affecting intuition at play, it might also influence our judgment about the second version of your thought experiment.

Firafix: Is that because the benefit that you or the child would enjoy would start only forty years into the future, and would then take a million years to be fully realized? So that, while in one sense the burden and the benefit would befall the same person, yet in another sense the relevant temporal segments of that person would be so far separated that it would almost be as if they belonged to different persons?

Kelvin: Also, the later segments of that person would only come into existence in one of the options. A generalized person-affecting perspective might hold that bringing additional time-segments of a person into existence is less desirable than making independently existing time-segments happier. Especially if the potential new time-segments are widely separated in time and linked to the already existing time-segments only via especially tenuous connections of personal identity.

Tessius: Ok, how about this. Third version of my thought experiment. You either have an excellent normal human life, where all days are pretty good; or else you have a life that consists of alternating good days and bad days. You have one of your enormously fantastic posthuman utopian days, then you have N bad days; and it repeats like that. What is the greatest number of N for which you would prefer this future to a future that is just an excellent normal human life?

Kelvin: How bad would the bad days be?

Tessius: Maybe we could say a bad day would be one such that if you took the excellent normal human life with only good days, and you spliced in a bad day between every good day in that life, you would then be indifferent between having that life and dying immediately? And maybe we could stipulate that the choice you make here would start to affect your life only a decade from now, to reduce any possibly irrational fear-of-imminent-bad effect.

Firafix: Maybe *N* equals 10, or 20?

Kelvin: I predict that once you actually experience one of those fantastic posthuman utopian days, your opinion would change. You would become willing to endure a much larger number of bad human days in order to have another one of those superdays.

Tessius: Hey, look at that dog! He's taking a leak on a tombstone.

Dog owner: Fido! *Bad* dog! *Bad* dog!

Tessius: One shudders to think what unwitting offenses we may be committing on an average day.

Dog owner: I'm so sorry. Every time we walk this graveyard, he runs off and does that. Always the same stone! "Nospmit". Unusual name. I really hope that's not one of your ancestors?

Firafix: No, we're just here for a stroll.

Tessius: The weather bureau has forecast rain, so Fido's misdeed will hopefully soon be absolved.

Dog owner [*looking skywards*]: Yes it does look about to come down.

Tessius: Do you notice something about Nospmit's grave—is it not uncommonly thriving?

Dog owner: There's a lot of greenery there.

Tessius: More so than around the other graves?

Dog owner: I believe so. There are even blackberries growing there… I actually picked a couple, but I'm not sure I should eat them.

Tessius: Whatever ill might have gone into the soil, it appears that something good has come out of it. We must believe in the possibility of redemptive processes.

Dog owner: You think so?

Tessius: We must hope. Maybe you could try to think kind thoughts while you eat them?

Dog owner: Okey dokey. [*Eats berries.*]

Dog: [*Growls.*]

Dog owner: Alright, here you go. [*Gives dog biscuit to Fido.*]

Firafix: He seems like a nice dog.

Dog owner: Love him to bits, but you know how they are. I just felt a drop—we better get going. Have a nice day. Fido, come here!

Tessius: Goodbye, sir.

Firafix: Bye, Fido.

Kelvin: Maybe we should make our way indoors?

Firafix: I want to show you guys something first. A couple of the tombstones... Here is one.

Mr. M. SVENSSON
1940 - 2021

Ate meals 114,238 times
Went to bathroom 152,771 times
Showered or bathed 28,934 times
Made love 11,213 times

Tessius: 11,213...

Firafix: There is just something—I don't know, summarizing a whole life like that... So cold and bloodless, yet poignant somehow. Like, is this what our lives amount to?

Kelvin: I'd say the numbers are bogus.

Firafix: Perhaps he was an accountant obsessed with quantifying his own life?

Kelvin: Who kept an exact tally of the number of meals he ate while he was a toddler?

Firafix: Perhaps his parents kept track when he was little. Maybe he came from an accounting family.

Tessius: Don't you think that 11,213… rather strains credulity… particularly for an accountant who was obsessed with logging the number of times he went to the bathroom?

Firafix: Well, the exact numbers don't matter! You guys are so…

Tessius: So?

Firafix: So—well, you know.

Tessius: Sagacious? Truth-loving? Epistemically virtuous?

Firafix: Not one of those.

Tessius: Wise?

Firafix: No, it starts with an "A".

Tessius: Accurate?

Firafix: No, the second letter is "N".

Tessius: An—, an—, … Angelic?

Firafix: The third letter is "A".

Tessius: Anagogic?

Firafix: Nope. The fourth and final letter is "L".

Tessius: Analytical!

Firafix: Well, something along those lines.

Kelvin: There's another row of text.

Firafix: Where?

Kelvin: Down there, hidden by the grass.

Tessius: [*Kneels down.*] Let's see… [*reads*] *"Died 2 times, once when Mary left and once from thyroid cancer."*

Firafix: Oh. I never saw that. That's so sad.

Tessius: It does cast things in a different light.

Firafix: I can't believe all these people have died. They were once alive, just like we are. They were once little girls with springy legs and little boys with red apple cheeks. They were mothers and fathers, and then they were grandparents.

[*Kelvin lays an arm around Firafix's neck.*]

Firafix: I feel this breeze coming from the past. I sometimes feel closer to these people than to many of my contemporaries.

Tessius: The dead don't have the same ability to annoy us.

Firafix: Well, that's probably part of it. But I also feel sorry for them. They don't have any more life left.

Tessius: Yes, it is sad.

Kelvin: Yes. They died before daybreak.

[*Silence.*]

Tessius: Still… 11,213 is not *bad.*

Firafix: I guess.

Kelvin: It's starting to get dark. We should get a move on if we are to make it to the poetry slam in time.

Firafix: I want to show you one more tombstone first. It's over in that corner… Here, look at this one. It doesn't have a name or a date.

Having practiced my entire life
I have now finally mastered
The art of stillness

Tessius: Not only is he doing nothing very well, he's doing it in Carrara marble.

Firafix: Do you think it was a Zen master?

Tessius: Either that or a lazy bum.

Firafix: Well, I guess it's the same now.

Kelvin: Look at that weeping willow, in the rain.

Firafix: It is beautiful.

Tessius: I'm reminded of these lines:

"into the hopper goes
the queen of hearts
two-faced treachery
the extra ace
sighs and jubilations
the tangled neurons of the scholarly head
thorns and twigs
the little louse and the fat king
all that was born goes into the hopper"

And here's another one:

"Not much longer
Already can be heard
the murmured rumor
of the final fall
fall
fall
fall
into the great turbine.

If we weren't so busy making ends meet
If we weren't so busy
If we weren't so busy doing nothing
If we weren't so busy whistling a ditty
Might we not wonder:
How do the blades turn?
And what do they power?"

Firafix: That's almost like an exam question... I wonder, if there were an exam, how we would do?

Kelvin: You mean if we sat the exam for the lecture series?

Firafix: No, I mean generally. Like, if our existence were an exam. It's not, but...

Tessius: Are you sure about that?

Firafix: Well, no. I suppose it could be.

Tessius: I reckon we might do best to adopt that as a working hypothesis. *Something* is being examined—though whether it is us, or whether we are the examiners, or whether we are being judged based on our judgments, or some combination...

Kelvin: We're in the dark.

Firafix: So we must have faith.

Kelvin: Basically.

Tessius: Yes.

Carnival

Kelvin: We're late, but we might still be able to catch the end.

Firafix: We can head down that alley, then cut across the College courtyard. That should be the fastest way.

Tessius: Whoa the carnival is in full swing. Let's see if we can plow through the throngs.

Drunk student #1: I'm pulling the rope sideways. What are you doing?

Drunk student #2: Earning to give.

Drunk student #1: But you have a butler!

Drunk student #2: To prevent burnout.

Drunk student #1: Man.

Drunk student #3: His dad owns a warship!

Drunk student #2: Oh, the man-o'-war? The fart won't sell it. Hey, let's go down to the wharf and spray-paint Greenpeace on it.

Drunk students #1 and *#3*: Yeah! [*The three students depart.*]

Tessius: Tomorrow's leaders.

Drunk student #4: Hey, are you a unicorn?

Firafix: Do I look like one?

Drunk student #4: One whose horn's been ground up for aphrodisiacs!

Drunk students (several): Ha, ha, ha.

Drunk student #4: I'd still pay for a ride. Five dollars?

Drunk student #5: You gonna pawn your shoes?

Drunk students (several): Ha, ha, ha.

Drunk student #4: C'mon!

Firafix: No thanks.

Drunk student #4: Six dollars!

Kelvin: Back off.

Drunk student #4: I think the groom here is getting a bit uppity.

Firafix: Let's keep moving.

Tessius: Freshmen: another wrinkle to the theodicy problem.

Firafix: It's just around the corner…

Crowd of students [*singing loudly*]:

> *… and the clouds take the shape of* [redacted]
> *ola, olé, oja!*
> *here we are just getting by*
> *but in utopia we'll get high*
> *and the blunts will be rolled on a virgin's thigh*
> *ola, olé, oja!*
> *on the roof of my palace*

I'll swig from a golden chalice
While I stroke my collosal [redacted]…

Kelvin: We should have brought umbrellas.

Tessius: And ear plugs.

Firafix: We made it… It seems it's still going on. Let's duck inside.

Poetry slam

Poet:

…
Poetic kinetic paper wings
Illusion infusion
Dance glance and moon swoon
Iridescent crescent
Tender surrender
World unfurled
Ephemeral cathedral
Showcase
Staircase
Interface
Embrace grace

Audience: [*Wild cheering.*]

Emcee: Thank you. That was absolutely wonderful. So powerful.

I feel that we are all visitors and travelers on this planet, but our next poet tonight is also a voyager in a more literal sense, having just recently flown in. What a wondrous thing that is—birds of aluminum that crisscross the sky with people in their bellies. But you know what's still more wondrous? That we can use words and imagination to travel not only to distant countries, but even to places that no longer exist, or that never existed. We can travel to moments and possibilities. And there are no security pat-downs or immigration checks on these journeys of the imagination, no lost luggage. Your ticket is a willing heart and your visa is an open mind. But I am still grateful when the aluminum birds bring us visitors such as our guest tonight: Walter Diego.

Please, bring your hands together and stomp your feet, and send some good vibes as he makes his way to the mic!

Audience: [*Clapping, whooping.*]

Walter Diego [*with heavy accent*]: Thank you. Your words were so true, but I must take issue with one thing you said, about there being no lost luggage in the travels of the mind. When one reaches the more advanced years, the airlines sometimes get a bit sloppy in that regard. [*Audience laughter.*]

The poem that I will read is called "Old Seamen". Now I did bring a little carry-on just in case, now where did I put it? [*Fumbles in pockets.*] Oh, here! [*Takes out and unfolds a piece of paper; clears throat.*]

Kelvin [*whispers*]: It's Professor Bostrom.

Firafix: What?

Kelvin: Look—the beard and the eyebrows are fake.

Tessius: Well tickle me pink, the cat's whiskers!

Walter Diego [*reads*]:

> OLD SEAMEN
> and once more to sea!
> this ancient schooner
> yet has another sea-ride
> in its rotting planks
> the mast can still hold loft a sail!
> and these my weathered hands
> though less dexterous than afore
> have the more experience
> know their way with hook and tackle
> and soon will heave unto the deck
> a thrashing tuna
> and with a grin the inveterate ocean-robber
> will stuff a pipe
> and regard the final trophy by his side
> and then the horizon and the salty waves
> which he will leave to younger men—

before laying up for good at port
to sit by the pane and foghorns' bellow

Emcee: Thank you Diego, and thank you all for coming tonight. And I hope to see you all next time!

Summer air

Tessius: A thrashing tuna! Indeed.

Kelvin: Good idea to wear a disguise.

Tessius: So sassy. Firafix, are you blushing?

Firafix: I would if I could.

Tessius: He looked spiff though when he took his jacket off, in that black turtleneck. Admit it—that rail-thin look! Like a dagger penetrating the rib cage and stabbing the heart. —Verily, you are blushing!

Firafix: Nay!

Kelvin: Well, the rain has cleared up.

Promoter: Epic bash alert! Trance-continental express departing for next-level delirium! Pulsar strobes, supernova beats. Surrender to the pandemonium! The party singularity! Is happening! Tonight!

Tessius: [*Takes flier.*] Look, dj teddybjörn is doing a gig. You guys wanna go? It's down at the docks.

Kelvin: I don't think so.

Tessius: This one could be really good. Let's go dancing!

Kelvin: Nah.

Firafix: C'mon Kelvin, let's have some fun. It's a special night tonight.

Kelvin: How so?

Firafix: We make it so.

Tessius [*chanting*]: Go, go, go, go

Kelvin: Okay.

Firafix: Yay!

Tessius: Hey, take some of this.

Kelvin: What is it—is it legit?

Tessius: Pure quality.

Firafix: Are you sure?

Tessius: Trusted source. Somebody from the chemistry department.

Kelvin: I don't know…

Tessius: H-index above 70.

Kelvin: I guess it's alright then.

Tessius: Here's one. And for you—how many?

Firafix: About ten.

Tessius: Here you go.

Firafix: Thanks.

Kelvin: Thank you.

Tessius: Ahhh, breathe the air! It is…

Kelvin: It is.

Firafix: It's summer.

*

[*Ed. note: And they all lived happily ever after. Including Pignolius, Heißerhof, and Svensson (reconstituted), and… well, yes, even the Dean, the Registrar, and Nospmit.*]

ENDNOTES

1 Gates (2017)

2 Musk (2023)

3 Hanson (2021)

4 "This is about the wonderful Land of Cockaigne" (c. 1458), lines 19–22

5 Pleij (2003), p. 3

6 Hesiod (c. 700 B.C.), lines 132–9

7 Keynes (1930)

8 Keynes spoke in terms of "the standard of life in progressive countries one hundred years hence", which he predicted "will be between four and eight times as high as it is to-day" (ibid., pp. 364–5). From the context, we can interpret him as making, primarily, a claim about rising labor productivity, and, secondarily, a claim that workers who are much more productive will choose to work much less, thus taking out much of their productivity gains in increased leisure rather than increased income. (Keynes didn't define what he meant by "progressive countries", so the geographic scope of his claim is unclear.)

9 Ibid., p. 366

10 In the US; Bergeaud et al. (2016). In the UK the numbers are somewhat smaller. Strictly speaking, what Keynes predicted was that "assuming no important wars and no important increase in population, the economic problem may be solved, or be at least within sight of solution, within a hundred years." (Keynes, 1930, pp. 365–6).

11 In the UK and US; International Labour Organization (2023).

12 In the UK, the expected portion of remaining waking life spent at work for a 20-year-old male declined from slightly under 40% in 1931 to a bit over 20% in 2011; Crafts (2022), p. 818.

13 Or perhaps it will turn out he didn't even get the timing wrong. AI has been making rapid progress recently.

14 Note that the marginal value of income is still decreasing in this picture: going from $1,000,000 to $2,000,000 does not provide a welfare lift one thousand times as large as going from $1,000 to $2,000, even if the second million bought you five extra years of healthy life instead of a larger summer house. But if wages continue to climb at the right pace, then incentives to work could in principle remain constant even if the gaps between rungs of the well-being ladder grow wider as we move higher up on the income ladder.

15 Shulman & Bostrom (2021); Bostrom & Shulman (2022)

16 Keynes himself alludes to this possibility: "Needs . . . which satisfy the desire for superiority, may indeed be insatiable; for the higher the general level, the higher still are they."; Keynes (1930), p. 365.

17 Hirsch (1977)

18 It may be a defect of contemporary economics that it often ignores (in practice if not in theory) the great extent to which consumption is driven by positional status concerns. This means that we may overestimate the welfare gains of increasing average levels of consumption, especially at higher income levels where most of our basic noncompetitive needs are catered for.

19 In the limit, the costs to the employer of hiring the human, training them, supervising them, etc. would make it not worth hiring them at *any* salary level—not even at one cent an hour. And for the employee, in the limit, the extra metabolic expenditure from working would be greater than the calories that could be brought with the salary.

20 Cf. also Trammell & Korinek (2023).

21 In this simplified model, the outcome does not depend on whether the robots are owned by humans or operate as independent economic agents who sell their labor at market rates. In either case, the long-term equilibrium is one in which the robots receive bare subsistence income from their labor.

22 Cf. Malthus (1803), p. 14.

23 Parfit (1986), p. 148

24 After the plague, "such a shortage of labourers ensued that the humble turned up their noses at employment, and could scarcely be persuaded to serve the eminent for triple wages", per *Historia Roffensis;* (Dene, 1348, p. 70).

25 Cf. Clark (2007), p. 5.

26 Shakespeare (1607), 1.1.11

27 Klein Goldewijk et al. (2017). All the numbers are very imprecise.

28 Hare (2017)

29 Bostrom (2007, 2014a)

30 Royal Commission of Inquiry into Children's Employment (1842)

31 Welles (1949)

32 Cf. Hanson (2008); Bostrom (2014b).

33 If there are secure property rights, and a large fraction of total wealth is held by clans that internally regulate reproduction or practice primogeniture or similar non-diluting intergenerational wealth transfer protocols, average income could remain substantially higher than subsistence.

34 Scott Alexander's riff on this, "Meditations on Moloch", is worth a read; Alexander (2014).

35 Cf. McMahan (1981).

36 And, if he is right, there will be more people to cite him—see how marvelously bibliometrics works as a measure of academic merit.

37 Eventually, in this scenario, pigeons would evolve to lose many of the capacities they needed to fend for themselves in the wild. A selection gradient on their reproduction would need to be applied to preserve or enhance those traits—or to develop new traits that would make pigeon lives more eudaemonic.

38 Apologies for any offense this comparison may cause—to either party.

39 Johnston & Janiga (1995), pp. 190–205

40 Cf. Bostrom (2009), p. 48.

41 Heilbroner (1995), p. 8

42 See Kugel (2007), pp. 54–6.

43 Mummert et al. (2011)

44 Kremer (1993); Hanson (2000). The figures are unavoidably very inexact, but I think qualitatively correct.

45 "I thank God every morning that I'm not in charge of the Roman Empire." Pignolius seems to be quoting Goethe (1808), lines 2093–4. Or the other way around?

46 Because I have done so elsewhere: Shulman & Bostrom (2021); Bostrom & Shulman (2022).

47 See Killingsworth (2021) and Bartels (2015), though it's worth noting that the heritability of subjective well-being could also run via channels other than native temperament, such as health.

48 Ord (2021)

49 Ord (2020), pp. 232, 411

50 Bostrom (2014b), pp. 101–3; see references therein.

51 See e.g. Olson (2018), Hanson et al. (2021), and Armstrong & Sandberg (2013).

52 Bostrom (2022)

53 Cf. Bostrom (2013).

54 As envisaged by Eric Drexler (1986, 1992)

55 Freitas (2022)

56 Cf. Pearce (1995).

57 Finnveden et al. (2022)

58 Cf. Bostrom (2019).

59 Bostrom & Shulman (2022), pp. 8–10

60 Bostrom (2003a)

61 Jackall (1988)

62 Mitchell (1966)

63 Mitchell (1967)

64 Jefferson (1826), p. 1517

65 Not to be conflated with the infamous Club of Rome 1972 report *Limits to Growth*; Meadows et al. (1972)! Hirsch (1977) deserves credit for being perhaps the first to develop this theme at length, although the basic idea had been noticed before (e.g. by Phillip H. Wicksteed and others).

66 Shakespeare (1604), 2.2.121–126

67 The *observable universe* is finite, as is the *accessible universe*; but that does not mean that the universe is finite. If it's infinite, then it's a safe bet that there are extraterrestrial civilizations out there—in fact, infinitely many of them.

68 A pandigital number is an integer in a given base that includes each digit at least once. Thus, in base 10, the smallest pandigital number is 1,023,456,789.

69 For example, Kass (2003), Sandel (2007), and Habermas (2003) have each voiced general ethical objections to genetic engineering of humans, though it is not clear that even these arch-bioconservatives would assert that there is an absolute prohibition in all possible circumstances.

70 Shulman & Bostrom (2021); Bostrom (2022)

71 Warren (1997); Kagan (2019)

72 See e.g. Chalmers (2022) for more discussion.

73 Meme about future jobs. Orphan work.

74 Chaitin's omega represents the probability that a particular universal prefix-free Turing machine halts when given a random input.

The number of possible human visual experiences is not necessarily the same as the number of possible human visual stimuli. Two distinct stimuli will often produce the same experience, as many details of the incoming activation pattern are discarded at higher levels of processing. On the other hand, a single visual stimulus is capable of triggering more than one visual experience, since the brain can interpret it phenomenally in multiple different ways. For example, the same visual stimulus of a Necker cube can produce at least two distinct visual experiences depending on how the input is mentally projected into 3D space.

It seems safe to say, however, that the number of possible visual experiences is very large. It would only take a few hundred elements or features that can independently vary and be co-present in a visual experience for this to be the case, and for it therefore to be impossible for even a mature civilization to generate all possible human visual experiences by enumeration. If we can denote the independent qualitative features of a human visual experience, we could then adapt the construction in the text by using a coding scheme that maps these experiential elements to the initial digits of a decimal number, and then the argument given in the text should work.

75 E.g. Proverbs 16:27 (The Living Bible)

76 Proverbs 16:27 (World English Bible)

77 Keynes (1930), p. 368

78 Skidelsky & Skidelsky (2012)

79 Posner (2012)

80 O'Shea (2019); Gall & Smith (2019); "Nine months for lotto millionaire" (2006)

81 McFadden (2019)

82 Patel (2021)

83 Engraving of Michelangelo Buonarrotti. Photograph by Gustav Schauer. Wellcome Collection.

84 Photograph of Michael Carroll. Copyright by Albanpix. Reprinted with permission.

85 Estimate, based on data from Hatfield (2002), pp. 229, 337–43. Hatfield argues that conversions of Renaissance fortunes to modern currencies based on the value of gold (such as this one) are "misleading, as gold was much rarer, and therefore much more valuable, during the Renaissance than it is now." p. XXII. (In his old age, Michelangelo made big gifts to his servants and charitable causes.)

86 Posner (2012)

87 Cf. Fox (2005).

88 Catalano et al. (2011)

89 Hetschko et al. (2014)

90 See, e.g., Leino-Arjas et al. (1999).

91 Holder (2012): "Whether based on self-reports or estimates from their parents, 90% or more of the children are typically rated as happy [in Canada and India]".

92 Artists, despite having low average income and high unemployment, report significantly higher job satisfaction compared to other professions (Bille et al., 2013).

93 Janotík (2016)

94 E.g. Blanchflower (2020). For a critique of the claim that happiness follows a U-shaped trajectory across the lifecycle, see Kratz & Brüderl (2021).

95 Illustration by Elena Samokysh-Sudkovskaya. From *Eugene Onegin* (p. 7) by A. Pushkin, 1918, Petrograd: Golike & Wilborg.

96 Ertz (1943), p. 137

97 Farmer & Sundberg (1986)

98 Marx & Engels (1846), p. 47

99 The numbers are made up; maybe it should be 80%? At any rate, I think we've come a very substantial part of the way toward post-scarcity utopia—not because there does not remain pockets of destitution in developed countries or because there is not a lot more neat stuff that it would be nice if we could afford, but because it seems plausible that the

most basic things (such as avoiding famines) are the most important, and, for the time being, knock on wood, we are doing *relatively* well on those. Still, serious material deprivation remains quite widespread even in some wealthy countries; and the rest of the world obviously remains further away from anything that could be described as a post-scarcity condition.

100 What about preserving the ecosystem in an unmodified condition? Imagine that some technologically advanced civilization arrived on Earth and was now pondering how to manage things. Imagine they said: "The most important thing is to preserve the ecosystem in its natural splendor. In particular, the predator populations must be preserved: the psychopath killers, the fascist goons, the despotic death squads—while we could so easily deflect them onto more wholesome paths, with a little nudge here and maybe some gentle police action there, we must scrupulously avoid any such interference, so that they can continue to prey on the weaker or more peaceful groups and keep them on their toes. Furthermore, we find that human nature expresses itself differently in different environments; so we must ensure that there continue to be… slums, concentration camps, battlefields, besieged cities, famines, and all the rest. What a tragedy if this rich natural diversity were replaced with a monoculture of healthy, happy, well-fed people living in peace and harmony." Unless somehow there are bigger cosmic or theological considerations at play than we can see, this would be appallingly callous. Yet is not this the attitude that many humans today take with respect to the animal kingdom? (those, that is, who care about nature at all, other than simply as an instrumentally useful resource for human exploitation).

101 Gates (2017)

102 Sahlins (1972), pp. 17, 56

103 Freitas (1999, 2003)

104 You would think that the task of *giving* a lecture would have been automated quite some time ago, or at any rate mass-produced by replaying recordings to multiple audiences. Yet the live lecture is still going strong, not just in academia but on the corporate keynote circuit too. Perhaps most professors and public speakers are best conceived of as DJs (of themes, ideas, and pep), in which case our earlier remarks apply.

105 Communicating *philosophical* ideas may be particularly tricky in this regard. Much of the craft and practice of contemporary academic philosophy is about trying to prevent and correct misinterpretations. A noble aspiration, which I support, but also one that seems intractable

enough to keep the profession occupied for some time. People are still working on explicating Aristotle.

I wonder if we are going about things in the right way? Like thus: first we paint the idea; then we put nails in to hold the paint in place; then we put little screws in each nail to ensure they stay put; then we add little drops of superglue on the screws to prevent them from unwinding. And yet, as soon as we launch the construction, it all comes off. Maybe we need to add some tiny nails to keep the glue from flaking?

106 "AI complete"—meaning that if we knew how to do that, we'd also know how to create at least human-level artificial general intelligence.

107 Hoffer (1954), p. 151

108 "Ask yourself whether you are happy, and you cease to be so. The only chance is to treat, not happiness, but some end external to it, as the purpose of life." (Mill, 1873, p. 147).

109 L. (1848), p. 2. (This line, which seems to have first appeared in 1848 in an article in a New Orleans newspaper signed only "L.", is widely misattributed to Nathaniel Hawthorne, probably because of misleading formatting in an 1891 book; see O'Toole, 2014.)

110 Bostrom (2008a)

111 Ibid.

112 Bostrom & Ord (2006)

113 Bostrom (2008a)

114 Cf. Suits (1978).

115 Nozick (1974), p. 42

116 Nozick (1989, 2000)

117 Sandberg & Bostrom (2008)

118 See Bostrom (2006) for more on this.

119 Cf. Barnes (1991), Chalmers (1996), and Bostrom (2006).

120 Cf. Bostrom (2006).

121 Giedd et al. (2015), pp. 44–5

122 Kaplan et al. (2020)

123 Bostrom (1997)

124 Bostrom (2019)

125 Bostrom (2013)

126 See Suits (1978) for a notable exception.

127 E.g. Egan (1994)

128 I do not assert that these are the *only* value concepts that could be undermined or disturbed in a plastic world. For example, one could likewise ask what happens to naturalness, the beauty of wilderness, reverence, spontaneity, authenticity, virtue, dignity-as-a-quality (Kolnai 1976), duty, fate (and so on and so forth) in a plastic world; and these questions may be more or less relevant to how much we should be looking forward to such a utopia. I would tentatively contend, however, that conditional on purposefulness, interestingness, fulfillment, richness, and meaning not imposing any insuperable obstacle to the construction of an attractive AI-powered utopia, then none of these other potential value concepts is likely to do so either.

129 When I say "perfection", I speak a bit loosely—we don't need to assume *literal* perfection, a condition in which there is no possible change that would improve things even ever so slightly. So let's say that by "perfection" we mean a condition *close enough* to strict perfection that the matters covered in this lecture series become a pertinent concern.

130 Schopenhauer (1851), p. 16

131 Schopenhauer (1818), p. 312. The only respite, according to Schopenhauer, lies in either a radical self-abnegation of the will or in absorption in some intellectual activity / disinterested contemplation that lets us, for a while, to become spectators of the world rather than participants.

132 Cf. Fisher (1993), p. 396.

133 While I use the term "objective attribution" I don't intend to make any strong claim about the metaphysics of this property.

134 Though note that not all the functionality of negatively valenced mental states can be straightforwardly expressed in terms of behavioral response propensities. Negative experiences also shape our learning processes and our personality development.

135 This is especially so if we also banished Boredom's sisters, Habituation and Adaptation.

136 *Ecce Homo*, painting by Elías García Martínez. Attempted restoration by Cecilia Giménez. Copyright 2012 by Centro de Estudios Borjanos. Reprinted with permission.

137 Nozick (1989), p. 106

138 Egan (1994), pp. 251–2

139 Yudkowsky (2009). Actually, Yudkowsky seems to be using the term "fun" to mean something like "finally valuable activity". I, however, want to separate out different elements that might contribute to making a life finally valuable, and I use the term "interestingness" to refer to only one of these elements—as distinct from, for example, "pleasure" or "meaning".

140 Yudkowsky (2008)

141 On this point, see Fischer (1994), in response to Williams (1973).

142 Mill (1873), p. 149

143 Etinson (2017)

144 Mill (1873), pp. 151–3

145 Cf. Stace (1944).

146 We could also compare the rut-avoidance hypothesis to one explanation of why we tend to habituate more to pleasures than to pains.

Imagine a person who happens to find some particular random stimulus—such as a purple hexagon—unusually painful. One day he comes across a picture of a purple hexagon, finds the experience aversive, so he looks away and gets on with his life. This person wasted 5 seconds. Compare this to somebody who finds a purple hexagon unusually *pleasurable.* One day she comes across a purple hexagon and is deluged with a flood of pleasure. Unless this eventually gets boring—unless the pleasure subsides—this second person would be at risk of wasting her entire life looking at the purple hexagon.

This asymmetry in the behavioral consequences between misfiring pleasure and misfiring pain (for most random stimuli that might accidentally become capable of triggering an abnormal amount of either), could explain why we tend to habituate more to pleasurable stimuli than to painful stimuli—it could be an evolved safeguard against noise in the function that matches positive and negative reinforcement signals to the adaptive value of different stimuli or situations. (I owe this point to Carl Shulman.)

147 Our psyches may be constituted in such a way that there are also levers that control the overall balance between intrinsified goals and more elemental regulators of behavior, so that the relative influence of these different types of motivation can be dynamically adjusted according to cues such as nutrient availability, social situation, life stage, etc. The mechanistic metaphor can only be taken so far.

148 E.g. Ord (2020); MacAskill (2022)

149 Hanson (2021)

150 Lewis (1989)

151 Yudkowsky (2007)

152 I say this *could* be the case. It *need* not be the case, however, since a country might be a global outlier in many different dimensions other than its degree of internal diversity. (Consider a catalog of 1,000 different wallpapers, each with its own distinctive textures, patterns, or motifs. One of these, a red wallpaper, has the greatest internal diversity because of how complex and variegated its pattern is. But suppose that out of the 1,000 wallpapers, 999 are red, and 1 is blue. The blue wallpaper is uniformly blue with no texture—it has a very low degree of internal diversity—yet it contributes more to the diversity of the catalog than the red wallpaper with the greatest internal diversity.)

153 It is also possible to approach it experimentally, which I recommend. Here's how: We will assume that you find the present book interesting. You then keep buying more copies and observe how the total amount of interestingness varies as the number of copies in your possession increases from 1 to 100. Plot result on graph paper and do a regression analysis. (For greater accuracy, you can try buying 200 copies, or even 1,000.)

154 In this example, interestingness corresponds to information. We don't gain any new information when we receive a copy of a document we already have. But in general we can't equate interestingness with information, at least not if we use the standard information theoretic formalism. In that formalism, the maximally information-dense book is one filled with random characters (Shannon, 1948); yet such a book, far from being maximally interesting, is maximally boring. (Interestingness would be closer to something like "theoretically significant or intrinsically important information".)

155 Setting aside special cases like some of Andy Warhol's portraits of Marilyn Monroe, where what is interesting is the fact of repetition itself. Or cases like a misprinted stamp, which might be interesting not only because it is unique but also because there are many other identical stamps from which it differs only very slightly.

156 Note that we are focusing on *non-instrumental* contributions here—how much better or more interesting a particular life makes the world directly by virtue of that life's internal characteristics. A life that consists solely of pressing a button that creates billions of super-interesting galaxies far away would be causally responsible for a huge amount of interestingness; but this is not the kind of contribution I have in

mind here. Think instead of the way that a book contributes to the interestingness of a library.

157 I am not advocating average utilitarianism, only using it to illustrate how the two questions could have different answers.

158 Bostrom (2002)

159 Cf. Bostrom (2011). You could have a conception of yourself not as a particular concrete individual but instead as the set of all identical copies—or, even more broadly, as the set of all systems throughout the multiverse that are sufficiently similar to the token of you that is present in this lecture hall (or that are decision-theoretically correlated with the decision algorithm implemented by this token). On this view, the total impact of "your" actions (i.e. of the set of you-copies) would be infinite in a Big World.

However, unless you construe yourself very broadly—so broadly as to encompass within your self concept many other of the human individuals here on Earth—this aggregate you would still be responsible for a tiny percentage of total interestingness in the world.

As for the absolute amount of interestingness that the aggregate "you" would be contributing, this would depend on just how broadly you defined your self concept. If the aggregate you consists only of all the individuals throughout the multiverse who live lives identical to the life of your present token, then this aggregate you—although it would have infinite causal impacts on the world—would seem to contribute only a negligible amount to the total interestingness of the world, since the world would contain infinitely many other people arbitrarily similar to you and living arbitrarily similar lives to yours. If, however, the aggregate you consists of individuals who live a variety of different lives—and who are linked only in a looser sense, for example by having similar utility functions and correlated decision algorithms, then the aggregate you may actually contribute a substantial amount in absolute terms to the total interestingness of the world.

160 Note that large numbers here work against us in our quest for making an individually humanly-noticeable interestingness contribution in *two* ways.

First, because the more people there are, the more likely it is that there exist other people who are similar to you, which reduces your contributed interestingness in absolute terms (for the same reason that adding a second or a third copy of the same book to a library contributes less interestingness to the library than adding the first copy).

Second, because the greater the number of individuals who contribute a given amount of interestingness to the world, the less humanly-noticeable each such contribution is when toting up the world's overall interestingness (just like your tax contributions make a less noticeable difference to the total tax revenues of your country if you live in India or China than if you live in Tuvalu).

161 Bostrom (2003b)

162 A couple of other ways in which we might conceivably contend for a significant amount of cosmic interestingness: (a) if we have some special designation by a superior being, so that what happens to us has much greater ramifications or importance in a higher realm than what happens to other people; or (b) if we construe our identity in a more encompassing manner than is normally done, so that we are able to count toward "our" external interestingness not only what is done by us construed as individual four-dimensional spacetime worms but everything that is done by some larger class of beings (e.g. all the agents whose decisions are sufficiently strongly correlated with ours).

On the former possibility, I have not much to say here, except that it might seem unlikely that, if there are many humanlike creatures, we would have been specially singled out ahead of all the rest. But if we have, then perhaps whatever special designation we have now we would continue to have in utopia, so that we wouldn't need to fear a reduction in this putative component of our well-being?

On the second possibility, it is true that somebody with a sufficiently expansive conception of self could think of himself as deserving a lot of credit. In the limiting case, if you think that you are one with the universe, then presumably you hold that your life is very interesting indeed. On this conception, it is possible that you would lose some interestingness by transitioning to a utopian condition too early. If human conditions are generally ephemeral on cosmological timescales, while utopian conditions last for maybe trillions of years, then it could be that the total amount of interestingness in the cosmos gains more from human conditions lasting a bit longer. This could be true even if any time slice of a utopian condition is more interesting than any time slice of a human condition—for we could have a situation like that in the Solburg & Lunaburg example, where making one part of the whole more interesting decreases the interestingness of the whole.

163 Hirsch et al. (2015)

164 Bostrom (2008a, 2008b)

165 Ecclesiastes 3:5 (King James Version)

166 Cf. Wilkenfeld (2013).

167 There is an argument that, in the standard picture, the existential safety consideration dominates for impersonal aggregative consequentialist ethical views; whereas, for person-affecting ethical views, both safety and speed are important (Bostrom 2003a). However, even for impersonal views, which focus on existential risk minimization, it is a further (difficult) question which policies would best promote this objective in practice. The original argument itself is consistent with the view that we ought to advance with all possible haste along all technological and economic paths, for example in order to win a race against foul competitors or to minimize various kinds of "state risk" (Bostrom 2014b, chapter 14).

168 Bostrom (2008b)

169 Cf. Parfit (1984).

170 Bostrom (2005)

171 The most obvious justification for preference modifications in a rational actor model would be if our default time preferences are hyperbolic. But in richer and more psychologically realistic models of human decision-making, there may well be additional normative justifications for adjusting aspects of the ways we relate to the passage of time and to our own futures.

172 Ecclesiastes 3:6 (King James Version)

173 Althaus & Gloor (2019)

174 Bostrom & Shulman (2022); Bostrom (2022)

175 Huxley (1946), p. x

176 Huxley (1962), pp. 308–9

177 See Aquinas (1274), p. 601.

178 1 Corinthians 13:12 (King James Version)

179 Feinberg (1980), p. 265

180 Ibid., 271

181 Ibid.

182 Ibid., 273

183 Cf. Tomasik (2014).

184 Oishi & Westgate (2022), p. 794

185 Proust (1913), pp. 118–9

186 We can refer to this putative value as "active experience", although I mean to be agnostic here as to whether the value is grounded in the phenomenal experiences accompanying goal-directed activity or in the activity itself—in many cases, those would anyway be difficult to separate. But I think we can distinguish, at least in a rough-and-ready manner, a life that is tilted toward active experience from one that leans mostly or entirely on passive experience.

It is possible, if we examine "passive experience" very closely, that we should find that it, too, essentially involves aspects of seeking and striving. For example, consider somebody who is engaged in a disinterested contemplation of a sunset—they are passively perceiving the warm flowering colors and taking in the beauty of the spectacle. Yet the processing going on in their visual system and in other parts of their brain—which is necessary for this "passive experience" to occur (and in which, in fact, it might *consist*)—may essentially involve computational processes that have an "aim-ful" or purposive nature. For instance, early visual processing may in some sense be "aiming" to extract contours from the incoming retinal projections.

Not wishing to take a stand on this issue here, what I would say is that even if some aiming suffuses all experience, we can still make out a difference corresponding to the distinction I wish to draw. In this case, the difference would not be between strictly aimless versus aimful experience, but between experience that has only this kind of rudimentary micro-aimfulness and experience that has the more fully articulated aimfulness characteristic of longer-range goal-oriented (or mission-oriented) activity.

187 Cf. Suits (1978) and subsequent literature on this. See also Danaher (2019).

188 Bostrom & Ord (2006)

189 "Tout le malheur des hommes vient d'une seule chose, qui est de ne savoir pas demeurer en repos dans une chambre."; Pascal (1670), pp. 48–9.

190 Freitas (1999, 2003)

191 Bostrom (2014b), pp. 30–6

192 Hall (1993)

193 It is a separate question whether we could *postpone*, even if not permanently avoid, obsolescence by going all-in on transhumanist enhancements. Robin Hanson (2016), for example, has argued that whole brain emulations will be the first form of machine intelligence to

fully rival biological brains, and that this will usher in an era in which human uploads have an edge over artificial intelligences on many tasks. Although biological humans would at that point be outcompeted by the uploads, the uploads would hold their own for maybe hundreds of years of subjective time—which might correspond to a year or two of calendar time—before they are finally surpassed in all tasks by AIs. I think such a scenario looks fairly unlikely, although I don't think we can completely rule it out. But in any case, this book is focused on the longer term. Even Professor Hanson believes that uploads will eventually be outcompeted.

194 Bostrom (2004)

195 Metz (2013), p. 235

196 Cf. e.g. Cohen (2011).

197 Metz (2013), p. 224

198 Though angels *are* often depicted as performing tasks for God (e.g. Milton 1674; Alighieri 1321). Maybe this contributes to making their lives more meaningful?

199 Ord (2021). By the way, when I speak of somebody having "the mass of a galaxy": this wouldn't necessarily mean that their girth would swell to 100,000 light years in diameter, but rather that we might want our minds to have the kind of computational and memory capacity that would require a galaxy's worth of resources to instantiate. Perhaps the best way to use those resources would involve concentrating them into a smaller volume.

200 Metz (2013), p. 235

201 Tooby & Cosmides (1996), pp. 133–6

202 Weber (1905, 1919); Henrich (2020)

203 I would remain agnostic here as to the specifics of the causal explanation, e.g. whether any degree of group selection was involved or the degree to which the faculties in question were specifically shaped to function in this way by genetic selection versus being a result of commonly shared features of human culture and upbringing during ontogeny.

204 In reality, there are of course complex interactions between lust and love, but the simplistic model presented may serve to illustrate the point.

205 Cf. Nozick (1981)'s account.

206 Cf. Frankl (1946).

207 Nietzsche (1889), pp. 60–1

208 Shulman & Bostrom (2021), p. 316

209 Though he might be content to let these ethics continue to inhabit the basement, as moral codes for the masses. Nietzsche was unapologetically elitist and far more concerned with the peaks of human achievement and flourishing than with its average level or its lowest troughs. Perhaps this is another corollary of his axiom—it would be so much harder, he may have reckoned, to affirm existence as valuable if one places substantial weight on the welfare of the downtrodden and disadvantaged multitudes.

210 Camus (1942)

211 Taylor (1970), p. 26

212 Stevenson (1892), p. 203

213 Bacon (1626), p. 35

214 Hepburn (1966), p. 128; Wolf (2010)

215 Metz (2013). This is also so in e.g. Robert Audi's account; Audi (2005).

216 Adams (2010)

217 Stace (1944) used a grasscounter example to illustrate the point that not all truths are interesting. Rawls (1971), Stace's student, raised the hypothetical in a discussion of the good rather than the interesting.

218 It is true that the field of grass-counting is not yet degree-granting, but I think a plausible path could be staked out: some research grant funding opportunities, a dedicated peer-reviewed journal or two, perhaps a scholarly society and an annual conference—and we'd be well up and running.

219 The relevant feature of this payoff is not that it would involve being transported to a new place or being transformed into a new type of being. What is key, rather, is that the stakes in the meaning-giving purpose are extrinsic to the episode upon which the meaning is bestowed. To have meaning bestowed upon the travails of one's own ordinary life, one therefore needs something that is external to this ordinary life or that involves important values that transcend it.

220 But investors in karma coins beware. Scammers are legion.

221 Shulman & Bostrom (2021)

222 Narveson (1973), p. 80

BIBLIOGRAPHY

Adams, R. M. 2010. "Comment". In S. Wolf, *Meaning in Life and Why it Matters* (pp. 75–84). Princeton: Princeton University Press.

Alexander, S. 2014. "Meditations On Moloch". *Slate Star Codex* (blog), July 30th. https://slatestarcodex.com/2014/07/30/meditations-on-moloch/

Alighieri, D. [1321] 1921. *The Divine Comedy*. Cambridge: Harvard University Press.

Althaus, D., & Gloor, L. 2019. "Reducing Risks of Astronomical Suffering: A Neglected Priority". Unpublished manuscript. https://longtermrisk.org/reducing-risks-of-astronomical-suffering-a-neglected-priority/

Aquinas, T. [1274] 1947. *Summa Theologica* (Fathers of the English Dominican Province, Trans.; Vol. 1). New York: Benziger Brothers.

Armstrong, S., & Sandberg, A. 2013. "Eternity in six hours: Intergalactic spreading of intelligent life and sharpening the Fermi paradox". *Acta Astronautica, 89*, 1–13.

Audi, R. 2005. "Intrinsic Value and Meaningful Life". *Philosophical Papers, 34*(3), 331–355.

Bacon, F. [1626] 1915. *New Atlantis* (A. B. Gough, Ed.). Oxford: Clarendon Press.

Bartels, M. 2015. "Genetics of Wellbeing and Its Components Satisfaction with Life, Happiness, and Quality of Life: A Review and Meta-analysis of Heritability Studies". *Behavior Genetics, 45*(2), 137–156.

Bergeaud, A., Cette, G., & Lecat, R. 2016. "Productivity Trends in Advanced Countries between 1890 and 2012". *Review of Income and Wealth, 62*(3), 420–444.

Bille, T., et al. 2013. "Happiness in the arts—International evidence on artists' job satisfaction." *Economics Letters, 121*(1), 15–18.

Blanchflower, D. G. 2021. "Is Happiness U-shaped Everywhere? Age and Subjective Well-Being in 145 Countries". *Journal of Population Economics, 34*(2), 575–624.

Bostrom, N. 1997. "Predictions from Philosophy?". Unpublished manuscript. https://nickbostrom.com/old/predict

Bostrom, N. 2002. "Self-Locating Belief in Big Worlds: Cosmology's Missing Link to Observation". *The Journal of Philosophy, 99*(12), 607–623.

Bostrom, N. 2003a. "Astronomical Waste: The Opportunity Cost of Delayed Technological Development". *Utilitas, 15*(3), 308–314.

Bostrom, N. 2003b. "Are We Living in a Computer Simulation?". *The Philosophical Quarterly, 53*(211), 243–255.

Bostrom, N. 2004. "The Future of Human Evolution". In C. Tandy (Ed.), *Death and Anti-Death: Two Hundred Years After Kant, Fifty Years After Turing* (Vol. 2, pp. 339–371). Palo Alto, California: Ria University Press.

Bostrom, N. 2005. "The Fable of the Dragon Tyrant". *Journal of Medical Ethics, 31*(5), 273–277.

Bostrom, N. 2006. "Quantity of Experience: Brain-Duplication and Degrees of Consciousness". *Minds and Machines, 16*(2), 185–200.

Bostrom, N. 2007. "Technological Revolutions: Ethics and Policy in the Dark". In N. M. de S. Cameron & M. E. Mitchell (Eds.), *Nanoscale: Issues and Perspectives for the Nano Century* (pp. 129-152). Hoboken, NJ: John Wiley & Sons.

Bostrom, N. 2008a. "Letter from Utopia". *Studies in Ethics, Law, and Technology, 2*(1), 1–7.

Bostrom, N. 2008b. "Why I Want to be a Posthuman When I Grow Up". In B. Gordijn & R. Chadwick (Eds.), *Medical Enhancement and Posthumanity*. New York: Springer.

Bostrom, N. (2009). "The Future of Humanity". *Geopolitics, History, and International Relations, 1*(2), 41–78.

Bostrom, N. 2011. "Infinite Ethics". *Analysis and Metaphysics, 10*, 9–59.

Bostrom, N. 2013. "Existential Risk Prevention as Global Priority". *Global Policy, 4*(1), 15–31.

Bostrom, N. 2014a. "Crucial Considerations and Wise Philanthropy". *Good Done Right* conference, All Souls College, University of Oxford, July 9th. https://www.nickbostrom.com/lectures/crucial-considerations

Bostrom, N. 2014b. *Superintelligence: Paths, Dangers, Strategies*. Oxford: Oxford University Press.

Bostrom, N. 2019. "The Vulnerable World Hypothesis". *Global Policy, 10*(4), 455–476.

Bostrom, N. 2022. "Base Camp for Mt. Ethics". Unpublished manuscript. https://nickbostrom.com/papers/mountethics.pdf

Bostrom, N., & Ord, T. 2006. "The Reversal Test: Eliminating Status Quo Bias in Applied Ethics". *Ethics, 116*(4), 656–679.

Bostrom, N., & Shulman, C. 2022. "Propositions Concerning Digital Minds and Society". Unpublished manuscript. https://nickbostrom.com/propositions.pdf

Camus, A. [1942] 2013. *The Myth of Sisyphus* (J. O'Brien, Trans.). London: Penguin.

Catalano, R., Goldman-Mellor, S., Saxton, K., Margerison-Zilko, C., Subbaraman, M., LeWinn, K., & Anderson, E. 2011. "The Health Effects of Economic Decline". *Annual Review of Public Health, 32*, 431–450.

Chalmers, D. J. 2022. *Reality+: Virtual Worlds and the Problems of Philosophy.* London: Allen Lane.

Clark, G. 2007. *A Farewell to Alms: A Brief Economic History of the World.* Princeton: Princeton University Press.

Cohen, G. A. 2011. "Rescuing Conservatism: A Defense of Existing Value". In R. J. Wallace, R. Kumar, & S. Freeman (Eds.), *Reasons and Recognition: Essays on the Philosophy of T.M. Scanlon* (pp. 143–174). Oxford: Oxford University Press.

Crafts, N. 2022. "The 15-Hour Week: Keynes's Prediction Revisited". *Economica, 89*(356), 815–829.

Danaher, J. 2019. *Automation and Utopia: Human Flourishing in a World without Work.* Cambridge: Harvard University Press.

Dene, W. [1348] 1994. "The Plague Seen from Rochester". In R. Horrox (Ed.), *The Black Death* (pp. 70-73). Manchester: Manchester University Press.

Drexler, K. E. 1986. *Engines of Creation: The Coming Era of Nanotechnology.* Garden City, NY: Anchor Books.

Drexler, K. E. 1992. *Nanosystems: Molecular Machinery, Manufacturing, and Computation.* Hoboken, NJ: John Wiley & Sons.

Egan, G. 1994. *Permutation City.* New York: HarperCollins.

Ertz, S. 1943. *Anger in the Sky, a novel.* New York: Harper & Brothers.

Etinson, A. 2017. "Is a Life Without Struggle Worth Living?". *The New York Times,* October 2.

Farmer, R., & Sundberg, N. D. 1986. "Boredom Proneness–The Development and Correlates of a New Scale". *Journal of Personality Assessment, 50*(1), 4–17.

Feinberg, J. 1980. "Absurd Self-Fulfillment". In P. van Inwagen (Ed.), *Time and Cause: Essays Presented to Richard Taylor* (pp. 255–281). London: D. Reidel.

Finnveden, L., Riedel, C. J., & Shulman, C. 2022. "Artificial General Intelligence and Lock-In". Unpublished manuscript. https://lukasfinnveden.substack.com/p/agi-and-lock-in

Fischer, J. M. 1994. "Why Immortality Is Not So Bad". *International Journal of Philosophical Studies, 2*(2), 257–270.

Fisher, C. D. 1993. "Boredom at Work: A Neglected Concept". *Human Relations, 46*(3), 395–417.

Fox, K. 2005. *Watching the English: The Hidden Rules of English Behaviour.* London: Hodder and Stoughton.

Frankl, V. E. [1946] 1984. *Man's Search for Meaning: An Introduction to Logotherapy* (I. Lasch, Trans.). New York: Simon & Schuster.

Freitas, R. A., Jr. 1999. *Nanomedicine, Volume I: Basic Capabilities.* Austin, Texas: Landes Bioscience.

Freitas, R. A., Jr. 2003. *Nanomedicine, Volume IIA: Biocompatibility.* Austin, Texas: Landes Bioscience.

Freitas, R. A., Jr. 2022. *Cryostasis Revival: The Recovery of Cryonics Patients through Nanomedicine*. Scottsdale, AZ: Alcor Life Extension Foundation.

Gates, B. 2017. "What if people run out of things to do?" *Gates Notes* (blog), May 22. https://www.gatesnotes.com/Homo-Deus

Gall, C., & Smith, L. 2019. "Lotto Lout Mickey Carroll Is a Skint Coalman in a Rented Flat but Is 'Happier Now'". *Daily Mirror*, 15 November.

Giedd, J. N., et al. 2015. "Child Psychiatry Branch of the National Institute of Mental Health Longitudinal Structural Magnetic Resonance Imaging Study of Human Brain Development". *Neuropsychopharmacology, 40*(1), 43–49.

Goethe, J. W. von. 1808. *Faust. Eine Tragödie.* Tübingen: J. G. Cotta.

Habermas, J. 2003. *The Future of Human Nature*. Cambridge: Polity Press.

Hall, J. S. 1993. "Utility Fog: A Universal Physical Substance". In *Vision 21: Interdisciplinary Science and Engineering in the Era of Cyberspace*, 115-126.

Hanson, R. 2000. "Long-Term Growth As A Sequence of Exponential Modes". Unpublished Manuscript. https://mason.gmu.edu/~rhanson/longgrow.pdf

Hanson, R. 2008. "Abstract/Distant Future Bias". *Overcoming Bias* (blog), November 26. https://www.overcomingbias.com/p/abstractdistanthtml

Hanson, R. 2016. *The Age of Em: Work, Love, and Life when Robots Rule the Earth.* Oxford: Oxford University Press.

Hanson, R. 2021. "'Long Reflection' Is Crazy Bad Idea". *Overcoming Bias* (blog), October 3. https://www.overcomingbias.com/p/long-reflection-is-crazy-bad-ideahtml

Hanson, R., Martin, D., McCarter, C., & Paulson, J. 2021. "If Loud Aliens Explain Human Earliness, Quiet Aliens Are Also Rare". *The Astrophysical Journal, 922*(2), 182.

Hare, B. 2017. "Survival of the Friendliest: Homo sapiens Evolved via Selection for Prosociality". *Annual Review of Psychology, 68*, 155–186.

Hatfield, R. 2002. *The Wealth of Michelangelo*. Rome: Edizioni di Storia e Letteratura.

Heilbroner, R. 1995. *Visions of the Future: The Distant Past, Yesterday, Today, and Tomorrow*. Oxford: Oxford University Press.

Henrich, J. 2020. *The Weirdest People in the World: How the West Became Psychologically Peculiar and Particularly Prosperous*. London: Penguin.

Hepburn, R. W. 1966. "Questions about the Meaning of Life". *Religious Studies, 1*(2), 125–140.

Hesiod. [c. 700 B.C.] 1993. *Works and Days and Theogony* (S. Lombardo, Trans.). Indianapolis: Hackett Publishing.

Hetschko, C., Knabe, A., & Schöb, R. 2014. "Changing Identity: Retiring From Unemployment". *The Economic Journal, 124*(575), 149–166.

Hoffer, E. 1954. *The Passionate State of Mind, and Other Aphorisms*. New York: Harper & Row.

Hirsch, D., et al. 2015. "More Ties Than We Thought". *PeerJ Computer Science, 1*, e2.

Hirsch, F. [1977] 2005. *Social Limits to Growth*. London: Routledge.

Holder, M. D. 2012. *Happiness in Children: Measurement, Correlates and Enhancement of Positive Subjective Well-Being*. New York: Springer.

Homer. [c. 700 B.C.] 1900. *The Odyssey* (S. Butler, Trans.). A. C. Fifield: London.

Huxley, A. 1946. Foreword to *Brave New World*. New York: Harper & Row.

Huxley, A. 1962. *Island*. New York: Harper & Row.

International Labour Organization. 2023. "Statistics on Working Time". ILOSTAT, 25 April. https://ilostat.ilo.org/topics/working-time/

Jackall, R. [1988] 2010. *Moral Mazes: The World of Corporate Managers*. Oxford: Oxford University Press.

Janotík, T. 2016. "Empirical Analysis of Life Satisfaction in Female Benedictine Monasteries in Germany". *Revue Économique*, *67*(1), 143–165.

Jefferson, T. [1826] 1984. Letter to Roger C. Weightman. In M. D. Peterson (Ed.), *Thomas Jefferson: Writings*. New York: The Library of America.

Johnston, R. F., & Janiga, M. 1995. *Feral Pigeons*. Oxford: Oxford University Press.

Kagan, S. 2019. *How to Count Animals, More Or Less*. Oxford: Oxford University Press.

Kaplan, J., et al. 2020. "Scaling Laws for Neural Language Models" (arXiv:2001.08361). *arXiv*. https://doi.org/10.48550/arXiv.2001.08361

Kass, L. R. 2003. "Ageless Bodies, Happy Souls: Biotechnology and the Pursuit of Perfection". *The New Atlantis*, *1*, 9–28.

Keynes, J. M. [1930] 1931. "Economic Possibilities for Our Grandchildren". In *Essays in Persuasion*. London: Macmillan.

Killingsworth, M. A. 2021. "Experienced Well-Being Rises With Income, Even Above $75,000 per Year". *Proceedings of the National Academy of Sciences*, *118*(4).

Klein Goldewijk, K., et al. 2017. "Anthropogenic land use estimates for the Holocene – HYDE 3.2". *Earth System Science Data*, *9*(2), 927–953.

Kremer, M. 1993. "Population Growth and Technological Change: One Million B.C. to 1990". *The Quarterly Journal of Economics*, *108*(3), 681–716.

Kolnai, A. 1976. "Dignity". *Philosophy*, *51*(197), 251–271.

Kratz, F., & Brüderl, J. 2021. "The Age Trajectory of Happiness". *PsyArXiv*. https://doi.org/10.31234/osf.io/d8f2z

Kugel, J. L. 2007. *How to Read the Bible: A Guide to Scripture, Then and Now*. New York: Free Press.

L. 1848. "A Chapter of Definitions". *The Daily Crescent*, June 23.

Leino-Arjas, P., et al. 1999. "Predictors and Consequences of Unemployment Among Construction Workers: Prospective Cohort Study". *BMJ*, *319*(7210), 600–605.

Lewis, D. 1989. "Dispositional Theories of Value". *Proceedings of the Aristotelian Society, Supplementary Volumes*, *63*, 113–137.

MacAskill, W. 2022. *What We Owe The Future: A Million-Year View*. London: Oneworld Publications.

Malthus, T. R. [1803] 2018. *An Essay on the Principle of Population: The 1803 Edition* (S. C. Stimson, Ed.). New Haven: Yale University Press.

Marx, K., & Engels, F. [1846] 1976. *The German Ideology*. In W. Lough, C. Dutt, W. Lough, & C. P. Magill (Trans.), *Karl Marx/Friedrich Engels Collected Works* (Vol. 5). London: Lawrence & Wishart.

McFadden, B. 2019. "Lotto lout Michael Carroll is working as a £10-an-hour LUMBERJACK after being left penniless when he blew his £10 million jackpot on drug-fuelled parties". *The Daily Mail Online*, February 11. https://www.dailymail.co.uk/news/article-6689797/Lotto-lout-Michael-Carroll-working-10-hour-LUMBERJACK.html

McMahan, J. 1981. "Problems of Population Theory". *Ethics*, *92*(1), 96–127.

Meadows, D. H., et al. 1972. *The Limits to Growth: A Report for the Club of Rome's Project on the Predicament of Mankind*. New York: Universe Books.

Metz, T. 2013. *Meaning in Life: An Analytic Study*. Oxford: Oxford University Press.

Mill, J. S. [1873] 1981. *Autobiography*. In J. M. Robson & J. Stillinger (Eds.), *Collected Works of John Stuart Mill* (Vol. 1). Toronto: University of Toronto Press.

Milton, J. [1667] 1900. *Paradise Lost*. In H. C. Beeching (Ed.), *The Poetical Works of John Milton*. London: Clarendon Press.

Mitchell, J. [1966] 2014. "Both Sides, Now" [Song]. All Access Entertainment.

Mitchell, J. [1967] 1974. "Conversation with Gene Shay" [Interview]. In "Transcript: A Joni Mitchell Retrospective–93.3 WMMR". January 29. https://jonimitchell.com/library/view.cfm?id=604

Mummert, A., Esche, E., Robinson, J., & Armelagos, G. J. 2011. "Stature and robusticity during the agricultural transition: Evidence from the bioarchaeological record". *Economics & Human Biology*, *9*(3), 284–301.

Musk, E. 2023. "Broad subject interview with @DavidFaber" [Interview]. Twitter, May 16. https://twitter.com/i/spaces/1RDxlavQqaRKL_

Narveson, J. 1973. "Moral Problems of Population". *The Monist*, *57*(1), 62–86.

Nozick, R. 1974. *Anarchy, State, and Utopia*. New York: Basic Books.

Nozick, R. 1981. *Philosophical Explanations*. Cambridge: Harvard University Press.

Nozick, R. 1989. *The Examined Life: Philosophical Meditations*. New York: Simon and Schuster.

Nozick, R. 2000. "The Pursuit of Happiness". *Forbes ASAP*, October 2.

Nietzsche, F. [1889] 1988. *Götzen-Dämmerung oder Wie man mit dem Hammer philosophirt*. In *Friedrich Nietzsche: Sämtliche Werke Kritische Studienausgabe in 15 Einzelbänden* (Vol. 6). Berlin: de Gruyter.

"Nine months for Lotto millionaire". 2006. *BBC News*. http://news.bbc.co.uk/2/hi/uk_news/england/norfolk/4699870.stm

Oishi, S., & Westgate, E. C. 2022. "A psychologically rich life: Beyond happiness and meaning". *Psychological Review, 129*(4), 790–811.

Olson, S. J. 2018. "Expanding Cosmological Civilizations on the Back of an Envelope" (arXiv:1805.06329). *arXiv*. https://doi.org/10.48550/arXiv.1805.06329

Ord, T. 2020. *The Precipice: Existential Risk and the Future of Humanity*. London: Bloomsbury Publishing.

Ord, T. 2021. "The Edges of Our Universe" (arXiv:2104.01191). *arXiv*. https://doi.org/10.48550/arXiv.2104.01191

O'Shea, G. 2019. "Rangers Daft Lotto Lout Cuts Logs & Delivers Coal for £10 After Blowing £10M". *The Scottish Sun*, February 11.

O'Toole, G. 2014. "Happiness Is A Butterfly, Which When Pursued, Seems Always Just Beyond Your Grasp". *Quote Investigator* (blog), April 17. https://quoteinvestigator.com/2014/04/17/butterfly/

Parfit, D. 1984. *Reasons and Persons*. Oxford: Oxford University Press.

Parfit, D. 1986. "Overpopulation and the Quality of Life". In P. Singer (Ed.), *Applied Ethics* (pp. 145–158). Oxford: Oxford University Press.

Pascal, B. [1670] 1897. *Pensées de Pascal* (E. Havet, Ed.). Paris: Libraire C. Delagrave.

Patel, B. 2021. "Infamous Lottery Winner Michael Carroll Remarries His Ex-wife". *The Daily Mail Online*, October 12. https://www.dailymail.co.uk/news/article-10078187/Infamous-lottery-winner-Michael-Carroll-remarries-ex-wife.html

Pearce, D. 1995. *The Hedonistic Imperative*. https://www.hedweb.com/

Pleij, H. 2003. *Dreaming of Cockaigne: Medieval Fantasies of the Perfect Life* (D. Webb, Trans.). New York: Columbia University Press.

Posner, R. A. 2012."Working 9 to 12". *The New York Times*, August 17.

Proust, M. [1913] 1956. *Swann's Way* (C. K. S. Moncrieff, Trans.). New York: Random House.

Pushkin, A. [1833] 1918. *Eugene Onegin*. Petrograd: Golike & Wilborg.

Rawls, J. 1971. *A Theory of Justice*. Cambridge: Harvard University Press.

Royal Commission of Inquiry into Children's Employment. 1842. *The Condition and Treatment of the Children Employed in the Mines and Collieries of the United Kingdom*. London: William Strange.

Sahlins, M. 1972. *Stone Age Economics*. Chicago: Aldine-Atherton.

Sandberg, A. & Bostrom, N. 2008. "Whole Brain Emulation: A Roadmap". Technical Report #2008-3, Future of Humanity Institute, Oxford University. http://www.fhi.ox.ac.uk/brain-emulation-roadmap-report.pdf

Sandel, M. 2007. *The Case Against Perfection*. Cambridge: Harvard University Press.

Schopenhauer, A. [1851] 1901. "The Wisdom of Life". In T. B. Saunders (Trans.), *Essays of Arthur Schopenhauer*. New York: A. L. Burt.

Schopenhauer, A. [1818] 1969. *The World as Will and Representation* (E. F. J. Payne, Trans.; Vol. 1). New York: Dover Publications.

Shakespeare, W. [1604] 1898. *Measure for Measure*. London: Bliss & Sands.

Shakespeare, W. [1607] 1918. *The Tragedy of Macbeth* (C. M. Lewis, Ed.). New Haven: Yale University Press.

Shannon, C. E. 1948. "A Mathematical Theory of Communication". *The Bell System Technical Journal, 27*(3), 379–423.

Shulman, C., & Bostrom, N. 2021. "Sharing the World with Digital Minds". In S. Clarke, H. Zohny, & J. Savulescu (Eds.), *Rethinking Moral Status* (pp. 306–326). Oxford: Oxford University Press.

Skidelsky, E., & Skidelsky, R. 2012. *How Much is Enough?: Money and the Good Life*. London: Penguin.

Stace, W. T. 1944. "Interestingness". *Philosophy, 19*(74), 233–241.

Stevenson, R. L. 1892. "The Lantern-Bearers". In *Across the Plains*. London: Thomas Nelson and Sons.

Suits, B. [1978] 2005. *The Grasshopper: Games, Life and Utopia*. Plymouth: Broadview Press.

Taylor, R. [1970] 2010. "The Meaning of Life". In D. Benatar (Ed.), *Life, Death, & Meaning: Key Philosophical Readings on the Big Questions* (pp. 21-30). Rowman & Littlefield.

Tomasik, B. 2014. "Is There Suffering in Fundamental Physics?". *Essays on Reducing Suffering* (blog), August 17. https://reducing-suffering.org/is-there-suffering-in-fundamental-physics/

Tooby, J., & Cosmides, L. 1996. "Friendship and the Banker's Paradox: Other pathways to the Evolution of Adaptations for Altruism". *Proceedings of the British Academy, 88*, 119–143.

"This is about the wonderful Land of Cockaigne". [c. 1458] 2003. In H. Pleij, *Dreaming of Cockaigne: Medieval Fantasies of the Perfect Life* (D. Webb, Trans.). New York: Columbia University Press.

Trammell, P., & Korinek, A. 2023. "Economic Growth under Transformative AI" (Working Paper 31815). *National Bureau of Economic Research*. https://doi.org/10.3386/w31815

Vollset, S. E., et al. 2020. "Fertility, mortality, migration, and population scenarios for 195 countries and territories from 2017 to 2100: A forecasting analysis for the Global Burden of Disease Study". *The Lancet*, 396 (10258), 1285–1306.

Warren, M. A. 1997. *Moral Status: Obligations to Persons and Other Living Things*. Oxford: Oxford University Press.

Weber, M. [1905] 2001. *The Protestant Ethic and the Spirit of Capitalism* (T. Parsons, Trans.). London: Routledge.

Weber, M. [1919] 2004. "Science as a Vocation". In D. Owen & T. Strong (Eds.), & R. Livingstone (Trans.), *The Vocation Lectures*. Cambridge: Hackett Publishing.

Welles, O. (Director). 1949. *The Third Man* [Film]. London Films.

Wilkenfeld, D. A. 2013. "Understanding as Representation Manipulability". *Synthese, 190*(6), 997–1016.

Williams, B. 1973. "The Makropulos Case: Reflections on the Tedium of Immortality". In B. Williams, *Problems of the Self: Philosophical Papers 1956–1972* (pp. 82–100). Cambridge: Cambridge University Press.

Wolf, S. 2010. *Meaning in Life and Why It Matters*. Princeton: Princeton University Press.

Yudkowsky, E. 2007. "Your Strength as a Rationalist". *Less Wrong* (blog), August 10. https://www.lesswrong.com/posts/5JDkW4MYXit2CquLs/your-strength-as-a-rationalist

Yudkowsky, E. 2008. "Complex Novelty". *Less Wrong* (blog), December 20. https://www.lesswrong.com/posts/aEdqh3KPerBNYvoWe/complex-novelty

Yudkowsky, E. 2009. "The Fun Theory Sequence". *Less Wrong* (blog), January 25. https://www.lesswrong.com/posts/K4aGvLnHvYgX9pZHS/the-fun-theory-sequence

INDEX

V

W

Y

ACKNOWLEDGMENTS

I am grateful to many more people than I can mention for enabling this work, by creating the material or intellectual preconditions for its coming into existence. But for their more direct contributions, I want to explicitly thank Guive Assadi, Emily Campbell, Richard Yetter Chappell, Will Hammond, Guy Kahane, Anton Korinek, Matthew van der Merwe, Thaddeus Metz, Geoffrey Miller, Toby Newberry, Carl Shulman, ChatGPT-4, Claude 2, Tanya Singh, and Jan-Erik Strasser for extensive comments on the manuscript; Gilbert N. Morris, Eric Neyman, Toby Ord, David Pearce, and Anders Sandberg for helping to answer some particular questions; Liz Hudson, Frances Key Phillips, Sam Blake, and Pascal Porcheron for help with copyediting; Gwen Bradford, Stephen Campbell, Dale Dorsey, Nick Fletcher, Thomas Hurka, Antti Kauppinen, Eden Lin, Michael Prinzing, Aaron Smuts, Michael Steger, Louise Sundararajan, Lars Svendsen, Valerie Tiberius, and Susan Wolf for in-depth expert conversations on specific topics that helped me learn and refine preliminary ideas; Matthew van der Merwe and Toby Newberry for project management; Guive Assadi for research assistance; and the team at Ideapress for their uncommon dispatch.

Whether it was all worth it?